T0219868

Lecture Notes in Computer Science 10968

Commenced Publication in 1973
Founding and Former Series Editors:
Gerhard Goos, Juris Hartmanis, and Jan van Leeuwen

More information about this series at http://www.springer.com/series/7409

Francis Y. L. Chin · C. L. Philip Chen
Latifur Khan · Kisung Lee
Liang-Jie Zhang (Eds.)

Big Data – BigData 2018

7th International Congress
Held as Part of the Services Conference Federation, SCF 2018
Seattle, WA, USA, June 25–30, 2018
Proceedings

 Springer

Editors
Francis Y. L. Chin
The University of Hong Kong
Hong Kong
Hong Kong, Special Administrative
   Region of China

C. L. Philip Chen ⓘ
University of Macau
Macao
Macao, Special Administrative
   Region of China

Latifur Khan
The University of Texas at Dallas
Richardson, TX
USA

Kisung Lee ⓘ
Louisiana State University
Baton Rouge
USA

Liang-Jie Zhang
Kingdee International Software Group
   Company Limited
Shenzhen
China

ISSN 0302-9743   ISSN 1611-3349   (electronic)
Lecture Notes in Computer Science
ISBN 978-3-319-94300-8   ISBN 978-3-319-94301-5   (eBook)
https://doi.org/10.1007/978-3-319-94301-5

Library of Congress Control Number: 2018947341

LNCS Sublibrary: SL3 – Information Systems and Applications, incl. Internet/Web, and HCI

Printed on acid-free paper

This Springer imprint is published by the registered company Springer International Publishing AG
part of Springer Nature
The registered company address is: Gewerbestrasse 11, 6330 Cham, Switzerland

Preface

The 2018 International Congress on Big Data (BigData Congress 2018) aimed to provide an international forum that formally explores various business insights of all kinds of value-added "services." Big data is a key enabler of exploring business insights and the economics of services.

This volume presents the papers accepted for the BigData Congress 2018, held in Seattle, USA, during June 25–30. The main topics of the 2018 BigData Congress included but were not limited to: big data architecture, big data modeling, big data as a service, big data for vertical industries (government, health care, etc.), big data analytics, big data toolkits, big data open platforms, economic analysis, big data for enterprise transformation, big data in business performance management, big data for business model innovations and analytics, big data in enterprise management models and practices, big data in government management models and practices, and big data in smart planet solutions.

We accepted 32 papers, including 22 full papers and ten short papers. Each was reviewed and selected by at least three independent members of the BigData Congress 2018 international Program Committee. We are pleased to thank the authors whose submissions and participation made this conference possible. We also want to express our thanks to the Organizing Committee and Program Committee members for their dedication in helping to organize the conference and reviewing the submissions. We owe special thanks to the keynote speakers for their impressive speeches.

May 2018

Kisung Lee
Francis Y. L. Chin
C. L. Philip Chen
Latifur Khan
Liang-Jie Zhang

Organization

General Chair

Shengzhong Feng   Shenzhen Institutes of Advanced Technology,
Chinese Academy of Sciences, China

Program Chairs

Francis Y. L. Chin   The University of Hong Kong, SAR China
C. L. Philip Chen   University of Macau, SAR China
Latifur Khan   University of Texas – Dallas, USA

Program Vice Chair

Kisung Lee   Louisiana State University, USA

Application and Industry Track Chair

Keke Chen   Wright State University, USA

Big Data for Social Networking Track Chair

Bin Wu   Beijing University of Posts and Telecommunications,
China

Big Data Management and Systems Track Chair

Qi Zhang   IBM T.J. Watson Research Center, USA

Big Data Search, Mining, and Visualization Track Chair

Noseong Park   University of North Carolina at Charlotte, USA

Short Paper Track Chair

Liqiang Wang   Central Florida University, USA
Qishi Wu   New Jersey Institute of Technology, USA

Publicity Chair

Raju Vatsavai   North Carolina State University, USA

Services Conference Federation (SCF 2018)

General Chairs

Calton Pu	Georgia Tech, USA
Wu Chou	Essenlix Corporation, USA

Program Chair

Liang-Jie Zhang	Kingdee International Software Group Co., Ltd., China

Finance Chair

Min Luo	Huawei, USA

Panel Chair

Stephan Reiff-Marganiec	University of Leicester, UK

Tutorial Chair

Carlos A Fonseca	IBM T.J. Watson Research Center, USA

Industry Exhibit and International Affairs Chair

Zhixiong Chen	Mercy College, USA

Operations Committee

Huan Chen (Chair)	Kingdee, China
Jing Zeng	Tsinghua University, China
Yishuang Ning	Tsinghua University, China
Sheng He	Tsinghua University, China
Cheng Li	Tsinghua University, China

Steering Committee

Calton Pu	Georgia Tech, USA
Liang-Jie Zhang (Chair)	Kingdee International Software Group Co., Ltd., China

Program Committee

Pelin Angin	Middle East Technical University, Turkey
Valentina Emilia Balas	Aurel Vlaicu University of Arad, Romania
Alfredo Cuzzocrea	ICAR-CNR and University of Calabria, Italy

Alfredo Cuzzocrea	University of Trieste, Italy
Peter Baumann	Jacobs University, Germany
Wagner Meira Jr.	UFMG, Brazil
Florin Rusu	University of California, Merced, USA
Xumin Liu	Rochester Institute of Technology, USA
Elizabeth Liddy	University of Syracuse, USA
Qi Yu	Rochester Institute of Technology, USA
Daniel Grosu	Wayne State University, USA
Jaroslaw Szlichta	University of Ontario, Canada
Wolfgang Nejdl	Institut für Verteilte Systeme, Germany
Amjad Gawanmeh	Khalifa University, UAE
Jun Shen	University of Wollongong, Australia
Latifur Khan	University of Texas at Dallas, USA
Anastasios Gounaris	Aristotle University of Thessaloniki, Greece
Ningfang Mi	Northeastern University, China
Hatem Ltaief	KAUST, Saudi Arabia
Xiang Zhao	National University of Defense, China
Lijun Chang	University of New South Wales, Australia
Haopeng Chen	Shanghai Jiao Tong University, China
Luca Cagliero	Politecnico di Torino, Italy
Marcio K. Oikawa	UFABC, Brazil
Albert Lam	Hong Kong Baptist University, SAR China
Yan Bai	University of Washington Tacoma, USA
Hailong Sun	Beihang University, China
Jiuyun Xu	China University of Petroleum, China
Ernesto Damiani	University of Milan, Italy
Joao E. Ferreira	USP, Brazil
Jianwu Wang	UMBC, USA
Alfredo Goldman	USP-IME, Brazil
Raju Vatsavai	NCSU, USA
Jian Zhang	Louisiana State University, USA
Jian Wang	Wuhan University, China
Jian-Yun Nie	Univeristy of Montreal, Canada
Ludovico Boratto	University of Cagliari, Italy
Marios Dikaiakos	University of Cyprus, Cyprus
Luiz Angelo Steffenel	University of Reims, France
Haytham ElGhazel	Polytech Lyon, France
Xiaofang Zhou	University of Queensland, Australia
Yasuhiko Kanamasa	Fujitsu, Japan
Makoto Yui	Treasure Data, USA
Harald Kornmayer	DHBW Mannheim, Germany
Amy Apon	Clemson University, USA
Guangming Tan	Chinese Academy of Sciences, China
Weiguo Liu	Shandong University, China
Chi Zhou	Shenzhen University, China
Yanjie Wei	Chinese Academy of Sciences, China
Yong Zhang	Chinese Academy of Sciences, China

Contents

Short Paper Track: BigData Analysis

Short Paper Track: BigData Modeling

Research Track: BigData Modeling

Time Series Similarity Search Based on Positive and Negative Query

Jimin Wang$^{(\boxtimes)}$, Qi Liu$^{(\boxtimes)}$, and Pengcheng Zhang$^{(\boxtimes)}$

College of Computer and Information, Hohai University, Nanjing 211100, China
{wangjimin, pchzhang}@hhu.edu.cn, 1034188416@qq.com

Abstract. Traditional time series similarity search, based on relevance feedback, combines initial, positive and negative relevant series directly to create new query sequence for the next search; it can't make full use of the negative relevant sequence, even results in inaccurate query results due to excessive adjustment of the query sequence in some cases. In this paper, time series similarity search based on separate relevance feedback is proposed, each round of query includes positive query and negative query, and combines the results of them to generate the query results of each round. For one data sequence, positive query evaluates its similarity to the initial and positive relevant sequences, and negative query evaluates it's similarity to the negative relevant sequences. The final similar sequences should be not only close to positive relevant series but also far away from negative relevant series. The experiments on UCR data sets showed that, compared with the retrieval method without feedback and the commonly used feedback algorithm the proposed method can improve accuracy of similarity search on some data sets.

Keywords: Time series similarity search · Relevance feedback
Positive query · Negative query

1 Introduction

Data mining is the process of extracting knowledge from massive data. In reality, most of the data are time series, so it has important theoretical and practical significance to mine potential useful knowledge from the time series data [1]. Time series data mining mainly includes classification, clustering, sequence pattern matching, similarity search and prediction. In many time, similarity search is the important foundation of the others, and it was proposed by Agrawal in 1993 [2] to find similar pattern for the given pattern in time series. Similarity search can help us to make useful decisions, for example, we can find a similar sales pattern in the sales records of various commodities to make a sales strategy [3], and we can forecast the natural disaster by searching similar precursor of natural disasters [4].

The traditional time series similarity search extracts data features firstly to reduce the data dimension [5–7], and then building index for them [8, 9]. Finally, based on similarity measure function [10–13], the sequences similar to the query sequence are retrieved in the index structure and displayed to the users. However, in the beginning of search, users usually can't describe the query sequence clearly, so once search is unable

© Springer International Publishing AG, part of Springer Nature 2018
F. Y. L. Chin et al. (Eds.): BIGDATA 2018, LNCS 10968, pp. 3–16, 2018.
https://doi.org/10.1007/978-3-319-94301-5_1

to find the suitable similar sequences for users. Feedback-based strategy allows users to express their satisfaction or dissatisfaction on the query results, and make multiple queries to improve the query accuracy and users' satisfaction.

Feedback technology was applied in information retrieval at first, and Keogh introduced it into the time series data mining in the literature [14], user can set different weights indicating the similarity or dissimilarity degree for the query result sequences and return them to search system, then the new query sequence for next query is generated by the feedback sequences by some strategy. Time series similarity search based on relevance feedback diversification was proposed in Literature [15], MMR [16] was used on the feedback sequences to ensure the diversity of the query results, and then new query sequence was generated by the feedback sequences.

The above time series similarity search methods combine initial, positive and negative relevant sequences directly to create new query series for the next query, it's easy to make the query sequence change too much causing the worse query results. For lack of query topics, Wang [17] proposed a negative relevance feedback method in text retrieval, only the negative relevant feedback vectors and query vector were used to do the next query. Peltonen [18] proposed a negative feedback information retrieval system based on machine learning, allowing users to make positive and negative relevance feedback directly by the interactive visual interface. Studies had shown that making full use of negative feedback sequence can improve the retrieval accuracy.

In this paper, we propose a time series similarity search method based on positive and negative relevance feedback sequences separately. The contributions of this paper are summarized as follows:

- A novel similarity search method based on relevance feedback for time series is proposed. By combining positive query and negative query, the characteristics hidden in positive and negative relevant sequences are easily extracted. Considering the multiple categories of negative relevant sequences, two strategies, single positive relevance feedback model and multi-positive relevance feedback model, can be used during negative query. The proposed method can improve the accuracy of the similarity search.
- The proposed method is validated by a set of dedicated experiments. The experimental data are included in UCR archive [22], and experiments show that, compared with non-feedback and traditional feedback methods, the proposed method can improve the accuracy of the query on some data sets of UCR archive.

The rest of the paper is organized as follows. In Sect. 2, we review the relevance feedback strategy based on vector model. In Sect. 3, we describe our proposed time series similarity search based on positive and negative query. Experiments are analyzed and discussed in Sect. 4. We conclude this paper and discuss our future work in Sect. 5.

2   Relevance Feedback Strategy Based on Vector Model

Relevance feedback is one of query extension technologies which have become one of the key technologies to improve the recall and precision in information retrieval. Relevance feedback based on vector model is generally implemented by query modification which is proposed by Van Rijsbergen [19]. For the query results, user can label the similar vectors(called positive relevance feedback vectors) or dissimilar vectors(called negative relevance feedback vectors), query vector will be modified according to the original, positive and negative relevance feedback vectors, and then used to search similar vectors again, until user is satisfied with the results, or give up the search.

2.1   Rocchio Algorithm

The classical feedback algorithm based on vector model was proposed by Rocchio in the SMART system [20], new query vector is generated following Eq. 1

$$\overrightarrow{q_{new}} = \alpha \times \vec{q} + \beta \times \frac{1}{|D_r|} \sum_{\overrightarrow{d_j} \in D_r} \overrightarrow{d_j} - \gamma \times \frac{1}{|D_{nr}|} \sum_{\overrightarrow{d_j} \in D_{nr}} \overrightarrow{d_j} \tag{1}$$

Where, $\overrightarrow{q_{new}}$ is the new query vector, $\vec{q}$ is the original query vector, D_r is the collection of positive relevant document vectors, D_{nr} is the collection of negative relevant document vectors, α, β and γ are the weights to control their impact on new query vector. The optimal weight values can be assessed by knowledge of the data, or experimentally.

2.2   Ide dec-hi Algorithm

Ide [21] improved the Rocchio algorithm by using the most dissimilar negative relevant vector instead of the average of negative relevant vectors, the new query vector is generated following Eq. 2.

$$\overrightarrow{q_{new}} = \alpha \times \vec{q} + \beta \times \frac{1}{|D_r|} \sum_{\overrightarrow{d_j} \in D_r} \overrightarrow{d_j} - \gamma \times \max_{\overrightarrow{d_j} \in D_{nr}} \overrightarrow{d_j} \tag{2}$$

Where, $\max_{\overrightarrow{d_j} \in D_{nr}} \overrightarrow{d_j}$ is the most dissimilar negative relevant vector. Both of the above feedback algorithms rely heavily on positive relevance feedback, but if the query topics are few so that the positive relevance vectors are few or no, the above methods will be difficult to work.

2.3   Negative Relevance Feedback Algorithm

Wang [17] proposed a negative relevance feedback algorithm for this extreme case in document retrieval. User labels the negative relevant vectors, and search system establishes a positive query (the query vector is the original query vector) and a negative query (the query vector is composed of the negative relevant vectors), finally, the similarity is obtained by combining them following Eq. 3.

$$S_{combine}(Q, D) = S(Q, D) - \beta \times S(Q_{neg}, D) \qquad (3)$$

Where, Q is the original query vector, D is the document vector to be checked, Q_{neg} is the negative query vector generated from negative vectors by some strategy, $S(Q, D)$ is the similarity measure of the positive query, $S(Q_{neg}, D)$ is the similarity measure of the negative query, $S_{combine}(Q, D)$ is the final similarity measure. β controls the impact of $S(Q_{neg}, D)$ to the final similarity measure.

3   Time Series Similarity Search Based on Positive and Negative Query

In this paper, we introduce the negative feedback strategy in time series similarity search, and propose a time series similarity search method based on positive and negative query. For the query results, user labels the positive and negative relevant sequences, and then search system makes positive query and negative query, finally combines the results of positive query and negative query to get final similar sequences, the above steps will be executed repeatedly until user is satisfied with the query results or ignores the search. This method mainly includes: query sequence modification, positive and negative query, the combination of positive and negative query.

3.1   Query Sequence Modification

Time series similarity search can also be seemed as a kind of information retrieval. In this paper, new query sequence is generated following Eq. 4 based on the Rocchio algorithm. Negative relevant sequences are used to do negative query and do not participate in the modification, so γ is set to be 0.

$$q_{new} = \alpha \times q + \beta \times \frac{1}{|S_{PR}|} \sum_{s_j \in S_{PR}} s_j \qquad (4)$$

Where, q_{new} is new query sequence for next query, q is the original query sequence, S_{PR} is the set of positive relevance sequences.

3.2   Positive and Negative Query

During the query, we perform positive query and negative query for every sequence s in data sequences. Positive query computes the similarity between s and q_{new}, negative query computes the similarity between s and the positive relevant sequences.

Positive Query. For every sequence s in data sequences, some similarity measure, such as Euclidean distance, will be used to compute the similarity, identified by $Sim(q_{new}, s)$, between s and q_{new}.

Negative Query. The main problem to be solved in our algorithm is to determine the similarity, identified by $Sim(q_{neg}, s)$, between s and the negative relevance sequences. We present two strategies to solve this problem, one is the single negative relevance feedback model, the other is the multi-negative relevance feedback model.

Single Negative Relevance Feedback Model. All the negative relevance sequences were combined to generate the single average sequence q_{neg} by Eq. 5.

$$q_{neg} = \sum_{s_j \in S_{NR}} s_j * w_j \tag{5}$$

Where, S_{NR} is the set of negative relevance sequences, and $\sum_{j=1}^{|S_{NR}|} w_j = 1$. This strategy treats all the negative relevance sequences as the same category, and use the mean sequence represent the characteristics of most negative relevance sequences. The weight w_j can be set according to the dissimilarity, specified by the user's subjective setting, between the query sequence and each negative relevance sequence. For example, user can ranks all the negative feedback sequences by the dissimilarity subjectively as s_1, s_2, s_3, ..., s_n, $n = |S_{NR}|$, then we can set $w_j = (n - j + 1) / \sum_{j=1}^{|S_{NR}|} j$. If user can't rank the negative feedback sequences, then we can generate the single average sequence q_{neg} by Eq. 6.

$$q_{neg} = \frac{1}{|S_{NR}|} \sum_{s_j \in S_{NR}} s_j \tag{6}$$

That is, $w_j = \frac{1}{|S_{NR}|}$.

Multi-negative Relevance Feedback Model. The negative relevance sequences may belong to multiple categories. Single negative feedback model can't reflect the difference of the negative relevance sequences between different categories. The multi-negative relevance model clusters the negative relevance sequences into n clusters, and merges the sequences in each cluster to generate the mean sequence q_{neg_i}, $i = 1, 2, 3,$ ..., n, $n = |S_{PR}|$ to represent the characteristics of each cluster independently. For every sequence s in data sequences, some similarity measure, such as Euclidean distance, will be used to compute the similarity, identified by $Sim(q_{neg\_i}, s)$, between s and q_{neg_i}. We get the similarity between s and negative feedback sequences by Eq. 7.

$$Sim(q_{neg}, s) = F(Sim(q_{neg\_i}, s)) \tag{7}$$

Where, F may be *MAX*, *MIN* or *AVG* operation. *MAX*($Sim(q_{neg\_i}, s)$) means the max similarity between s and n representative sequences is chosen to represent the final similarity. *MIN*($Sim(q_{neg\_i}, s)$) means the min similarity between s and n representative sequences is chosen to represent the final similarity. *AVG*($Sim(q_{neg\_i}, s)$) means the average similarity between s and n representative sequences is chosen to represent the final similarity.

3.3   The Combination of Positive and Negative Query

In our method, the final query results should be similar to the query sequence, but also are dissimilar to the negative relevant sequences, so the final similarity measure following Eq. 8.

$$S_c(q_{new}, s) = Sim(q_{new}, s) - \lambda \times Sim(q_{neg}, s) \qquad (8)$$

$Sim(q_{new}, s)$ measures the similarity between q_{new} and the checked sequence s in positive query. $Sim(q_{neg}, s)$ measures the similarity between the negative relevance sequences and the checked sequence s in negative query, λ controls the magnitude of dissimilarity. $Sc(q_{new}, s)$ represents the final similarity degree between q_{new} and s. It can be seen that to make $Sc(q_{new}, s)$ high, either $Sim(q_{new}, s)$ is high or $Sim(q_{neg}, s)$ is low, that is, either the similarity between q_{new} and s is high or the similarity between q_{neg} and s is low.

4   Experiments and Analysis

4.1   Experimental Data

In this paper, we use 17 sets of the UCR data [22] to make experiment, and all the sequences in each data set have been labeled category. The information of each data set used in our experiments is shown in Table 1.

Table 1. Data set

No.	Data set	Number of categories	Size
1	ChlorineConcentration	3	3840
2	StarLightCurves	3	8326
3	Two_Patterns	4	4000
4	Trace	4	100
5	OliveOil	4	30
6	Haptics	5	308
7	Beef	5	30
8	OSUleaf	6	242
9	Synthetic_control	6	300
10	InlineSkate	7	550
11	Fish	7	175
12	Lighting7	7	73
13	MedicalImages	10	760
14	SwedishLeaf	15	625
15	WordsSynonyms	25	638
16	Adiac	37	391
17	50words	50	455

4.2   Method

Five experiments are performed on each data set, that is similarity searching based on no feedback, Rocchio algorithm, Ide dec-hi algorithm, single negative relevance feedback model (labeled SNRF) and multi-negative relevance feedback model (labeled MNRF). Euclidean distance is used as the distance measure.

Usually the initial query sequence can express most of user's intent, the query sequence modification only do a few fine-tuning. In order to ensure that the modified query sequence dose not deviate from the original sequence largely, the value of α of Eq. 4 is fixed to 1, and the other parameters increment from 0 to 1 by step 0.1 to find the optimal value in our experiments. In our experiments, the negative relevant sequences are clustered according to the category of sequence directly in multi-negative relevance feedback model, and in single negative relevance feedback model, Eq. 6 is used to generate the representative sequence.

In order to validate our proposed single negative and multi-negative relevance feedback similarity search model, kNN and leave-one-out cross validation are performed. P-R value, which is the integration of recall and precision, is used to assess the quality of the query, P represents the accuracy rate and R represents the recall rate. For each query sequence, kNN is performed to find r, which is called recall number, relevant sequences. P-R value of data set for r_j, noted as PR_{rj}, is calculated by the following Eq. 9.

$$PR_{rj} = \frac{1}{N_q} \sum_{i=1}^{N_q} \frac{r_j}{k_i} \qquad (9)$$

where N_q is the number of query sequences, k_i is the nearest neighbor number for the i-th query sequence to find r_j relevant sequences. The P-R average value of data set is the average of all PR_{rj}, where $r_j= 1,2,\ldots,10$. In the similarity search with feedback, for each query sequence, we carried out two times feedback, the optimal P-R average value and the corresponding parameters will be retained.

4.3   Discussion and Comparison

Figure 1 shows the precision-recall curve (according to the optimal P-R average value shown in Tables 2, 3, 4, 5 and 6) of some data sets, it can be seen that, precision-recall of the methods based on feedback are obviously better than that of no feedback. When the number of recall is small, the precision of SNRF and MNRF are similar to that of Rocchio and Ide dec-hi. However, when the number of recall increases, the precision of SNRF and MNRF is better than that of Rocchio and Ide dec-hi.

Tables 2, 3, 4, 5 and 6 show the optimal P-R average value of the methods used in our experiments, and the corresponding parameters. Figure 2 shows the P-R average value of 17 data sets. It can be seen that, compared to the no feedback, the similarity search with feedback can obviously improve the accuracy. The results of single-negative feedback model and multi-negative feedback model are all better than that of Rocchio and Ide dec-hi except the results on Two_Patterns and 50words, but the results of five methods on Two_Patterns and 50words are very close. There are some

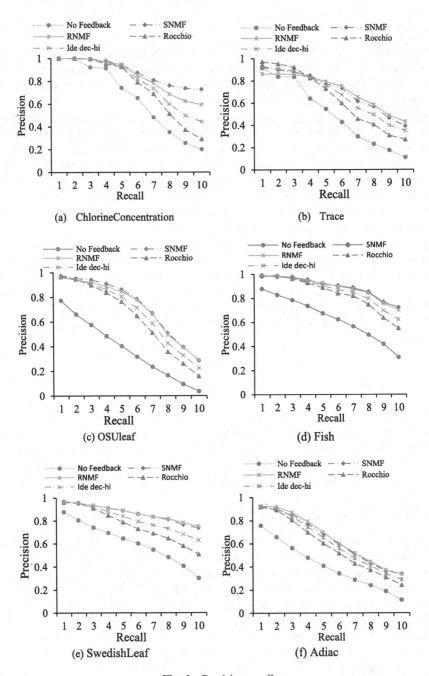

Fig. 1. Precision-recall

Table 2. P-R of no feedback model

No.	Name of data sets	P-R value
1	ChlorineConcentration	0.651
2	StarLightCurves	0.829
3	Two_Patterns	0.869
4	Trace	0.501
5	OliveOil	0.669
6	Haptics	0.242
7	Beef	0.259
8	OSUleaf	0.375
9	Synthetic_control	0.817
10	InlineSkate	0.205
11	Fish	0.631
12	Lighting7	0.403
13	MedicalImages	0.553
14	SwedishLeaf	0.611
15	WordsSynonyms	0.501
16	Adiac	0.404
17	50words	0.479

Table 3. P-R of single negative relevance feedback model

No.	Name of data sets	P-R value	β	λ
1	ChlorineConcentration	0.881	0.7	1
2	StarLightCurves	0.927	0.8	1
3	Two_Patterns	0.959	0.9	1
4	Trace	0.698	0.9	0.7
5	OliveOil	0.872	0.7	0.5
6	Haptics	0.565	0.8	1
7	Beef	0.470	0.8	0.9
8	OSUleaf	0.728	0.8	0.8
9	Synthetic_control	0.962	0.6	0.3
10	InlineSkate	0.453	0.7	0.6
11	Fish	0.893	0.9	0.8
12	Lighting7	0.685	0.9	0.8
13	MedicalImages	0.805	0.9	0.9
14	SwedishLeaf	0.866	0.8	0.8
15	WordsSynonyms	0.711	0.9	0.7
16	Adiac	0.630	0.7	0.5
17	50words	0.662	0.9	0.7

Table 4. P-R of multi-negative relevance feedback model

No.	Name of data sets	P-R value	β	λ
1	ChlorineConcentration	0.844	0.7	0.3
2	StarLightCurves	0.915	0.9	1
3	Two_Patterns	0.953	0.7	1
4	Trace	0.700	0.9	0.1
5	OliveOil	0.871	0.8	0.5
6	Haptics	0.540	0.9	0.5
7	Beef	0.449	0.8	0.2
8	OSUleaf	0.717	0.7	0.3
9	Synthetic_control	0.967	0.8	0.5
10	InlineSkate	0.450	0.9	0.1
11	Fish	0.888	0.7	0.5
12	Lighting7	0.699	0.9	0.1
13	MedicalImages	0.817	0.8	0.3
14	SwedishLeaf	0.871	0.9	0.7
15	WordsSynonyms	0.725	0.8	0.1
16	Adiac	0.648	0.9	0.5
17	50words	0.685	0.8	0.1

Table 5. P-R of feedback based on Rocchio

No.	Name of data sets	P-R value	β	γ
1	ChlorineConcentration	0.752	0.7	0.7
2	StarLightCurves	0.872	0.9	0.7
3	Two_Patterns	0.973	0.8	0.6
4	Trace	0.643	0.9	0.5
5	OliveOil	0.815	0.8	0.5
6	Haptics	0.479	0.7	0.3
7	Beef	0.395	0.8	0.6
8	OSUleaf	0.635	0.7	0.3
9	Synthetic_control	0.932	0.9	0.5
10	InlineSkate	0.403	0.7	0.3
11	Fish	0.834	0.9	0.4
12	Lighting7	0.620	0.7	0.7
13	MedicalImages	0.716	1	0.4
14	SwedishLeaf	0.761	0.8	0.6
15	WordsSynonyms	0.686	0.9	0.8
16	Adiac	0.578	0.8	0.5
17	50words	0.659	0.9	0.4

Table 6. P-R average value of feedback based on Ide dec-hi

No.	Name of data sets	P-R value	β	γ
1	ChlorineConcentration	0.752	0.7	0.7
2	StarLightCurves	0.893	0.9	0.6
3	Two_Patterns	0.991	0.9	0.7
4	Trace	0.677	0.6	0.4
5	OliveOil	0.843	0.7	0.4
6	Haptics	0.530	0.6	0.4
7	Beef	0.432	0.8	0.6
8	OSUleaf	0.696	0.8	0.5
9	Synthetic_control	0.947	0.6	0.5
10	InlineSkate	0.426	0.7	0.2
11	Fish	0.871	0.9	0.4
12	Lighting7	0.659	0.8	0.5
13	MedicalImages	0.761	1	0.3
14	SwedishLeaf	0.816	0.9	0.7
15	WordsSynonyms	0.712	0.9	1
16	Adiac	0.622	0.8	0.6
17	50words	0.673	1	0.2

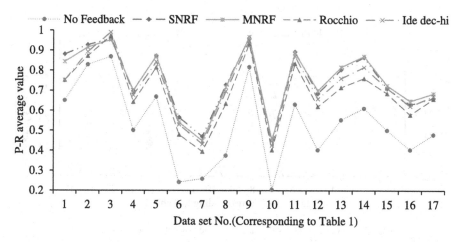

Fig. 2. P-R average value of data sets

differences between single negative feedback model and multi-negative feedback model, at the confidence level of 0.05, the Wilcoxon symbol rank sum test shows $p = 0.9758$ (bilateral test), indicating that for the 17 data sets, single negative feedback model and multi-negative feedback model performs basically the same.

However, in Table 7 and Fig. 3, it can been seen that when the number of categories is relatively small, the single negative relevance feedback model is better than the multi-negative relevance feedback model. When the categories is more, the multi-

Table 7. P-R difference value between SNF and MNF

No.	Name of data sets	P-R difference value (P-R of SNRF - P-R of MNRF)	Number of Categories
1	ChlorineConcentration	0.376	3
2	StarLightCurves	0.120	3
3	Two_Patterns	0.058	4
4	Trace	−0.023	4
5	OliveOil	0.002	4
6	Haptics	0.251	5
7	Beef	0.206	5
8	OSUleaf	0.114	6
9	Synthetic_control	−0.044	6
10	InlineSkate	0.033	7
11	Fish	0.052	7
12	Lighting7	−0.140	7
13	MedicalImages	−0.123	10
14	SwedishLeaf	−0.049	15
15	WordsSynonyms	−0.143	25
16	Adiac	−0.186	37
17	50words	−0.234	50

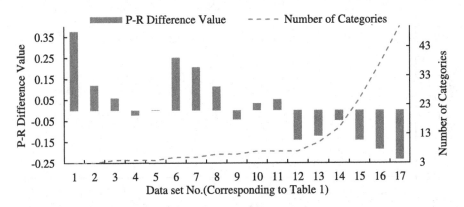

Fig. 3. P-R difference value and Number of categories of data set

negative relevance feedback model is better than single negative relevance feedback model, it maybe because the single average sequence q_{neg} can't represent the characteristics of the negative relevance sequences fully when the negative relevance sequences are more dispersed, that is, the number of categories is more.

5 Conclusion

Based on relevance feedback, we proposed a time series similarity search method which fully utilizes the positive and negative relevance sequences. The positive relevant sequences and negative relevant sequences are used to search similar sequences independently, after that, the similarity degree of positive query and negative query are combined to get the final similarity. Experiments show that the proposed method can improve the accuracy of the query on some data sets of UCR archive. Furtherly, we can study a more principled strategy to use the negative feedback sequences to conduct constrained query.

Acknowledgment. This research is supported by the Key Technologies Research and Development Program of China (2015BAB07B01), the National Natural Science Foundation of China (No. 61572171).

References

1. Wang, Y., Xu, C.: Data mining technology. Electron. Technol. Softw. Eng. **2015**(8), 204–205 (2015)
2. Agrawal, R., Faloutsos, C., Swami, A.: Efficient similarity search in sequence databases. In: Lomet, D.B. (ed.) FODO 1993. LNCS, vol. 730, pp. 69–84. Springer, Heidelberg (1993). https://doi.org/10.1007/3-540-57301-1_5
3. Luo, H.: Based on gray-ARIMA financial time series intelligent hybrid forecasting. Finan. Econ. Theory Pract. **35**(2), 27–34 (2014)
4. Zhu, Y., Li, S., Fan, Q.: Prediction of hydrological time series based on wavelet neural network. J. Shandong Univ. (Eng. Sci.) **41**(4), 119–124 (2011)
5. Li, Z., Guo, J., Hui, X.: Based on the common principal component of the multivariate time series dimensionality reduction method. Control Decis. **2013**(4), 531–536 (2013)
6. Li, H.: Research on feature representation and similarity measurement in time series data mining. Dalian University of Technology (2012)
7. Xiao, R.: Study on dimensionality reduction and similarity matching of uncertain time series. Donghua University (2014)
8. Li, Z., Zhang, F., Li, K.: A multi-time series index structure supporting DTW distance. J. Softw. **25**(3), 560–575 (2015)
9. Dai, K.: Research on time series query method based on linear hash index. Softw. Eng. **19**(8), 1–8 (2016)
10. Zhang, Q., Zhao, Z.: Z tree: an index structure for high-dimensional data. Comput. Eng. **33**(15), 49–51 (2007)
11. Xiao, R., Liu, G.: Research on time series similarity measure and clustering based on trend. Appl. Res. Comput. **31**(9), 2600–2605 (2014)
12. Zhang, H., Li, Z., Sun, Y.: New time series similarity measure method. Comput. Eng. Des. **35**(4), 1279–1284 (2014)
13. Goldin, D.Q., Millstein, T.D., Kutlu, A.: Bounded similarity querying for time-series data. Inf. Comput. **194**(2), 203–241 (2004)
14. Keogh, E.J., Pazzani, M.J.: An enhanced representation of time series which allows fast and accurate classification, clustering and relevance feedback. In: 4th International Conference of Knowledge Discovery and Data Mining, pp. 27–31. ACM Press, New York (1998)

15. Eravci, B., Ferhatosmanoglu, H.: Diversity based relevance feedback for time series search. Proc. VLDB Endow. **7**(2), 109–120 (2013)
16. Carbonell, J., Goldstein, J.: The use of MMR, diversity-based reranking for reordering documents and producing summaries. In: 21st Annual International ACM SIGIR Conference on Research and Development in Information Retrieval, Melbourne, pp. 335–336. ACM (1998)
17. Wang, X., Fang, H., Zhai, C.X.: A study of methods for negative relevance feedback. In: International ACM SIGIR Conference on Research and Development in Information Retrieval, Singapore, pp. 219–226 (2008)
18. Peltonen, J., Strahl, J., Floréen, P.: Negative relevance feedback for exploratory search with visual interactive intent modeling. In: 22nd International Conference on Intelligent User Interfaces, Raphael Resort, Limassol, pp. 149–159 (2017)
19. Van Rijsbergen, C.J.: A new theoretical framework for information retrieval. ACM SIGIR Forum **21**(1–2), 23–29 (1986)
20. Rocchio, J.J.: Relevance feedback in information retrieval. In: Salton, G. (ed.) The SMART System: Experiments in Automatic Document Processing, pp. 313–323. Prentice-Hall, Englewood Cliffs (1972)
21. Ide, E.: New experiments in relevance feedback. In: The SMART System: Experiments in Automatic Document Processing, pp. 337–354. Prentice-Hall (2000)
22. The UCR Time Series Classification Archive. http://www.cs.ucr.edu/~eamonn/\time_series_data/

Inter-Category Distribution Enhanced Feature Extraction for Efficient Text Classification

Yuming Wang[1,2]([✉]), Jun Huang[1], Yun Liu[1], Lai Tu[1], and Ling Liu[2]

[1] School of Electronic Information and Communications, Huazhong University
of Science and Technology, Wuhan 430074, Hubei, China
ymwang@mail.hust.edu.cn
[2] School of Computer Science, College of Computing,
Georgia Institute of Technology, Atlanta, GA 30332, USA

Abstract. Text data is one of the dominating data types in Big Data driven services and applications. The performance of text classification largely depends on the quality of feature extraction over the text corpus. For supervised learning over text documents, the TF-IDF (Term Frequency-Inverse Document Frequency) weighting factor is one of the most frequently used features in text classification. In this paper, we address two known limitations of TF-IDF based feature extraction method: First, the conventional TF-IDF weighting factor lacks of consideration about the synonymous relationship between feature terms. Second, for big corpus with large number of text documents and large number of feature terms, the computational complexity of text classification increases with the dimensionality of the feature space. We address these problems by introducing an optimization technique based on the Inter-Category Distributions (ICD) of terms and the Inter-Category Distributions of documents. We call this new weighting factor TF-IDF-ICD, namely TF-IDF with Inter-Category Distributions. To further enhance the effectiveness of our TF-IDF-ICD method, we describe a TF-IDF-ICD threshold based Dimensionality Reduction (DR) optimization. We test the text classifier with a corpus of 10,000 articles. The evaluation results show that the proposed TF-IDF-ICD based text classification method outperforms the conventional TF-IDF based classification solution by 7.84% at only about 43.19% of the training time used by the conventional TF-IDF based text classification methods.

Keywords: TF-IDF · Feature extraction · Text classification
Inter-Category Distribution (ICD) · Dimensionality reduction

1 Introduction

Text classification aims to categorize a document into one of the predefined class categories, denoted by $\mathcal{Y}$, where a document is represented in the form of bag

F. Y. L. Chin et al. (Eds.): BIGDATA 2018, LNCS 10968, pp. 17–25, 2018.
https://doi.org/10.1007/978-3-319-94301-5_2

of words $\mathcal{X}$, denoted as a feature vector $x \in \mathbb{R}^{d \times 1}$ with d unique terms [1]. Text classification algorithms have been applied successfully in many big data driven text applications and services, such as spam filtering [2], tagging online news [3], social media analysis [4], bioscience [5] and chat bot [6]. Statistical learning based methods, such as Support Vector Machine (SVM) and its family of algorithms, are widely used for text classification [7–9].

The typical workflow of statistical learning based text classification system consists of four core components: text preprocessing, feature extraction, classifier training and classification [10]. Research [11] has shown that the sophistication of the feature selection process is critical to the effectiveness and efficiency of the text classifier.

Traditional text feature extraction methods use TF-IDF algorithm [12] to extract text features and transform the corpus of N text documents with a vocabulary of d terms into the d dimensional vector space of N TF-IDF feature vectors. By representing each text document as a feature vector in a high dimensional vector space, a statistical feature based classifier, such as SVM, can be used for text classification through supervised learning. Several known limitations exist for the TF-IDF based feature extraction method. For example, the conventional TF-IDF weighting factor pays no attention to synonyms and semantic relation between synonyms. Moreover, when dealing with large corpus with tens of thousands documents and terms, the feature vector space has very high dimensionality and is highly skewed and sparse. As a result, the computation complexity increases dramatically for large corpus.

To address these limitations, we propose an optimized feature extraction method for text classifications by defining a new weighting factor based on the Inter-Category Distribution of terms and documents, called TF-IDF-ICD. Based on this new weighting factor, we introduce a Dimensional Reduction (DR) technique based on TF-IDF-ICD weighting factor, which confines the dimensionality of the feature vector space by limiting the number of *Feature Terms* to only the most critical features, namely those term features with high TF-IDF-ICD values. Our TF-IDF-ICD optimized text classification system consists of three steps: (1) By combining with the text preprocessing, the synonym merging, and converging the statistic weight of synonym, the TF-IDF-ICD factors of all terms to every single document in the corpus are calculated and the result is represented as a team feature vector to represent the corresponding document. (2) By filtering out the low weight terms and selecting only those high weighted terms as the *Feature Terms*, we reduce the dimensions of the feature vector space for the given corpus. (3) After performing feature extraction in the first two steps, the resulting document feature vectors are fed into an SVM classifier to train the classifier. We evaluate the performance of the proposed optimization method for text classification using a well known corpus of 10, 000 text documents. The experimental results show that our approach outperforms the conventional TF-IDF solution by 7.84% at only about 43.19% of the original training time, demonstrating the effectiveness of the proposed method for achieving better trade-off between classification accuracy and computational complexity.

2   Related Work

Text data is one of the dominating data types in Big data driven applications and services. A wide variety of techniques have been designed and researched for text classification. Document representation and feature selection are two of the fundamental core tasks in text classification.

The TF-IDF based method is one of the most popular methods for document representation. Given a corpus of N documents with d terms as the vocabulary, we transform each document in the corpus into a d-dimensional feature vector x based on TF-IDF weighting factor. Many feature selection methods have been proposed and compared in [13]. An important function in feature selection is to reduce the dimensionality of the term feature space through feature transformation methods, which create a new and smaller set of features as a function of the original set of features. A typical example of such a feature transformation method is Latent Semantic Indexing (LSI), and its probabilistic variant PLSA [14]. Once the feature space is determined, the document vectors are fed into a chosen text classifier. As noted in [15], text data is ideally suited for SVM classification because of the sparse high-dimensional nature of text.

Most of the conventional TF-IDF based methods lack of consideration on semantic features, such as synonyms and their relations. Several independent research efforts have been engaged to involve semantic features. For example, Huang et al. [16] proposed an improved method based on the feature term vectors extracted from TF-IDF, and analyzed the semantic similarity of feature terms with external dictionaries. Zhu et al. [17] proposed a method using word2vec model to calculate the similarity between words and words, which solves the problem that low-frequency words with high class discrimination are ignored by statistical weights. Qu et al. [18] considered the relations between the feature and the class, showing the problem of the traditional TF-IDF algorithms that ignore the inter-category distribution of feature words.

Our approach is inspired by these existing efforts. Concretely, we propose three optimizations to enhance the efficiency of TF-IDF based feature extraction and text classification: (1) We propose to merge synonyms in text preprocessing step. (2) We propose to compute the inter-category distribution of terms and the inter-category distribution of documents and use the product of these two ICD vectors to define the TF-IDF-ICD weighting factor as an optimization for feature extraction. (3) We introduce a tunable threshold control knob to perform dimension reduction on the TF-IDF-ICD vector space. Our experiments show the effectiveness of our approach compared to the existing methods.

3   Feature Extraction Optimization

3.1   Text Preprocessing with Synonym Fusion

The text preprocessing is the first step. We remove the text format and symbols and perform word segmentation and semantic labeling of the text documents using the open source HanLP [19] tools. Less significant terms are then filtered out for simplicity and noise cancellation. In this work, adverbs, locative words, number words, and auxiliary words etc, are regarded as less significant terms. Even though some solutions [20] remove the name of persons, places and organizations during text preprocessing. We argue that the terms such as the name of person have strong reliance on categories, such as the relation between "Taylor Swift" and music category, between "Yao Ming" and the sports category. Thus, we keep such terms instead of removing them. Our experiments show that the classification accuracy is higher by retaining these special terms.

Synonyms may dilute the statistical feature of a document. We propose to combine synonyms as one term for feature calculation. Synonyms in the text documents are identified according to the dictionary, such as the "HIT-SCIR Synonym Dictionary (Extended Edition)" [21], which contains $77,343$ terms organized according to a five-level classification structure to form a tree structure, and each leaf node represents a group of terms. Three possible relationships are identified among these terms: "=" stands for synonym, "#" stands for relevant terms and "@" stands for independent terms. We replace the terms marked with "=" with the first term of the group of terms and sum the term frequency so as to achieve the purpose of synonym fusion.

3.2   Text Feature Extraction with TF-IDF-ICD

In order to involve semantic features into the text classification, we advocate to use a new weighting factor, the TF-IDF with Inter-Category Distributions of both terms and documents, to replace the conventional TF-IDF.

Let $\mathbb{D} = \{d_1, d_2, d_3, \cdots, d_i, \cdots, d_N\}$ be the corpus of N documents, where each document d_i belongs to a category $C_j(j < m)$. m is the total number of categories. A document $d_i = \{W_{i1}, W_{i2}, \cdots, W_{ik}\}$ has k terms, where W_{ik} is the k-th feature term of document d_i. The TF-IDF-ICD weighting factor is calculated using Eq. 1.

$$w(W_{ik}) = TF(W_{ik}) \times (IDF(W_{ik}) + \alpha \times ICD(W_{ik})) \tag{1}$$

where α ($\alpha \geq 0$) is a weighting factor used to tune the weight ratio of the inter-category distributions to the inverse document frequency.

Let $n_{d_i, W_{ik}}$ denote the raw count of term W_{ik} appearing in document d_i and $D_{W_{ik}}$ denote the number of documents that contain term W_{ik} in the corpus. We define the term frequency adjusted by the document length and the inverse document frequency in Eqs. 2 and 3 respectively.

$$TF(W_{ik}) = \frac{n_{d_i, W_{ik}}}{\sum_k n_{d_i, W_{ik}}} \tag{2}$$

$$IDF(W_{ik}) = \log \frac{N}{D_{W_{ik}} + 1} \tag{3}$$

Before defining the extended weighting factor $ICD(W_{ik})$, we first define two concepts: $ICDT(W_{ik})$ and $ICDD(W_{ik})$.

$ICDT(W_{ik})$ describes the occurrence weight of term W_{ik}. It represents the inter-category distribution for term W_{ik} and is defined in Eq. 4. A higher weight indicates that term W_{ik} has a higher category preference.

$$ICDT(W_{ik}) = \sqrt{\frac{n_{C_j, W_{ik}}}{n_{W_{ik}} + 1}} \qquad (4)$$

where $n_{C_j, W_{ik}}$ is the raw count of term W_{ik} appearing in category C_j and $n_{W_{ik}}$ is the raw count of term W_{ik} appearing in the corpus.

$ICDD(W_{ik})$ describes the occurrence weight of documents in which the term W_{ik} appears. It is the inter-category distribution for a document and is defined in Eq. 5. Similarly, a higher weight indicates that term W_{ik} has a higher category preference.

$$ICDD(W_{ik}) = \log \frac{(D_{Cj, W_{ik}} + 1) \times N}{D_{W_{ik}} + 1} \qquad (5)$$

where $D_{C_j, W_{ik}}$ is the number of documents that contain term W_{ik} in category C_j, and $D_{W_{ik}}$ is the number of documents that contain term W_{ik} in the corpus.

The extended weighting factor $ICD(W_{ik})$ is defined as a product of two inter-category distributions, the ICD for the term and the ICD for the document, in Eq. 6.

$$ICD(W_{ik}) = ICDT(W_{ik}) \times ICDD(W_{ik}) \qquad (6)$$

3.3 Dimensionality Reduction

After we transform all documents in the corpus into the feature vectors based on the new TF-IDF-ICD weighting factor, the next step is to perform the Dimensionality Reduction by limiting the number of *Feature Terms* to be used for text classification. Based on the TF-IDF-ICD value of every term in the text document and thus the corresponding feature vector, we only select the terms with the top $\mu\%$ highest TF-IDF-ICD value as the critical *Feature Terms*. This operation reduces the dimensionality of the feature vectors of the document corpus. The union set of the selected *Feature Terms* from all the documents forms a feature vector space that represents the corpus. Consequently, the TF-IDF-ICD values of those critical *Feature Terms* of a document is used to represent the document. Our experiments show that adequate reduction of the number of *Feature Terms* has little effect on the accuracy of classification, while can significantly improve the time complexity of the text classification.

3.4 Text Classification with TF-IDF-ICD and DR Optimization

After the optimized feature extraction with TF-IDF-ICD and the ICD-based dimensionality reduction, we feed the optimized feature vector into the text classifier. We use the linear SVM algorithm to train the classifier, and use the K-fold cross validation to evaluate the classification performance.

4   Experimental Evaluation

The corpus used in this paper are labeled news reports. We collect 10000 articles from *Sina News*. The articles are labeled with 10 categories and 1000 news in each category. In the experiments, we uses 10-fold cross-validation to assess the classification results.

Recall Sect. 3.2 and Eq. 1, the parameter α is used as the probability for contrasting ICD with TF-IDF. Another important parameter in our approach is the dimensionality reduction (DR) threshold parameter $\mu\%$. Both α and $\mu\%$ are the tunable parameters in the experiments. To determine the best value of α, we fix $\mu\%$ to 20% and vary α from 0 to 4 by increasing 0.1 each time. The result is shown in Fig. 1. The overall trend of F-Measure increases as α increases from 0 to 4. When α increases to 1.4, F-Measure reaches the maximum value of 0.9490. As α continues to increase from 1.5 to 4.0, F-Measure first experiences a slight decrease and then becomes relatively stable. In *Sina News* Corpus, when α is 1.4, the F-measure accuracy is increased by 0.0237, compared to the conventional TD-IDF method without using ICD weighting factor (i.e., $\alpha = 0$, F-Measure: 0.9253).

$\mu\%$ represents the portion of *Feature Terms*. A higher $\mu\%$ results in a larger dimension of feature space which results in higher computational complexity. We set the α value by 1.4, which produces the best F-Measure as shown in the previous experiments, and report the F-measure results by varying $\mu\%$ from 5% to 100% at 5% each time. Figure 2 shows the results.

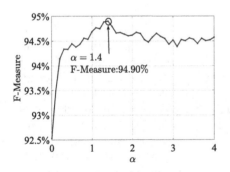

Fig. 1. F-Measure vs α ($\mu\% = 20\%$).

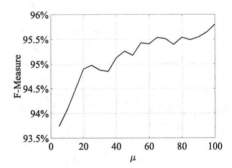

Fig. 2. F-Measure vs $\mu\%$ ($\alpha = 1.4$).

We observe that the F-Measure increases very fast as $\mu\%$ increases to 20%. This is because the feature size increases exponentially with the growth of $\mu\%$ and the increase ratio gets slower as $\mu\%$ continues to increase. Hence, $\mu\%$ is chosen to be 20% as a good trade-off setting between high accuracy and complexity.

We next compare the proposed TF-IDF-ICD method with and without DR ($\alpha = 1.4$, $\mu\% = 20\%$) with the traditional TF-IDF based method. To be fair, we also use the Dimensionality Reduction (DR) based on TF-IDF ($\mu\% = 20\%$) when comparing with our approach with DR. Table 1 shows the experimental results.

Table 1. Comparison between the improved method and TF-IDF.

Type	Precision		Recall		F-Measure	
Method	TF-IDF	TF-IDF-ICD	TF-IDF	TF-IDF-ICD	TF-IDF	TF-IDF-ICD
Value	94.04%	**95.80%**	93.67%	**95.81%**	93.87%	**95.80%**
Method	TF-IDF + DR	TF-IDF-ICD + DR	TF-IDF + DR	TF-IDF-ICD + DR	TF-IDF + DR	TF-IDF-ICD + DR
Value	88.25%	**94.94%**	85.90%	**94.86%**	87.06%	**94.90%**

As shown in the table, the proposed TF-IDF-ICD method has better accuracy performances than the traditional TF-IDF based method, no matter whether DR is used or not. Moreover, the accuracy performances of the proposed TF-IDF-ICD method is less affected by DR than the TF-IDF based method. In the case of using DR in both methods, the proposed method improves the F-Measure performance by 7.84% more than the TF-IDF based method does. The precision and recall performances are also improved by 6.69% and 8.96% respectively.

Using the dimension and training time of non-DR solution (i.e., $\mu\% = 100\%$) as the baseline, we also investigate the relative dimension of the vector space and the relative training time of the classifier. Figure 3 shows that the relative dimension increases with μ exponentially. When $\mu\% = 20\%$, the dimension is about $\frac{1}{20}$ of the non-DR value. Figure 4 shows that the relative training time reduces significantly with a smaller μ. When $\mu\% = 20\%$, the text classification using TF-IDF-ICD with DR reduces the training time to 43.19% of the time in non-DR solution.

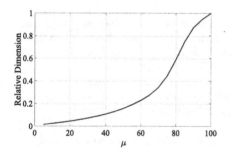
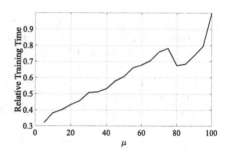

Fig. 3. Relative dimension vs $\mu\%$ ($\alpha = 1.4$, non-DR solution as baseline).

Fig. 4. Relative training time vs $\mu\%$ ($\alpha = 1.4$, non-DR solution as baseline).

It is worth noting that the ICD integration operation brings overhead of memory consumption and runtime increasing in document preprocessing. But based on our measurement in the experiment, the increased overhead is almost negligible compared to the original TF-IDF computation cost.

5   Conclusion

We have presented an efficient text classification scheme with three optimizations: (1) synonym fusion in the text preprocessing, (2) the enhanced feature extraction based on the Inter-Category Distributions (ICD) of both terms and documents, and (3) the dimensionality reduction based on the TF-IDF-ICD threshold based critical feature selection. We evaluate the performance of the proposed approach for text classification using a corpus of $10,000$ text documents. The experimental evaluation results ($\alpha = 1.4$, $\mu\% = 20\%$) show that our method outperforms the conventional TF-IDF based solution by 7.84% and spends only about 43.19% of the training time required by the TF-IDF based method.

Acknowledgement. The authors from Huazhong University of Science and Technology, Wuhan, China, are supported by the Chinese university Social sciences Data Center (CSDC) construction projects (2017–2018) from the Ministry of Education, China. The first author, Dr. Yuming Wang, is currently a visiting scholar at the School of Computer Science, Georgia Institute of Technology, funded by China Scholarship Council (CSC) for the visiting period of one year from December 2017 to December 2018. Prof. Ling Liu's research is partially supported by the USA National Science Foundation CISE grant 1564097 and an IBM faculty award. Any opinions, findings, and conclusions or recommendations expressed in this material are those of the author(s) and do not necessarily reflect the views of the funding agencies.

References

1. Joachims, T.: Learning to Classify Text Using Support Vector Machines: Methods, Theory and Algorithms, vol. 186. Kluwer Academic Publishers, Norwell (2002)
2. Almeida, T., Hidalgo, J.M.G., Silva, T.P.: Towards sms spam filtering: results under a new dataset. Int. J. Inf. Secur. Sci. **2**(1), 1–18 (2013)
3. Liu, S., Huang, K., Chai, J.: Research of news tagging based on word frequency statistics and user information. In: 2017 10th International Congress on Image and Signal Processing, BioMedical Engineering and Informatics (CISP-BMEI), pp. 1–5. IEEE (2017)
4. Ali, K., Dong, H., Bouguettaya, A., Erradi, A., Hadjidj, R.: Sentiment analysis as a service: a social media based sentiment analysis framework. In: 2017 IEEE International Conference on Web Services (ICWS), pp. 660–667. IEEE (2017)
5. Ramani, R.G., Jacob, S.G.: Benchmarking classification models for cancer prediction from gene expression data: a novel approach and new findings. Stud. Inf. Control **22**(2), 134–143 (2013)
6. Chu, Z., Gianvecchio, S., Wang, H., Jajodia, S.: Who is tweeting on Twitter: human, bot, or cyborg? In: Proceedings of the 26th Annual Computer Security Applications Conference, pp. 21–30. ACM (2010)
7. Yang, Y.: An evaluation of statistical approaches to text categorization. Inf. Retrieval **1**(1), 69–90 (1999)
8. Sebastiani, F.: Machine learning in automated text categorization. ACM Comput. Surv. **34**(1), 1–47 (2002)

9. Su, J.S., Bo-Feng, Z., Xin, X.: Advances in machine learning based text categorization. J. Softw. **7**, 1848–1859 (2006)
10. Aggarwal, C.C., Zhai, C. (eds.): Mining Text Data, 1st edn. Springer, New York (2012). https://doi.org/10.1007/978-1-4614-3223-4
11. Mladenić, D., Brank, J., Grobelnik, M., Milic-Frayling, N.: Feature selection using linear classifier weights: interaction with classification models. In: Proceedings of the 27th Annual International ACM SIGIR Conference on Research and Development in Information Retrieval, SIGIR 2004, pp. 234–241. ACM, New York (2004)
12. Salton, G., Yu, C.T.: On the construction of effective vocabularies for information retrieval. SIGIR Forum **9**(3), 48–60 (1973)
13. Yang, Y., Pedersen, J.O.: A comparative study on feature selection in text categorization. In: Proceedings of the Fourteenth International Conference on Machine Learning, ICML 1997, pp. 412–420. Morgan Kaufmann Publishers Inc., San Francisco (1997)
14. Hofmann, T.: Probabilistic latent semantic indexing. In: Proceedings of the 22nd Annual International ACM SIGIR Conference on Research and Development in Information Retrieval, SIGIR 1999, pp. 50–57. ACM, New York (1999)
15. Joachims, T.: A probabilistic analysis of the Rocchio algorithm with TFIDF for text categorization. In: Proceedings of the Fourteenth International Conference on Machine Learning, ICML 1997, pp. 143–151. Morgan Kaufmann Publishers Inc., San Francisco (1997)
16. Huang, C.H., Yin, J., Hou, F.: A text similarity measurement combining word semantic information with TF-IDF method. Chin. J. Comput. **34**, 856–864 (2011)
17. Zhu, L., Wang, G., Zou, X.: Improved information gain feature selection method for Chinese text classification based on word embedding. In: Proceedings of the 6th International Conference on Software and Computer Applications, pp. 72–76. ACM (2017)
18. Qu, S., Wang, S., Zou, Y.: Improvement of text feature selection method based on TFIDF. In: International Seminar on Future Information Technology and Management Engineering, FITME 2008, pp. 79–81. IEEE (2008)
19. HanLP: Han Language Processing (2014). https://github.com/hankcs/HanLP
20. Hua, X.L., Zhu, Q.M., Li, P.F.: Chinese text similarity method research by combining semantic analysis with statistics. Jisuanji Yingyong Yanjiu **29**(3), 833–836 (2012)
21. LTP-Cloud: Language Technology Platform Cloud (2017). https://www.ltp-cloud.com

Reversible Data Perturbation Techniques for Multi-level Privacy-Preserving Data Publication

Chao Li[(✉)], Balaji Palanisamy, and Prashant Krishnamurthy

University of Pittsburgh, Pittsburgh, USA
{chl205,bpalan,prashk}@pitt.edu

Abstract. The amount of digital data generated in the Big Data age is increasingly rapidly. Privacy-preserving data publishing techniques based on differential privacy through data perturbation provide a safe release of datasets such that sensitive information present in the dataset cannot be inferred from the published data. Existing privacy-preserving data publishing solutions have focused on publishing a single snapshot of the data with the assumption that all users of the data share the same level of privilege and access the data with a fixed privacy level. Thus, such schemes do not directly support data release in cases when data users have different levels of access on the published data. While a straight-forward approach of releasing a separate snapshot of the data for each possible data access level can allow multi-level access, it can result in a higher storage cost requiring separate storage space for each instance of the published data. In this paper, we develop a set of reversible data perturbation techniques for large bipartite association graphs that use perturbation keys to control the sequential generation of multiple snapshots of the data to offer multi-level access based on privacy levels. The proposed schemes enable multi-level data privacy, allowing selective de-perturbation of the published data when suitable access credentials are provided. We evaluate the techniques through extensive experiments on a large real-world association graph dataset and our experiments show that the proposed techniques are efficient, scalable and effectively support multi-level data privacy on the published data.

1 Introduction

Rapid advancements in data mining and data analytics techniques have made it possible to extract insights previously considered impossible. There is thus a significant incentive for collection and analysis of user information, some of which could be potentially sensitive and private [3,8]. Data Privacy is a crucial barrier to data analysis due to privacy risks [1,24]. Private data often arises in the form of associations between entities in real world such as medicines purchased by patients in a pharmacy store, films rated by users in a movie rating website or products purchased online by users. Such real-world associations are commonly represented as large, sparse bipartite graphs [5] with nodes representing the

F. Y. L. Chin et al. (Eds.): BIGDATA 2018, LNCS 10968, pp. 26–42, 2018.
https://doi.org/10.1007/978-3-319-94301-5_3

entities (e.g., patients and medicines) and edges representing the associations between them (e.g., medicines purchases by patients).

Privacy-preserving data disclosure schemes aim at publishing sensitive datasets such that the private information contained in the published data can not be inferred. These techniques primarily perturb the raw datasets to meet the privacy requirements before the data is published. While there have been several efforts on privacy-preserving data publishing in the past, most solutions have focused on publishing a single snapshot of a dataset that offers a fixed privacy level with the assumption that all users of the data share the same level of privilege to access the data [7,9,12,13,16,17,20–22]. Here a privacy level may directly correspond to an access privilege level. Hence, such schemes do not directly support data release in cases when data users have different levels of access on the published dataset.

In this paper, we develop a set of reversible data perturbation techniques for large bipartite association graphs that use perturbation keys to control the sequential generation of multiple snapshots of the perturbed data to offer multi-level access based on privacy levels. We assume that sensitive information in a dataset may arise either as: (i) an individual sensitive value indicating an individual's private information (e.g., did buyer 'Alice' purchase the medicine 'Citalo'?) or (ii) a statistical value representing some sensitive statistics about a group/sub-group of individuals (e.g., the total number of antidepressant medicines purchased by buyers in a given neighborhood represented by a zipcode). In many real-world scenarios, users of a dataset may have different privileges and may need to access the same dataset at different privacy/utility levels requiring multi-level access on the published data. For example, a data owner may prefer to share a sensitive dataset with a reputed data analytics team with a low degree of perturbation. However, the same data owner may release the dataset with a higher level of perturbation to a less privileged data analyst and the data may be disclosed to the general public with an even higher level perturbation to protect the data privacy further. The proposed reversible data perturbation schemes enable multi-level data access, allowing selective de-perturbation of the published dataset to reduce the degree of perturbation when suitable access credentials are provided. We evaluate the proposed techniques through extensive experiments on a large real-world association graph dataset. Our experiments show that the proposed techniques are efficient, scalable and effectively enable multi-level data privacy on the published data.

The rest of the paper is organized as follows. We introduce the key concepts and the differential privacy model in Sect. 2. The proposed reversible data perturbation techniques is discussed in Sect. 3. We present the experimental evaluation of the proposed techniques in Sect. 4. The related work is discussed in Sect. 5 and we finally conclude in Sect. 6.

2   Concepts and Models

In this section, we first present the general idea behind the multi-level privacy-preserving data publishing problem. We then introduce the graph representation of association data used in our work and briefly review the notion of differential privacy.

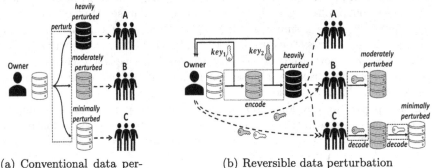

(a) Conventional data perturbation

(b) Reversible data perturbation

Fig. 1. Data perturbation schemes to support multi-level privacy-preserving data publication

2.1   Multi-level Data Access Using Reversible Data Perturbation

Privacy-preserving data publishing (PPDP) schemes are designed to prevent the inference of sensitive information in published datasets from data users accessing the published information. Dataset owners use privacy-preserving data publishing (PPDP) techniques to perturb their datasets prior to publishing. In many real-world scenarios, users of a dataset may have different privileges and may need to access the same dataset at different privacy/utility levels requiring multi-level access on the published data. The multi-level privacy-preserving data publishing requirement can be accomplished with a straight-forward approach of releasing a snapshot of the dataset for each privacy level using conventional data perturbation techniques such as differential privacy [7]. In the example shown in Fig. 1(a), the data owner is willing to share a minimally perturbed snapshot of the data with data user C and wants to share a moderately perturbed snapshot with data user B and a heavily perturbed snapshot with data user A. The perturbed snapshots for each of the data users can be generated by running a conventional data perturbation technique for each data user with the privacy parameters corresponding to the privilege level of the user. While this straight-forward approach achieves the multi-level data disclosure objective, it can however result in a higher storage cost requiring a replication of the dataset for each possible privacy level.

In this paper, we propose the concept of reversible data perturbation (Fig. 1(b)) that allows a data publisher to release data at multiple privacy levels using a single instance of the perturbed dataset. The reversible data perturbation approach consists of an encoding phase and a decoding phase. The dataset encoding is performed by the dataset owner. During the encoding process, the perturbed snapshots are first sequentially generated in order from the lowest privacy level with the least perturbation to the highest privacy level that requires a larger degree of perturbation. Between any two adjacent snapshots in the sequence, a perturbation key is used to pseudo-randomly generate the sequence of perturbation operations that transform the given snapshot into the next snapshot in the sequence.

In Fig. 1(b), we find that the data owner performs the encoding process to obtain the heavily perturbed dataset along with the two perturbation keys, key_1 and key_2, used in the process. After the encoding process, only the two keys need to be shared by the data owner and the heavily perturbed dataset is published to all users, namely users A, B and C. Later, when the data needs to be accessed at a privilege level, the data owner shares the relevant perturbation keys with the data user. In the example shown in Fig. 1(b), the data owner shares key_2 with data user B and key_1 and key_2 with data user C. Data user B can use key_2 to remove the noise and perturbation pseudo-randomly injected during the encoding process. Similarly, using key_1 and key_2, data user C can de-perturb the dataset further to obtain the minimally perturbed snapshot. Thus, in the reversible data perturbation approach, the data owner only creates one perturbed version of the dataset and uses perturbation keys that significantly reduce the storage cost associated with multi-level data sharing.

2.2   Bipartite Association Graphs

We use bipartite association graph datasets as the candidate data in the proposed reversible perturbation techniques. Several private data in real world arise in the form of associations between entities such as the drugs purchased by patients in a pharmacy store or the movies rated by viewers in a movie rating database or the products purchased by buyers in an online shopping website [4,10,11]. Such associations are best captured as bipartite association graphs with nodes representing the entities (e.g., drugs and patients) and the edges correspond to the associations between them (e.g., Patient Bob purchased the Insulin drug). A bipartite graph can be represented as $BG = (V, W, E)$, which consists of $m = |V|$ nodes of a first type, $n = |W|$ of a second type and a set of edges $E \subseteq V \times W$. Thus, a bipartite graph can represent a set of two-node pairings, where a two-node pairing (a, b) represents an edge between node $a \in V$ and node $b \in W$.

2.3   Differential Privacy

Next, we define the notion of Differential privacy that we use in the reversible data perturbation approach. Differential privacy [7] is a state-of-the-art privacy paradigm that makes conservative assumptions about the adversary's background knowledge and protects a single individual's information in a dataset by considering adjacent datasets which differ only in one record.

Definition 1 (Differential privacy [7]). *A randomized algorithm $\mathcal{A}$ guarantees ϵ-differential privacy if for all adjacent data sets D_1 and D_2 differing by at most one record, and for all possible results $S \subseteq Range(\mathcal{A})$, $Pr[\mathcal{A}(D_1) = S] \leq e^\epsilon \times Pr[\mathcal{A}(D_2) = S]$.*

The conventional (individual) differential privacy protects the inference of a single individual's information in a dataset. For example, in a bipartite graph representing the associations between drugs and patients, such a single individual's protection may correspond to the inference of the graph edge representing a patient (e.g., 'Alice') purchasing a drug (e.g.,'Citalo'). For the purpose of protecting sensitive information of a group of individuals (e.g., the total number of cancer medicines purchased by patients in a specific neighborhood), differential privacy can be further extended to support group data protection based on the notion of group differential privacy [19].

Definition 2 (Group differential privacy [19]). *A randomized algorithm $\mathcal{A}$ guarantees ϵ_g-group differential privacy if for all adjacent data sets D_1 and D_2 differing by at most one group $G_i \in G$, and for all possible results $S \subseteq Range(\mathcal{A})$, $Pr[\mathcal{A}(D_1) = S] \leq e^{\epsilon_g} \times Pr[\mathcal{A}(D_2) = S]$.*

Group differential privacy protects sensitive aggregate information about groups of records using higher noise injection and perturbation. When records of a dataset are grouped into larger groups, the transformed dataset will provide coarser aggregate information and the privacy offered by group differential privacy will be stronger. Therefore, by grouping the records of a dataset into multiple granularity levels, different privacy levels can be offered by implementing group differential privacy at different granularity levels in the dataset. In our work, we employ both individual and group differential privacy in the proposed reversible data perturbation process to provide multi-level disclosure of the data using a single instance of the perturbed dataset.

3 Reversible Data Perturbation Techniques

In this section, we present the details of the reversible data perturbation process and illustrate the encoding and decoding steps involved in the data perturbation.

3.1 Overview of Dataset Encoding Process

The overall encoding phase in the reversible data perturbation process can be viewed as a sequence of permutation and noise injection steps. Figure 3 illustrates the process with an example bipartite graph dataset where the original version of the bipartite graph is shown as snapshot S_0, which consists of eight left (patient) nodes denoted by *PID*, eight right (medicine) nodes denoted by *MID* and eleven edges representing which medicine was purchased by which patient. To protect group differential privacy, the bipartite graph is first partitioned into multiple levels of subgraphs representing different granularity levels based on the taxonomy tree shown in Fig. 2. If such a taxonomy tree is not available apriori, the dataset corresponding to the bipartite graph can be partitioned through a sequence of specializations based on granular subgraph generation techniques such as the one presented in [19]. In the Fig. 3 example, at level $L_{2\times2}$, both the left and right nodes are grouped into groups of two nodes and thus it generates sixteen subgraphs. At level $L_{4\times4}$, nodes are grouped into groups of four nodes and therefore it generates four subgraphs. At level $L_{8\times8}$, nodes are grouped into groups of eight nodes, which results in a single graph. Based on the partitioning, dataset owners can choose to make a low perturbed snapshot, S_1 at $L_{2\times2}$, a moderately perturbed snapshot, S_2 at $L_{4\times4}$ and a heavily perturbed snapshot, S_3 at $L_{8\times8}$.

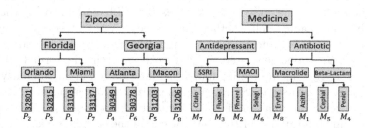

Fig. 2. Taxonomy trees

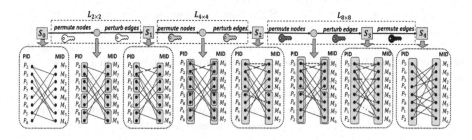

Fig. 3. Steps to encode a bipartite graph dataset

To generate each perturbed snapshot mentioned above, there is one step of (node) permutation followed by one step of (edge) perturbation. The purpose of node permutation is to ensure information generalization. For example, at snapshot S_0, left nodes P_2, P_3 and right nodes M_7, M_3 form a subgraph contained by $L_{2\times2}$, which has a single edge (P_2, M_3). Without node permutation, specific information in S_0, such as the edge (P_2, M_3) that indicates P_2 purchased M_3, can be viewed by users who only have privilege to view S_1 to learn generalized information about subgraphs at $L_{2\times2}$. In contrast, by permuting P_2, P_3 and also by permuting M_7, M_3 within their size-two groups, the label M_3 is swapped with M_7. Thus, instead of edge (P_2, M_3), a fake edge (P_2, M_7) indicating incorrect specific information is contained in S_1, whereas generalized information about the subgraph (e.g., one patient in Orlando purchased one SSRI medicine) is still maintained in S_1. This process is followed by the edge perturbation process which aims to prevent specific information to be inferred from the generalized information in the exposed snapshot. For example, after node shuffling, the subgraph between P_2, P_3 and M_3, M_7 shows generalized information that one patient in Orlando has purchased an SSRI medicine. It has four possibilities, namely (P_2, M_3), (P_2, M_7), (P_3, M_3) and (P_3, M_7). An adversary with some background knowledge may infer that (P_2, M_7), (P_3, M_3) and (P_3, M_7) cannot exist and therefore will be able to conclude the existence of edge (P_2, M_3) from the generalized information. To address this concern, edge perturbation can be used to perturb the edges of each subgraph based on randomized mechanisms (e.g., Laplace Mechanism [7]). In the example, users receiving S_1 can also view the injected edge (P_3, M_3) and thus learn that two patients in Orlando purchased SSRI medicine. Since the noise injection is based on differential privacy, the users viewing this data cannot tell whether the two edges are true edges or injected edges and therefore feel uncertain to conclude the existence of (P_2, M_3). After the two steps, S_1 can be generated, which reveals generalized information about subgraphs at $L_{2\times2}$ while protecting individual information in S_0. Similarly, after another round of two steps, S_2 reveals generalized information of $L_{4\times4}$ while protecting specific information of $L_{2\times2}$. Similarly, S_3 reveals $L_{8\times8}$ information while protecting information of $L_{4\times4}$. At the end of the encoding phase, if S_3 still contains sensitive information about $L_{8\times8}$ that the data owner is not willing to share to all possible users, edge permutation can be executed over S_3 to permute all the edges in S_3 so that the obtained S_4 is safe for disclosure.

To generate snapshot S_i from S_{i-1}, a *perturbation key* is used to first pseudo-randomly permute the two sides of nodes of each subgraph and then pseudo-randomly perturb edges within each subgraph. Also, edge permutation at the last step is pseudo-randomly implemented using a perturbation key. Next, we will show how to use perturbation keys to perform edge perturbation, node permutation and edge permutation steps in the reversible perturbation approach so that legitimate users can use perturbation keys to decode S_4 to any previous snapshots (e.g., S_0, S_1, S_2, S_3) containing finer information.

3.2   Reversible Edge Perturbation

For each subgraph, the reversible edge perturbation step first uses the perturbation key to pseudo-randomly sample a noise using the Laplace Mechanism [7]. When the sampled noise is positive, the procedures of noise injection and noise removal are performed as shown in Algorithms 1 and 2 respectively. During noise injection, the number of injected edges is sampled from Laplace pseudo-random value generator with mean 0, variance $\triangle f/\epsilon$ and seed K, where K is the key (line 1). Then, during each loop (line 3–12), two pseudo-random numbers are used to select one left node and one right node from the subgraph to form a new edge ne (line 4). If ne is not an existing edge, its selection will be confirmed (line 5–7); otherwise, this iteration will be skipped to avoid collision and this skipped index will be recorded into a list that will be attached with the key to be used during the decoding process later (line 8–10). The algorithm

---

Algorithm 1. Noise injection

Input : Bipartite graph $BG = (V, W, E)$, sensitivity $\triangle f$, budget ϵ, key K.
Output: Perturbed bipartite graph $\widetilde{BG}$.
1 $n = \lfloor LaplaceRandom(0, \triangle f/\epsilon, K) \rfloor$;
2 Initialize counter $c = 0$, index $i = 0$, new edge recorder $\overline{NE}$, skipped index recorder SI;
3 **while** $c < n$ **do**
4      $ne = (rand(2i, K) \bmod |V|, rand(2i + 1, K) \bmod |W|)$;
5   **if** $ne \notin E \cup \overline{NE}$ **then**
6          $\overline{NE} \leftarrow ne$; $c + +$;
7   **end**
8   **else**
9          $\overline{SI} \leftarrow i$;
10   **end**
11   i++;
12 **end**
13 $\widetilde{BG} = (V, W, E \cup \overline{NE})$;

Algorithm 2. Noise removal

Input : Perturbed bipartite graph $\widetilde{BG}$, sensitivity $\triangle f$, budget ϵ, key K,
           skipped index recorder SI.
Output: Bipartite graph BG.
1 $n = \lfloor LaplaceRandom(0, \triangle f/\epsilon, K) \rfloor$;
2 Initialize index $i = 0$;
3 **while** $i < n + |SI|$ && $i \notin SI$ **do**
4      $re = (rand(2i, K) \bmod |V|, rand(2i + 1, K) \bmod |W|)$;
5   Remove edge re from $\widetilde{BG}$;
6   i++;
7 **end**
8 $BG = \widetilde{BG}$;

---

complexity is $O(n)$. After this process, legitimate users can receive the perturbation key to reversibly remove the injected noise using Algorithm 2. With the same seed K, same n can be generated (line 1), which can then select and remove the same sequence of edges with assistance of SI (line 3–7). The complexity of this algorithm is $O(n + |SI|)$.

However, when noises are negative, instead of using $|SI|$ to record the skipped iterations, we need to record all removed edges using the perturbation key.

3.3 Reversible Node Permutation

For each subgraph, the reversible node permutation step uses the perturbation key to pseudo-randomly shuffle node labels (e.g., PID, MID) during the encoding phase and later uses the same key to recover their order during the decoding process. We

---

Algorithm 3. Node permutation: encoding

    Input   : Bipartite graph $BG = (V, W, E)$, key K.
    Output: Permuted bipartite graph $\overline{BG}$.
1 $R = PseudoRandom(K)$;
2 **for** $i = 0; i < |V|; i + +$ **do**
3 | Swap node $V[i]$ and node $V[R[i] \bmod |V|]$;
4 **end**
5 **for** $i = |V|; i < |V| + |W|; i + +$ **do**
6 | Swap node $W[i - |V|]$ and node $W[R[i] \bmod |W|]$;
7 **end**

Algorithm 4. Node permutation: decoding

    Input   : Permuted bipartite graph $\overline{BG} = (V, W, E)$, key K.
    Output: Bipartite graph BG.
1 $R = PseudoRandom(K)$;
2 **for** $i = |V| - 1; i \geq 0; i - -$ **do**
3 | Swap node $V[i]$ and node $V[R[i] \bmod |V|]$;
4 **end**
5 **for** $i = |V| + |W| - 1; i \geq |V|; i - -$ **do**
6 | Swap node $W[i - |V|]$ and node $W[R[i] \bmod |W|]$;
7 **end**

---

show its procedures in the encoding phase and decoding phase in Algorithms 3 and 4 respectively.

During the encoding phase, the perturbation key is used as a seed in the pseudo-random stream generator to generate a sequence of pseudo-random numbers denoted as R (line 1). Then, the first $|V|$ pseudo-random numbers in R are used to shuffle the left nodes in BG (line 2–4) while the pseudo-random numbers generated later, namely $|W|$ are used to shuffle the right nodes (line 5–7). Each pseudo-random number swaps two left (right) nodes. The first node is selected from top to bottom along with its position while the second node is pseudo-randomly selected by the pseudo-random number using modular arithmetic. At the end of Algorithm 3, both left nodes and right nodes are shuffled in a reversible manner. Later, during decoding phase, given the same key, the same R can be obtained (line 1). The same two groups of pseudo-random numbers in R are used to recover left nodes (line 2–4) and right nodes (line 5–7) respectively. The process within each loop is quite similar to that of the encoding process. However, instead of starting from top to bottom, during decoding process, the loops start from bottom to top with a reverse order so that operations implemented during encoding can be reversibly implemented during decoding, which results in the recovery of the original subgraph. Here, both the algorithms have a complexity of $O(|V| + |W|)$.

In Fig. 4, the labels of the nodes are permuted through reversible node permutation while the edges are permuted through reversible edge permutation (to be discussed later). In the example, Alice uses a perturbation key as a seed to the pseudo-random stream generator to get a sequence of pseudo-random numbers R. Then, the first six pseudo-random numbers (assumed to be $[35, 18, 46, 12, 27, 57]$) and second six pseudo-random numbers (assumed to be $[7, 18, 24, 29, 62, 67]$) in R are used to shuffle the left and right nodes of the bipartite graph respectively. Based on Algorithm 3, the first pseudo-random number $R_1 = 35$ swaps P_2 and P_3, followed by 18 swapping P_3 and P_6, 46 swapping P_4 and P_5, 12 swapping P_6 and P_5, 27 swapping P_4 and P_6, 57 swapping P_2 and P_4. As a result, left nodes in the left bipartite graph are permuted to the order in the right bipartite graph. Then, in Fig. 5, Bob gets the permutation key from Alice and uses the key as a seed and generates the same R as generated by Alice. Among the first six pseudo-random numbers $[35, 18, 46, 12, 27, 57]$, $R_6 = 57$ is first picked to swap P_2 and P_4, followed by 27 swapping P_4 and P_6, 12 to swapping P_6 and P_5 and so on. Therefore, the original order of the left nodes can be recovered.

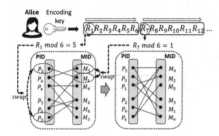

Fig. 4. Encoding

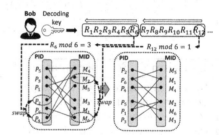

Fig. 5. Decoding

3.4   Reversible Edge Permutation

Edge permutation is implemented as the last step in the encoding phase and therefore it represents the first step during the decoding phase. The edges of the bipartite graph are represented using an adjacency matrix. For example, the adjacency matrix of the left bipartite graph in Fig. 4 is shown as the matrix E below, where the first row represents that P_2 is linked with M_1. Here, the edges can be shuffled by simply permuting the adjacency matrix.

$$E = \begin{bmatrix} 0 & 1 & 0 & 0 & 0 & 0 \\ 1 & 0 & 0 & 1 & 0 & 0 \\ 0 & 0 & 1 & 0 & 0 & 1 \\ 0 & 0 & 0 & 0 & 1 & 0 \\ 0 & 1 & 0 & 1 & 0 & 0 \\ 0 & 0 & 0 & 1 & 0 & 0 \end{bmatrix} \quad \overline{E} = \begin{bmatrix} 1 & 0 & 0 & 0 & 0 & 1 \\ 0 & 0 & 0 & 1 & 0 & 0 \\ 0 & 1 & 0 & 0 & 1 & 0 \\ 1 & 0 & 0 & 0 & 0 & 0 \\ 0 & 0 & 0 & 0 & 0 & 1 \\ 1 & 0 & 1 & 0 & 0 & 0 \end{bmatrix}$$

The encoding and decoding parts of the process are shown in Algorithms 5 and 6 respectively. Similar to node permutation, in both the algorithms, same

R can be obtained through the perturbation key (line 1). Then, we use the first $|V||W|$ pseudo-random numbers in R to perform $|V||W|$ rounds of swap operation (line 2–4). Each time, the first edge is selected based on a fixed order (top to bottom and left to right in Algorithm 5, right to left and bottom to top in Algorithm 6) and the second edge is pseudo-randomly selected by the pseudo-random number using modular arithmetic. In this way, by reversibly performing the swap operation during the decoding phase, the original order of the edges can be recovered. Here, the algorithms have a complexity of $O(|V||W|)$.

Algorithm 5. Edge permutation: encoding

Input   : Bipartite graph $BG = (V, W, E[|V|][|W|])$, key K.
Output: Permuted bipartite graph $\overline{BG}$.
1 $R = PseudoRandom(K)$;
2 **for** $i = 0; i < |V||W|; i + +$ **do**
3 $\quad$ Swap edge $E[\lfloor\frac{i}{|W|}\rfloor][i \bmod |W|]$ and edge
$\quad E[\lfloor\frac{R[i] \bmod |V||W|}{|W|}\rfloor][R[i] \bmod |V||W|) \bmod |W|]$;
4 **end**

Algorithm 6. Edge permutation: decoding

Input   : Permuted bipartite graph $\overline{BG} = (V, W, E[|V|][|W|])$, key K.
Output: Bipartite graph BG.
1 $R = PseudoRandom(K)$;
2 **for** $i = |V||W| - 1; i \geq 0; i - -$ **do**
3 $\quad$ Swap edge $E[\lfloor\frac{i}{|W|}\rfloor][i \bmod |W|]$ and edge
$\quad E[\lfloor\frac{R[i] \bmod |V||W|}{|W|}\rfloor][R[i] \bmod |V||W|) \bmod |W|]$;
4 **end**

In Fig. 4, if the first and second pseudo-random numbers generated by a key are 53 and 71, we first use 53 to swap $E[\lfloor\frac{0}{6}\rfloor][0 \bmod 6] = E[0][0]$ and $E[\lfloor\frac{53 \bmod 36}{6}\rfloor][(53 \bmod 36) \bmod 6] = E[2][5]$. Then, we use 71 to swap $E[0][1]$ and $E[5][5]$. By repeating this for all the 36 pseudo-random numbers, the adjacency matrix can be transformed as $\overline{E}$ to represent the right bipartite graph in Fig. 4.

4   Experimental Evaluation

In this section, we present the experimental study on the performance of the proposed reversible data perturbation techniques. We first briefly describe the experimental setup.

4.1   Experimental Setup

Our experiment setup was implemented in Java with an Intel Core i7 2.70 GHz 16 GB RAM PC. The bipartite graph dataset used in this work is the MovieLens dataset [10] which consists of 6,040 users (left nodes), 3,706 movies (right nodes) and 1,000,209 edges describing rating of movies made by users.

4.2 Experimental Results

The experimental results are organized into three parts. First, we evaluate the performance efficiency of the three key components of the reversible perturbation process separately, namely edge perturbation, node permutation and edge permutation. Then, we integrate the three components and evaluate the performance of the complete reversible data perturbation process. In our experiments, we generate three granularity levels and evaluate the time and space consumption for each granularity level during encoding and decoding phases. Finally, we evaluate the utility and privacy protection offered by the data perturbation process. We demonstrate that the noise injected for protecting differential privacy does not substantially reduce the utility of the data.

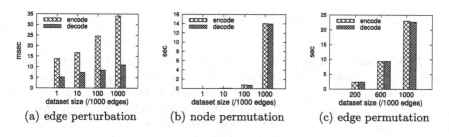

(a) edge perturbation   (b) node permutation   (c) edge permutation

Fig. 6. Algorithm performance

Algorithm Performance. The first set of experiments evaluates the performance efficiency of edge perturbation, node permutation and edge permutation separately. We evaluate the scalability of these algorithms by varying the size of dataset and we measure the time taken for their execution both in encoding and decoding phases. In Fig. 6(a), edge perturbation is evaluated. The dataset size is changed from one thousand edges to one million edges. Specifically, the one-million-edge dataset represents the entire MovieLens dataset. The results show that both noise injection (encode) and removal (decode) processes have significantly low time consumption cost and demonstrate high scalability. Even when the dataset size increases 1000 times, the time consumption increases only by a factor of 2. For a dataset with one million edges, the noise injection and removal processes cost only about 35 ms and 10 ms respectively. Here, compared with noise injection, the noise removal process usually has a lower time consumption. This is because the process of noise removal employs some meta data information attached to the perturbed dataset which significantly accelerates its speed of execution. Next, in Fig. 6(b), we evaluate the node permutation process with the same experiment setting. Unlike edge perturbation, although the time consumption of node permutation is significantly small for small datasets, it becomes acceptably larger for the one-million-edge dataset, which is about 14 s. Finally, in Fig. 6(c), we measure the time taken by the edge permutation process using dataset sizes that vary from 0.2 million edges to 1 million edges. The results show that the time taken by the process for the one-million-edge dataset is about 23 s, which is quite acceptable as the edge shuffling process is only required to be run once during the entire process.

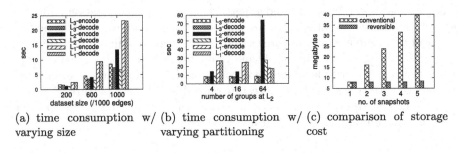

(a) time consumption w/ varying size

(b) time consumption w/ varying partitioning

(c) comparison of storage cost

Fig. 7. Multi-level performance

Multi-level Performance. The second set of experiments evaluate the performance of the multiple levels of perturbation during the process. In this part, we processed the dataset to generate three granularity levels of subgraphs, denoted as L_1, L_2 and L_3 respectively. We applied the partitioning algorithm proposed in [19] to generate the granularity levels. The algorithm runs several rounds of specializations and each specialization can partition a bipartite graph into four non-overlapping subgraphs. Therefore, after n rounds of specializations, the original bipartite graph has been partitioned to 4^n non-overlapped subgraphs. In this experiment, we use the MovieLens dataset and we consider the entire graph as level L_1, the 16 (4^2) subgraphs generated by two specializations as level L_2 and the 256 (4^4) subgraphs generated by four specializations as level L_3. For ease of understanding, L_1, L_2 and L_3 can be considered to roughly correspond with $L_{8\times8}$, $L_{4\times4}$ and $L_{2\times2}$ in the example of Fig. 3. In Fig. 7(a), we evaluate the encoding and decoding time for each granularity level when the dataset size is varied from 0.2 million edges to 1 million edges. As can be seen, as the dataset size increases, the time taken by all the three granularity levels also show a reasonable increase. Level L_3 needs to run edge perturbation and node permutation over the 256 subgraphs. Due to the very small subgraph size and the low sensitivity for protecting differential privacy for individual edges, L_3 has the lowest time consumption. At level L_2, although the number of subgraphs reduces to 16, the corresponding increase in subgraph size makes its time consumption higher than that of L_3 for large dataset size. Finally, the time consumption of level L_1 is dominated by edge permutation, which follows the same trend as shown in Fig. 6(c). In Fig. 7(b), we fix the dataset size as one million edges while changing the number of subgraphs at level L_2 from 16 to 4 and 64. This change at L_2, as shown in Fig. 7(b), has no influence on the results of L_3. The reduction from

16 to 4 makes an increase for both L_2 and L_1 while the increase from 16 to 64 makes results at L_2 significantly increased and results at L_1 obviously decreased. These results show that instead of the average size of subgraphs, the time consumptions of granularity levels are much more sensitive to the amount of the injected noises. Finally, in Fig. 7(c), we compare the storage cost required by conventional framework and reversible framework. Based on Fig. 3, three granularity levels can generate at most five snapshots. As can be seen, using the conventional framework, the storage cost is linearly increased with the number of generated snapshots as data owner needs to store all of them. However, the proposed reversible framework efficiently employs the use of perturbation keys to allow all snapshots to be recovered from a single published snapshot protected with the highest privacy level. Thus, the data owner only needs to store one snapshot. The size of the perturbation keys and the stored metadata for noise injection have little influence on the overall storage cost.

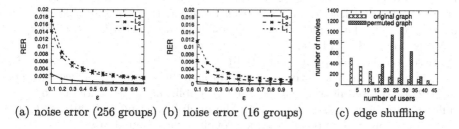

(a) noise error (256 groups) (b) noise error (16 groups)   (c) edge shuffling

Fig. 8. Utility and security

Utility and Security. In the final set of experiments, we evaluate the utility and privacy offered by the reversible multi-level data perturbation scheme. In the edge perturbation process, a large quantity of noise may have an impact on the utility of published dataset. To evaluate the data utility, we measured the relative error rate (RER) that represents the ratio of the sum of the error caused by the noise in each subgraph of a given level and the overall number of edges in the original bipartite graph. In Fig. 8(a), we measure RER of the three granularity levels when L_2 and L_3 have 16 and 256 subgraphs respectively. As can be seen, when the privacy budget ϵ is 1, RER of all the three levels is very small. Even if ϵ is decreased to a very strict value 0.1, the highest RER 0.017 appears at L_1, which is still acceptable. In Fig. 8(b), we reduce number of subgraphs at L_2 and L_3 to 4 and 16 respectively. The results show that this reduction makes RER of all the three levels lower. Finally, in Fig. 8(c), we evaluate the effectiveness of the edge permutation process by measuring the distribution of the degree of nodes before and after edge permutation. As can be seen, the distribution in the perturbed graph after edge permutation is substantially different than that of the original graph. This makes it harder to infer useful or true information after the edge permutation process.

5   Related Work

The problem of information disclosure has been studied extensively in the framework of statistical databases. Samarati and Sweeney [21,22] introduced the k-anonymity approach which has led to some new techniques and definitions such as l-diversity [17] and t-closeness [16]. There had been some work on anonymizing graph datasets with the goal of publishing statistical information without revealing information of individual records. Backstrom et al. [2] show that in fully censored graphs where identifiers are removed, a large enough known subgraph can be located in the overall graph with high probability. The safe grouping techniques proposed in [5] consider the scenario of retaining graph structure but aim at protecting privacy when labeled graphs are released. But, as mentioned earlier, these existing schemes have been focused on publishing a single instance of the perturbed dataset with a fixed privacy level without considering the requirements of multiple access levels. The key focus of this work is on developing a reversible data perturbation approach for bipartite association graph data that can facilitate the release of multiple levels of information using a single instance of the perturbed data, similar to the reversible location perturbation techniques recently proposed in [14,15].

Based on the concept of differential privacy [7], there had been many works focused on publishing sensitive datasets through differential privacy constraints [6,9,12,20,23]. Recent work had focused on publishing graph datasets through differential privacy constraints so that the published graph maintains as many structural properties as possible while providing the required privacy [20]. However, these existing schemes do not support multi-level access to the published dataset. The notion of group differential privacy and granular subgraph generation algorithms for bipartite graphs is recently introduced in [18,19]. This paper extends the work presented in [18,19] with a suite of reversible data perturbation techniques that provides a more scalable and cost-effective solution to releasing bipartite association graph data at multiple privacy levels using a single instance of the perturbed dataset.

6   Conclusion

Privacy-preserving data publishing techniques are critical for protecting sensitive information in published datasets. Existing solutions have focused on publishing a single snapshot of the perturbed dataset that offers a fixed privacy level with the assumption that all users of the data share the same privilege level to access it. In cases when data users have different levels of access on the published data, such schemes will require multiple snapshots corresponding to different privacy levels to be published and maintained, resulting in higher storage cost for the data. In this paper, we develop a set of reversible data perturbation techniques for large bipartite association graphs that use perturbation keys to control the sequential generation of multiple snapshots of the perturbed data to offer multi-level access to the data based on privacy levels. To support multi-level privacy, the proposed techniques require only a single snapshot of the data to be maintained which significantly reduces the storage cost. We evaluate the proposed reversible data perturbation techniques through experiments on a real large bipartite association graph dataset. The experiments demonstrate that the proposed techniques are scalable, effective and efficiently support multi-level data access using a single snapshot of the perturbed data.

References

1. Bigdata and future of privacy. https://epic.org/privacy/big-data/
2. Backstrom, L., et al.: Wherefore art thou r3579x?: anonymized social networks, hidden patterns, and structural steganography. In: WWW, pp. 181–190 (2007)
3. Batty, M.: Big data, smart cities and city planning. Dialog. Hum. Geogr. **3**(3), 274–279 (2013)
4. Celma, O.: Music Recommendation and Discovery in the Long Tail. Springer, Heidelberg (2010). https://doi.org/10.1007/978-3-642-13287-2
5. Cormode, G., et al.: Anonymizing bipartite graph data using safe groupings. Proc. VLDB Endow. **1**(1), 833–844 (2008)
6. Day, W.-Y., Li, N., Lyu, M.: Publishing graph degree distribution with node differential privacy. In: ICMD, pp. 123–138. ACM (2016)
7. Dwork, C., McSherry, F., Nissim, K., Smith, A.: Calibrating noise to sensitivity in private data analysis. In: Halevi, S., Rabin, T. (eds.) TCC 2006. LNCS, vol. 3876, pp. 265–284. Springer, Heidelberg (2006). https://doi.org/10.1007/11681878_14
8. Fan, W., Bifet, A.: Mining big data: current status, and forecast to the future. ACM SIGKDD Explor. Newsl. **14**(2), 1–5 (2013)
9. Friedman, A., Schuster, A.: Data mining with differential privacy. In: SIGKDD, pp. 493–502. ACM (2010)
10. Harper, F.M., Konstan, J.A.: The movielens datasets: history and context. ACM TIIS **5**(4), 19 (2016)
11. He, R., et al.: Ups and downs: modeling the visual evolution of fashion trends with one-class collaborative filtering. In: WWW, pp. 507–517 (2016)
12. Karwa, V., et al.: Private analysis of graph structure. Proc. VLDB Endow. **4**(11), 1146–1157 (2011)
13. Kasiviswanathan, S.P., Nissim, K., Raskhodnikova, S., Smith, A.: Analyzing graphs with node differential privacy. In: Sahai, A. (ed.) TCC 2013. LNCS, vol. 7785, pp. 457–476. Springer, Heidelberg (2013). https://doi.org/10.1007/978-3-642-36594-2_26
14. Li, C., Palanisamy, B.: De-anonymizable location cloaking for privacy-controlled mobile systems. Network and System Security. LNCS, vol. 9408, pp. 449–458. Springer, Cham (2015). https://doi.org/10.1007/978-3-319-25645-0_33
15. Li, C., Palanisamy, B.: ReverseCloak: protecting multi-level location privacy over road networks. In: ACM CIKM, pp. 673–682. ACM (2015)
16. Li, N., Li, T., Venkatasubramanian, S.: t-Closeness: privacy beyond k-anonymity and l-diversity. In: ICDE, pp. 106–115. IEEE (2007)
17. Machanavajjhala, A., et al.: l-diversity: privacy beyond k-anonymity. In: ICDE, p. 24. IEEE (2006)
18. Palanisamy, B., Li, C., Krishnamurthy, P.: Group differential privacy-preserving disclosure of multi-level association graphs. In: ICDCS, pp. 2587–2588. IEEE (2017)

19. Palanisamy, B., Li, C., Krishnamurthy, P.: Group privacy-aware disclosure of association graph data. In: IEEE Big Data (2017)
20. Sala, A., et al.: Sharing graphs using differentially private graph models. In: SIG-COMM, pp. 81–98. ACM (2011)
21. Samarati, P.: Protecting respondents identities in microdata release. IEEE Trans. Knowl. Data Eng. **13**(6), 1010–1027 (2001)
22. Sweeney, L.: k-anonymity: a model for protecting privacy. Int. J. Uncertain. Fuzziness Knowl.-Based Syst. **10**(05), 557–570 (2002)
23. Wang, Q., et al.: RescueDP: Real-time spatio-temporal crowd-sourced data publishing with differential privacy. In: INFOCOM, pp. 1–9. IEEE (2016)
24. Xindong, W., et al.: Data mining with big data. IEEE Trans. Knowl. Data Eng. **26**(1), 97–107 (2014)

Real-Time Analysis of Big Network Packet Streams by Learning the Likelihood of Trusted Sequences

John Yoon[1(✉)] and Michael DeBiase[1,2]

[1] Mercy College, Dobbs Ferry, NY 10522, USA
jyoon@mercy.edu, mdebiase3@mercymavericks.edu
[2] IT Department, Westchester County Government,
White Plains, NY 10601, USA

Abstract. Deep Packet Inspection (DPI) is a basic monitoring step for intrusion detection and prevention, where the sequences of packed packets are to be unpacked according to the layered network structure. DPI is performed against overwhelming network packet streams. By nature, network packet data is big data of real-time streaming. The DPI big data analysis, however are extremely expensive, likely to generate false positives, and less adaptive to previously unknown attacks. This paper presents a novel machine learning approach to multithreaded analysis for network traffic streams. The contribution of this paper includes (1) real-time packet data analysis, (2) learning the likelihood of trusted and untrusted packet sequences, and (3) improvement of adaptive detection against previous unknown intrusive attacks.

Keywords: Network traffic packet streams · Multithreading data analysis
Intrusion detection and prevention · Genetic algorithmic fitness

1 Introduction

Deep Packet Inspection (DPI) is a basic monitoring step in intrusion detection systems (IDSs), where the sequences of packed packets are to be unpacked according to the layered network structure. For example, IDSs aim at unpacking network packets for the IP or port number (which was packed at the network layer [1]) of source and destination computers to identify potential threats. They also unpack the transport layer protocols, such as TCP, UDP, TLS, HTTP, etc. to identify the session connection of malicious attacks. At the application layer, part of session data pieces (e.g., login and password credentials) are posed to access an application. In addition to the network structure data, payload contents may be fed into an application as needed. Both header data about network structures and payload contents will then be analyzed for the verification all the way up in the application layer [2–4].

DPI has been increasingly employed and visualized in IDSs. One of the widely used software packages is Wireshark [5]. Although Wireshark provides the filtering expression, it does not provide any additional analytic capabilities for massive real-time streaming data process. Automatic network packet filtering processes are embedded in

© Springer International Publishing AG, part of Springer Nature 2018
F. Y. L. Chin et al. (Eds.): BIGDATA 2018, LNCS 10968, pp. 43–56, 2018.
https://doi.org/10.1007/978-3-319-94301-5_4

routers and firewalls, which are either hardware devices or software packages. They determine whether network packets are accepted to proceed or deny. IP's are used to specify the block rules, and block rules are determined based on country code [6]. Once known, the state of active connections may or may not be stored to be used by a next step of DPI processes. We call them *stateful* inspection or *stateless* inspection depending on availability of the connection states. In stateful or stateless packet traffic inspection, denial or permission lead to non-negligible false-positives and false-negatives. Consider the following motivating examples:

Motivating Example 1 (Real-time vulnerability detection of packet streams). Fast incoming network packets are unpacked to identify their IP's. Network packets are unpacked at once for all network (session, transportation and network) layers, and the decision is made based on a rule set. Figure 1 illustrates this.

- Although denial of an IP can be made way before unpacking port numbers or session protocols, network equipment unpacks network packets for all layers of header data first, then followed by decision rules that are executed against unpacked header data. Note that the cells contain the data that can be captured by typical software packages, e.g., Wireshark [5], and IDSs determine the decision, which appears in the last column of the table below. The packet header data in the cells shaded are the data that does not need to be captured since the data in the cells with no shade are known to be enough for decision making. Consider the following packet sequences, which are all incoming.

Pckt Seq	Src IP	Dest IP	Protocol	Src Port	Dest Port	Action Result
p_1	92.168.3.2	172.16.2.1	TCP	72	25	Deny
p_2	96.121.7.1	145.5.7.9	HTTP	80	435	Permit
P_3	96.121.7.1	145.5.7.9	HTTP	217	512	Deny
P_4	96.121.7.1	145.5.7.9	UDP	217	512	Deny

Suppose that the rule set is defined in a firewall system: All packets requesting for HTTP protocol are permitted except the TCP request from 92.168.3.2.

Rule no	Direction	Src IP	Dest IP	Protocol	Src port	Dest port	Decision
1	In	92.168.3.2	172.16.2.1	Any	Any	Any	Deny
2	In	96.121.7.1	Any	UDP	Any	Any	Deny
3	In	Any	Any	HTTP	Any	435	Permit

According to the ruleset above (particularly Rule No 1), p_1 (or Packet Sequence 1) does not need to unpack session protocols and ports (highlighted) since Src IP is prohibited. p_4 does not need to unpack the port (highlighted) since IPs and protocols are known to be vulnerable due to Rule No 1. p_2 and p_3 needs to unpack for all header data to permit due to unless otherwise its vulnerability cannot be detected. p_2 is

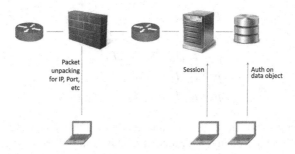

Fig. 1. Network packet analysis, consisting of (1) packet unpacking, and (2) access control. Note that all packet header data are unpacked at once and never to be referenced by applications.

permitted according to Rule No 3, while p_3 is denied according to Rule No 4. It is desired that if a vulnerability of packet is known, unpacking should stop.

Motivating Example 2 (Once attacked, more often unrecoverable). In this new technological era, once an infrastructure is attacked, its damage will be disastrous and its impacts will become paramount, and may be irrecoverable. Packets should be denied in advance if its vulnerability is highly likely. An early decision over real-time network packet streams is very hard, but it is undoubtedly an imperative task.

- Consider a packet, which is allowed to move into the systems all the way to an application system, such as a database storage through a web server. As illustrated in p_2 and p_5, they are continuous attempts: p_2 was permitted by network equipment, but p_5 was denied when the packet is to intrude into a database as a database administrator (DBA). Consider the following table, which illustrates packet sequences.

Pckt Seq	Src IP	Dest IP	Protocol	Src Port	Dest Port	...	Web Session	Database Session Role	Action Result
p_1	92.168.3.2	172.16.2.1	TCP	72	25				Deny
p_2	96.121.7.1	145.5.7.9	HTTP	80	435				Permit
p_3	96.121.7.1	145.5.7.9	HTTP	217	512				Deny
p_4	96.121.7.1	145.5.7.9	UDP	217	512				Deny
p_5	96.121.7.1	145.5.7.9	HTTP	80	435	...	GET	DBA	Deny

This table can be simplified as follows. I, T, P, S and R denote IP, Port, Protocol, Session data into a host server and session Role to an application software, respectively. Subscripts s, d denote source and destination, respectively. If subscript numbers are the same, they denote the same value. For example, I_{1s} and I_{1d} are the same IP: one for source and the other for destination.

$$P_1 : I_{1s}, I_{1d}$$
$$p_2 : I_{2s}, I_{2s}, T_{1s}, T_{2d}, P_1$$
$$p_3 : I_{2s}, I_{2s}, T_{3s}, T_{4d}, P_1 \tag{1}$$
$$p_4 : I_{2s}, I_{2s}, T_{3s}, T_{4d}$$
$$p_5 : I_{2s}, I_{2s}, T_{1s}, T_{2d}, P_1, S_1, R_1$$

If it is learned that p_5 is vulnerable and likely eventually be denied, a high tech solution will be able to deny p_2 in advance. This is illustrated in Fig. 1. No header data states of packets are propagated from network equipment to webservers and all the way to the backend storage, although the data states are useful to access control and authorization at backend servers.

To overcome the difficult task and to improve network monitoring systems as illustrated in the motivating examples, this paper proposes the following approaches:

- **Multithreading Packet Analysis.** As illustrated in Motivating Example 1, real-time packet streams are unpacked for each packet header data at each network layer [1]. We employ a multithreading packet analysis process to triage packet header data locally at a corresponding network layer and so some of packet header data is processed locally, reducing the time consumed for analysis/deep-learning time to the next threads.
- **Genetic Algorithmic Best-Fitting.** As illustrated in Motivating Example 2, we employ a genetic algorithmic approach to select the best fit to suspicious network packets, to provide the fitness score for the highest likelihood of prospective attacks. In addition to the best or the worst fits to known packet sequences, this paper takes into consideration the mutant packets which occur very rarely but non-negligibly [7].

The proposed packet analysis is illustrated in Fig. 2. Authorization requests are able to share the header data states of packets, and session contexts available from servers. Stateful DPI over multithreading architecture is proposed. The contribution of this paper includes (1) improving analysis time of the big data from network traffic packet

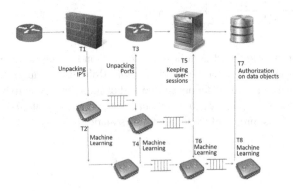

Fig. 2. Proposed packet analysis over multithreading and access control using states from packet header data. Note that multithreads perform to reduce bottlenecks

streams; (2) reducing false positives in fast and massive network packet processing; and (3) early detection of vulnerabilities.

The remainder of this paper is organized as follows: Sect. 2 describes background knowledge on (1) unpacking network traffic packets, and (2) genetic algorithms. Section 3 describes how multithreading architecture to speed up the intrusion detection process from the big data of network packet streaming. Section 4 describes a genetic algorithmic approach to identify a malicious sequence pattern of network packet streams. Section 5 describes our preliminary experiment results. Section 6 concludes our research work.

2 Background and Related Work

Network packets are analyzed by hardware, e.g., network equipment such as routers, firewall or switches, or by software. Both hardware and software network analyzers capture and log network traffic that passes over wired network or wireless network. Since network packets are fast streaming, the network analyzers unpack the packet streams to identify header data, and apply a filtering rule set to make a quick decision of either permit or denials of their access requests. The process of packet capturing and packet unpacking can be reasonably fast but not enough to deal with big packet streams, typical IDSs apply a ruleset to the header data. The ruleset is preset and its execution is against the entire header data of each packet.

Network packets can be monitored in GUI [5] or their vulnerabilities are exploited by command-line expression [8]. An extension of basic monitoring systems for network packets is a deep packet inspection (DPI). Since DIP is very expensive and its processing time is unacceptably slow, using a classification algorithm an improvement of DIP processing time is proposed. Network packets are classified to characterize the network traffic and to improve its analysis [3]. The use of regular expressions is another approach [2].

Its efforts are extended to a few other areas: DPI techniques are extended to a specific protocol such as HTTP and HTTPS over transport layer security (TLS) [4]. Session data particularly session IDs, which can be acquired in TLS, are kept to improve the revisited sessions [10], e.g., web services over TLS and universal secured socket layer (SSL). DPI techniques are also extended to take a few other issues into consideration. Privacy is a concern while DPI is performed. Since DPI is used to detect attacks and potential security breaches, a limited connection environment is proposed to preserve privacy during DPI [9]. However, this approach does not improve the quality of early detection of packet vulnerabilities.

To improve the detection accuracy of potential attacks or threats, and to identify unforeseen vulnerabilities in advance, machine learning and data mining techniques have been discussed. Scalable Programmable Packet Processing Platform (SP4) has been proposed and demonstrated to filter out unwanted traffic and detect DDoS attacks by using Support Vector Machine algorithm [11].

There are two major techniques in deep machine learning: Artificial neural network (ANN), and genetic algorithm. Genetic algorithm (GA) is an algorithm, introduced by John Holland at University of Michigan in 1970s, that mimics some of the processes observed in biological evolution and provides the steps to compute the biological evolutionary processes [12]. GA simulates the survival of the fittest among objects (or things or *solutions*) over the course of generation to generation evolutions. Similar to the evolution in the chromosome of DNA, we have (1) to define objects (or solutions) to represent the problems to solve. Those solutions are then (2) populated, from which a few solutions are (3) selected. Selection is made based on a fitness function, depending on application domains, which indicates the champions may produce off-spring with better fitness. Only those objects (or solutions) most successful in each selection will produce more offspring. Meanwhile, (4) GA operations can be applied as shown in Fig. 3.

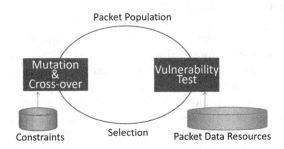

Fig. 3. General architecture of genetic algorithms

GA applies two major operations: mutation [13] and crossover [14]. These GA operators are applied to population of objects (or solutions) or their offspring. Similar to biological evolution process, the mutation or crossover operators take place randomly. When a mutation occurs, it occurs only in a small portion of an object (or solutions). For example, a sequence of DNA in unicellular bacteria may have 0.003 mutations per generation. The mutation rate depends on specious. Similarly, in DPI, the behavior of network attacks may change radically (say x days of attempt).

GA mutations can be a change/modification, an insertion of deletion that takes place at a smaller portion of a solution or an object. It is recommended that the rate of GA crossover is higher than the one for GA mutation. According to the recommendation [15, 16], the crossover rate is about 100 times the mutation rate. It is very likely that if there is active intrusion attacks on a particular IP and port, and if we see the sequence of a combination of attack tools, there are frequent crossover intrusion by starting with one tool and crossing over with attack tools.

Note that in Fig. 3, the data resource can be online real-time network intrusion data including historical data, which is downloaded by Python script in our experimentation. Constraints can be positive or negative impacts from internal and external sources. For example, mal-forecasted IP vulnerabilities, which is related with country, may have a negative impact. As such, constraints may be enforced to trigger a mutation.

Artificial neural network technique is applied to vulnerability test by constructing an attack tree [11]. Genetic algorithm has been applied to cryptanalysis [22]. Parameters, fitness functions and evolution processes for genetic algorithm are discussed to filter the traffic data and to reduce the complexity of intrusion detection [23, 24]. However, the GA approaches surveyed above do not satisfactorily forecast potential vulnerabilities from packet metadata and their stateful analysis.

3 Multithreading Analysis of Network Packet Streams

Multithreading is a mechanism of executing multiple threads of a process simultaneously. This mechanism is very useful to save the processing time when overwhelming amount of operations need to be performed together. It would be beneficial if overwhelming streaming data need to be processed at real time, especially if the streaming data processing time would create a major delay or a bottleneck for real-time processing.

Network traffic is streaming and bursty. DPI is needed over all levels of network layers from a network layer to an application layer for stateful analysis. There would be a bottleneck or an obstacle between unpacking network traffic streams and identifying attacks or analyzing packet payload (which may be very large or take longer to decrypt). For example, consider our college research network router, where the average data rate of incoming traffic and the average data rate for outgoing traffic are respectively 5 Mb/s and 7 Mb/s (as stored months or years, it is extremely a big data). The average total traffic is 12 Mb/s, and the average packet size is 150 bytes (in the range of 50 bytes and 400 bytes) and therefore there will be 80000 packets/s. The average time of a packet unpacking fully (fully for all levels: frame header data to port and payload header data) is 0.1 ms (on Intel 2.4 GHz CPUs), and the one for access control performed in Oracle Virtual Private Database (VPD on Intel 2.4 GHz CPUs) is 2 s. We call the former (1) *unpacking bottleneck*, and the latter (2) *stateful DPI bottleneck*. In this capacity, only 10000 packets can be unpacked per second, which means full unpacketing of packets is impossible, and even worse if TLS packets need to be decrypted. In this regards, inspections in most cases simply apply a rule set once as illustrated in Fig. 1.

However, Fig. 2 proposed in this paper shows that a multithreaded architecture will be able to reduce the bottleneck: multithreaded unpacking processes reduce the unpacking bottleneck by creating multiple queues, while multithreaded authentications reduce the stateful DPI bottleneck [25].

In overwhelmed network packet stream analysis, a multithreaded architecture is proposed to improve the big data process for unpacking network traffic at real-time. This paper proposes the following:

- Dynamic creation or removal of threads. Each CPU has a maximum capacity that can perform threads. The processing time of threads should not exceed or less a threshold λ than the CPU capacity. A detailed algorithm is provided in Fig. 4.
- Early decision of permit or deny for incoming packets. Part of early decision can be made by unpacking partial bytes of packets, which shows in the algorithm below, lines (9) and (12). Additional and more intelligent early decision will be addressed in the following section.

Algorithm 1. Dynamic Multithreading for Packet Unpacking

The following parameters are assumed: |c| denotes the number of CUPs; [C] denotes a list of CUPs that are unused and available; |P| denotes the number of packets P to consider; λ denotes a threshold that allows packets to wait, in # of packets; p[:n] and p[m:n] denotes the first n bytes of packet p and bytes from m to n bytes in p.

      Assume $|Q_i|$ denotes the number of packets being processed
      in the current queue i.
      Consider packets P incoming into this Packet Analysis in
      Q_i.

```
(1)
(2)   While |P| > 0
(3)   Begin
(4)      If |P| + |Qᵢ| > λ
(5)      Then If [C] is NOT empty
(6)           Then Create a new queue Q_{i+1}
(7)      For each first 40 bytes of packet p in P
(8)        Begin
(9)             Unpack p[40:]   // this unpacking for IP header data
(10)            Perform DPI on IP   // apply rules about IPs
(11)            Store IP's on the Global Heap
(12)            Unpack p[40:80] // this unpackinf for Port header
     data
(13)            Perform DPI on port // apply rules about ports
(14)            Store port's on the Global Heap
(15)            Perform DPI on protocol // apply rules about proto-
     col
(16)            Store Protocols on the Global Heap
(17)      If |Qᵢ| <= 0
(18)           Then Add Qᵢ to [C]
(19) End
```

Fig. 4. Algorithm 1: Dynamic multithreading for unpacking network traffics

- Propagation (or share) of the header data state of packets. This is shown in line (11), (14) and (16) in the algorithm below.
- Propagation (or share) of session context (login, role, etc.). This will be discussed in the next section.

Figure 5 illustrates the states of stacks of each CUP in the proposed multithreaded packet streams. Note that the global heaps in the algorithm above contain the states of packet header data for the network and transport layers, and the states of session contexts from the session and application layers. This figure and the algorithm shows how multithreading enables the big data of network traffic streams to be processed efficiently in network infrastructures. Lines (11), (14) and (16) show a part of global heap management in our multithreading for stateful packet analysis.

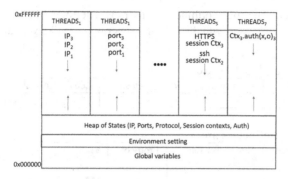

Fig. 5. Header data unpacked in network and transport layers, and state data extracted in session and application layers for multithreading

4   Using Genetic Algorithms to Learning Vulnerabilities

The goal of employing genetic algorithm (GA) is to generate both an optimal set of trusted packet sequences and those not trusted: (1) Packet sequences are learned to be trusted and therefore those packets are quickly determined to be permitted, while (2) those that will be untrusted are thus denied in advance. An ideal contribution of our GA approach is to forecast any vulnerability T3, T5 and T7 at T2. Similarly, GA results at T4 enable to skip or reduce the processes T5 and T7, and so on.

This section therefore takes the sequences of packets into consideration. A packet sequence PS can be defined over packets, p_i, as follows:

$$PS = p_1, p_2, \ldots, p_n \qquad (2)$$

For example back in Motivating Example 2, those five sequences can constitute a packet sequence, $PS = p_1,\ p_2,\ p_3,\ p_4,\ p_5$. The order in a packet sequence is very important in many cases. For example, a few control packets such as SYN and ACK packet come before TCP or UDP packet, and then FIN packet at last. However, it is not necessary that all packets in PS are related among themselves. For analysis purpose, PS can be simply a sequence of packets in a unit time period.

In what follows in this section, the fitness table and GA operations are characterized over network traffic packets.

4.1   Fitness Table

The parameters for the fitness functions are in three dimensions, (1) Geo-temporal factors, (2) port-relevant factors, and (3) user-relevant factors. The fitness table must represent the trustworthiness of network packets. Since it is well known that the trustworthiness of source IP's is determined dominantly by country [6]. The trustworthiness of source or destination ports is determined in part by its convention. For example, port numbers are designated to specific applications and protocols according to the network sorcery [26]. Any port access that does not follow the network sorcery are likely to attack. The trustworthiness of users is determined in part by their privileges. For example, accesses with the role of a super user are likely to attack if the other factors are met.

Geo-temporal factors are

Trusted IP	Untrusted IP	Time in src	Local time	•••
				•••

Port-relevant factors are

Conventional ports	Non-conventional ports	Protocol	•••
			•••

User-relevant factors are

Super user	Mid role	sub role	Casual role	•••
				•••

These three dimensional factors determine the fitness table, which is illustrated in Fig. 6. In the cube, each cell indicates a value (between −1 and +1) for three factors, one from each factor dimension, which is denoted by

$$-1 < \, = cell(g, t, u) < \, = 1 \qquad (3)$$

where g, t and u respectively denotes geo-temporal factors, port-relevant factors, and user-relevant factors.

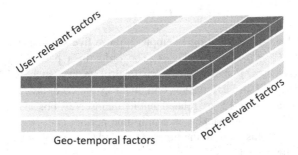

Fig. 6. Fitness cube of vulnerability factors

So, for given packet sequence P, which is constituted over n packets, the fitness function proposed in this paper is:

$$F(P) = \sum_{i=1}^{n} \omega_i cell_i(g, p, u) \qquad (4)$$

where ω denotes the weight for a cell value, $0 < \, = \omega \leq 1$. Hence, $-1 < \, = F(P) < \, = 1$.

4.2   GA Operations

As illustrated in Fig. 3, for any given pair of packet series, PS_i and PS_j, which contain respectively n and m packets, the following three operations are performed. Crossover and Selection operations are performed a lot more frequently than Mutation operation. Mutation operations take place at a rate of one to five percent of the frequency of selection or crossover operations.

One of the goals for using GA approaches is to learn the dichotomy of packet sequences. An example of packet sequence dichotomy is shown in Fig. 7(c). Three packet sequences, each sequence consists of 10 packet frames. The initial fitness of those packet sequences is closer to 0, which means they are neither fitness 1, i.e., trusted, nor -1, i.e., untrusted. As combination of GA operators is iterated multiple times, the fitness of each is converged to either 1 or -1. Some sequences can be converged early, depending of the values on the fitness cube.

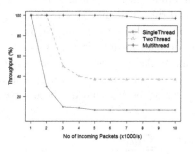
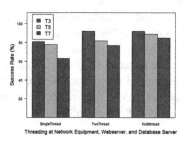

(a) Throughput of analysis   (b) Success rate w.r.t false positive

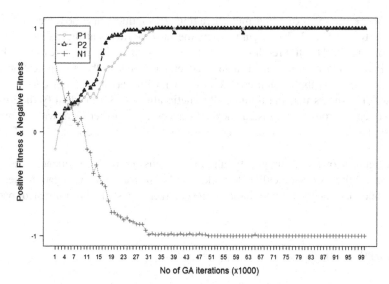

(c) Dichotomy of packet sequences by genetic algorithm:
Positive and negative fitness trained

Fig. 7. Dichotomy of packet sequences by genetic algorithm

5 Experimental Results

To evaluate performance of multithreading, packets are received from 4 different interfaces, including packets from a wifi router, on Intel 7i of 3.6 GHz 8 CPU core desktop. The total size of real-time network traffic packet streams for overall 10 h was 10 TB.

Of the experiments for several different features, Fig. 7(a) shows the throughput of
packet processing in percentage. The throughput of packet processing is defined as

$$Throughput = \frac{\sum unpacking\&analyzed}{\sum p_i} \tag{5}$$

Three cases are considered: (1) the case with no threading – simply called a single
thread; (2) the case of using two threads such as a thread for unpacking and a thread for
authentication at backend. (3) the case of multithreading: If additional CPUs are
allocated for multithreading, due to overwhelming packet delays, some packets are
simply permitted or denied by the first matched rule. This may lead to false negatives
and false positives. The figure shows that multithreaded packet analysis outperforms.

Figure 7(b) shows the success rate based on false positive in percentage. It shows
how much of packets are finally denied at the backend database server although the
packets are passed up to that point of packet traversal. Unfortunately, false negatives
are filtered out before they reach the webserver or the backend database server. The
success rate is defined as follows:

$$Success\,rate = \frac{\sum PS - \sum denied}{\sum PS} \tag{6}$$

One of the purposes of using multithreading is to reduce false positives in packet
inspection. Recall Fig. 2. This figure counts the false positives at the threads, T3, T5
and T7 of Fig. 2. The figure shows that there are more false positives at backend
servers. Part of the reasons found at the backend is because the authentication at a
database server takes the session contexts together with network header data.

Figure 7(c) shows that the training of genetic algorithms can identify the trusted
packet sequences from the untrusted packet sequences. This experiment was performed
on maximum 8 children of cross-over operations, with mutation of 0.1% of other
operations.

This figure shows that the genetic algorithm trains the packet sequences well and
therefore their fitness can quickly determine to either positive one or negative one. The
positive one and negative one indicate respectively trusted and untrusted packet
sequences.

6 Contribution

This paper proposed a multithreading architecture that can handle the big data of
network traffic packet streams and it also analyze the network packet streams at real-
time. The multithreaded network packet analysis propagates the network header data to

the global heap area, and shares them at each level of analysis thread. The data available at the global heap area can then be used for backend database access, and together control traffic with the session contexts that can be captured at the webserver.

This paper shows that the real-time big data has been efficiently handled by multithreaded analysis. A fitness of network traffic packet streams is well defined to apply to a genetic algorithm to forecast both trusted and untrusted packet sequences.

The contribution of this paper includes (1) improving analysis time of the big data from network traffic packet streams; (2) reducing false positives in fast and massive network packet processing; and (3) early decision of vulnerabilities.

The future work will be an optimization of load-balancing over multithreads. This paper assigns loads randomly to multithreads. That means that a job in a queue may be delayed much longer if the CPU is held by a job previously assigned. Knowing the data about the jobs loaded in queues, job assignments will be able to improve.

References

1. Cesare, S., Xiang, Y.: Classification of malware using structured control flow. In: Proceedings of 8th Australasian Symposium on Parallel and Distributed Computing, vol. 107 (2010)
2. Kumar, S., Dhamapurikar, S., Yu, F., Crowley, P., Turner, J.: Algorithms to accelerate multiple regular expressions matching for deep packet inspection. In: The ACM SIGCOMM Computer Communication Review, vol. 36, no. 4 (2006)
3. Cascarano, N., Ciminiera, L., Risso, F.: Improving cost and accuracy of DPI traffic classifiers. In: Proceedings of ACM Symposium on Applied Computing, Switzerland (2010)
4. Miura, R., Takano, Y., Miwa, S., Inoue, T.: GINTATE: scalable and extensible deep packet inspection system for encrypted network traffic. In: Proceedings of ACM Conference on SoICT, Viet Nam (2017)
5. Wireshark. http://www.wireshark.org. Accessed 15 Feb 2018
6. How to block traffic by country in the CSF firewall. https://www.liquidweb.com/kb/how-to-block-traffic-by-country-in-the-csf-firewall/. Accessed 15 Feb 2018
7. Caruccio, L., Deufemia, V., Polese, G.: Evolutionary mining of relaxed dependencies from big data collections. In: Proceedings of the 7th International Conference on Web Intelligence, Mining and Semantics (2017)
8. https://www.metasploit.com/. Accessed 22 Feb 2018
9. Fan, J., Guan, C., Ren, K., Cui, Y., Qiao, C.: SPABox: safeguarding privacy during deep packet inspection at a MiddleBox. IEEE/ACM Trans. Netw. 25, 3753–3766 (2017)
10. Lin, Z.: TLS session resumption: full-speed and secure. https://blog.cloudflare.com/tls-session-resumption-full-speed-and-secure/. Accessed 9 Mar (2018)
11. Gill, H., Lin, D., Sarna, L., Mead, R., Lee, K., Loo, B.: SP4: scalable programmable packet processing platform. In: ACM SIGCOMM Computer Communication Review, October 2012
12. Goldberg, D.: Genetic Algorithms in Search, Optimization, and Machine Learning. Addison-Wesley, Boston (1989)
13. Nareddy, S., Westover, E., Hillesland, K., Kim, W.: Genome dynamics in coevolved genomes: database management system for tracing mutations. In: Proceedings of the 5th ACM Conference on Bioinformatics, Computational Biology and Health Informatics, pp. 633–634 (2014)

14. Bogard, J.: A probabilistic functional crossover operator for genetic programming. In: Proceedings of the 12th Annual Conference on Genetic and Evolutionary Computation, pp. 925–931 (2010)
15. Stanhope, S.A., Daida, J.M.: Optimal mutation and crossover rates for a genetic algorithm operating in a dynamic environment. In: Porto, V.W., Saravanan, N., Waagen, D., Eiben, A. E. (eds.) EP 1998. LNCS, vol. 1447, pp. 693–702. Springer, Heidelberg (1998). https://doi.org/10.1007/BFb0040820
16. http://www.obitko.com/tutorials/genetic-algorithms/recommendations.php. Accessed 28 Jan 2016
17. LeFevre, J., Sankaranarayanan, J., Hacigumus, H., Tatemura, J., Polyzotis, N.: Towards a workload for evolutionary analytics. In: Proceedings of the 2nd Workshop on Data Analytics in the Cloud (2013)
18. Fan, W., Geerts, F., Cao, Y., Deng, T., Lu, P.: Querying big data by accessing small data. In: ACM Symposium of Principles of Database Systems, pp. 173–184 (2015)
19. http://eoddata.com/default.aspx. Accessed 29 Jan 2016
20. The R project for statistical computing. https://www.r-project.org/. Accessed 29 Jan 2016
21. Tumoyan, E., Kavchuk, D.: The method of optimizing the automatic vulnerability validation. In: Proceedings of the 5th International Conference on Security of Information and Networks (2012)
22. Bergmann, K., Scheidler, R., Jacob, C.: Cryptanalysis using genetic algorithms. In: Proceedings of the 10th Annual Conference on Genetic and Evolutionary Computation, July 2008
23. Hoque, M., Mukit, M., Bikas, M.: An implementation of intrusion detection system using genetic algorithm. Int. J. Netw. Secur. Appl. **4** (2012)
24. Hashemi, M., Muda, Z., Yassin, W.: Improving intrusion detection using genetic algorithm. Inf. Technol. J. **12**, 2167–2173 (2013)
25. Khan, A., Gleich, D., Pothen, A., Halappanavar, M.: A multithreaded algorithm for network alignment via approximate matching. In: Proceedings of the International Conference on High Performance Computing, Networking, Storage and Analysis (2012)
26. RFC Sourcebook, TCP/UDP ports. http://www.networksorcery.com. Accessed 1 Mar 2018

Forecasting Traffic Flow: Short Term, Long Term, and When It Rains

Hao Peng$^{(\boxtimes)}$ ⓘ, Santosh U. Bobadeⓘ, Michael E. Cotterellⓘ,
and John A. Miller

Department of Computer Science, University of Georgia, Athens, GA, USA
{penghga,santosh.bobade,mepcott}@uga.edu, jam@cs.uga.edu

Abstract. Forecasting is the art of taking available information of the past and attempting to make the best educated guesses of the ever unforeseen future. From the historical data, patterns can be observed and forecasting models have been developed to capture such patterns. This work focuses on forecasting traffic flow in major urban areas and freeways in the state of Georgia using large amounts of data collected from traffic sensors. Much of the existing work on traffic flow forecasting focuses on the immediate short terms. In addition to that, this work studies the forecasting powers of various models, including seasonal ARIMA, exponential smoothing and neural networks, for relatively long terms. A second experiment that incorporates precipitation data into forecasting models to better predict traffic flow in rainy weather is also conducted. Dynamic regression models and neural networks are used in this experiment. In both experiments, neural networks outperformed the others overall.

Keywords: Traffic flow forecasting · Big data analytics
Time series analysis · Seasonal ARIMA · Exponential smoothing
Neural networks · Dynamic regression

1 Introduction

Forecasting is an area of interest in most people's lives. People would like to know what the weather will be like this weekend in order to make appropriate plans. Investors are interested in knowing how well certain stocks will perform before purchasing or selling. Travelers would like to know what the traffic conditions will be like in order to make travel plans in advance. The desire to know the future, or make the best possible predictions of it, and the benefits it brings have driven scientists to develop various forecasting models and study the factors that may improve forecasting accuracies.

The advancements in computing hardware and big data frameworks have enabled scientists to perform big data analytics on increasing amounts of data. In recent years, large numbers of traffic sensors have been deployed throughout major urban areas and freeways in the United States to collect data such as

© Springer International Publishing AG, part of Springer Nature 2018
F. Y. L. Chin et al. (Eds.): BIGDATA 2018, LNCS 10968, pp. 57–71, 2018.
https://doi.org/10.1007/978-3-319-94301-5_5

traffic flow, speed, lane occupancy, etc. These increasing amounts of data allow scientists to develop, test and validate forecasting models in order to achieve higher forecasting accuracies, which is a very important task for improving traffic management and control systems. The data used in this study are traffic flow data (number of vehicles per hour) from Jan 1, 2013 to June 30, 2017 reported by traffic sensors deployed throughout the state of Georgia. These data may be accessed from the Georgia Department of Transportation website.[1] Various forecasting models, including seasonal ARIMA, exponential smoothing and neural networks, are trained, validated and evaluated on these data.

It is very intuitive that weather conditions such as rain affect traffic flow. There have been some work that studies the relationship between weather conditions and traffic flow, such as in [6,21,22]. Recent work in [27] used rainfall information to help predict traffic speed; yet surprisingly not as much research has been done to directly incorporate rainfall data into forecasting models to help predict traffic flow. Though a few such studies have emerged in the very recent years and are discussed in the Related Work section. In this work, the dynamic regression model is used to incorporate precipitation data as an exogenous explanatory variable that can be used in combination with the seasonal ARIMA and exponential smoothing while neural networks simply require an extra neuron in the input layer to take into account the rainfall data.

Much of the existing work on forecasting traffic also focuses on the immediate short term (e.g., forecast 15 min ahead into the future); typically, only a couple of months of data from relatively few sensors are utilized (e.g., in [5,17,18]). Though short term traffic forecasting is certainly an important task for traffic control systems such as smart traffic lights, relatively long term traffic forecasting can be beneficial to the general public for trip planning in advance, which may also lead to the development and improvements of related software/apps.

Therefore, in order to address these issues, this work aims to complement existing work in the literature with the following contributions: (1) evaluating various commonly used forecasting models on large amounts of traffic flow data covering a great number of locations and times; (2) studying the forecasting powers of the models for both short and relatively long terms, 24 h into the future in this case; and (3) examining the effects of the inclusion of rainfall data on short term traffic flow forecasting.

All implementations of the forecasting models used in this study are available in the SCALATION project, which is a Scala-based project for analytics, simulation and optimization freely available under an MIT License. For more information about this project, please visit cs.uga.edu/~jam/scalation.html.

The rest of this paper is organized as follows: Sect. 2 discusses the background of the forecasting models. Related Work is discussed in Sect. 3. Section 4 contains Evaluations, in which experimental setups and results are discussed. Finally, Conclusion and Future Work are in Sect. 5.

---

[1] http://www.dot.ga.gov/DS/Data.

2 Background

Various forecasting models have been developed throughout the years. Classical statistical models such as ARIMA [1] are still very capable and commonly used today. Machine Learning models such as Neural Networks (NN) are also gaining increasing popularity and attention due to recent progress made in deep learning. This section details a couple of forecasting models that are implemented in SCALATION and used in this study.

2.1 ARIMA Family of Models

The AutoRegressive Moving Average, or ARMA (p,q) model is a commonly used and classical statistical model that may be defined as

$$Y_t = \sum_{i=1}^{p} \phi_i Y_{t-i} + \sum_{i=1}^{q} \theta_i \epsilon_{t-i} + \epsilon_t , \tag{1}$$

where Y_t is the zero-mean transformed response variable of interest; p and q are the orders of the AutoRegressive(AR) and Moving Average (MA) components, respectively; ϕ's and θ's are the parameters of the AR and MA components, respectively;[2] and the error term ϵ_t is assumed to be independently and identically distributed (i.i.d.) from $\mathcal{N}(0, \sigma^2)$ for some constant variance σ^2. Techniques such as Box-Cox transformations can be applied to stabilize variance while differencing is usually done to stabilize the mean. By combining the ARMA model with differencing, a more general model known as the AutoRegressive Integrated Moving Average, or ARIMA (p,d,q) model, where d is the order of the differencing, is obtained.

The ARIMA model may be further improved if the data are known to contain seasonality, which is just a repeated pattern over a fixed period of time. This is definitely applicable for the traffic flow data, in which the seasonal period could be one week (e.g., we would expect similar traffic flow on the same road during the morning rush hours of this Monday and previous Mondays). A more general seasonal model may be defined as

$$Y_t = \sum_{i=1}^{p} \phi_i Y_{t-i} + \sum_{i=1}^{q} \theta_i \epsilon_{t-i} + \sum_{i=1}^{P} \Phi_i Y_{t-is} + \sum_{i=1}^{Q} \Theta_i \epsilon_{t-is} + \epsilon_t , \tag{2}$$

where s is the seasonal period; P and Q are the orders of the seasonal AR and MA components, respectively; and Φ's and Θ's are the parameters of the seasonal AR and MA components, respectively. Seasonal differencing may also be applied as necessary, and this type of model is know as the Seasonal ARIMA, or simply SARIMA $(p,d,q) \times (P,D,Q)_s$ model, where D is the order of the seasonal difference.

---

[2] In some definitions, these parameters may be located in opposite sides of the equation and therefore have the opposite signs.

In this study, the SARIMA$(1, 0, 1) \times (0, 1, 1)_{120}$ model with weekly seasonality of 24 hours/day $\times$ 5 weekdays/week = 120 hours/week is chosen since it was used in several related studies including [18, 24, 28]. Alternatively, SARIMA models may be selected based on a scoring function in an automated fashion as described in [12]. This automated order selection process is also implemented in SCALATION. The SARIMA models selected using the automated process based on the AICc criterion as recommended in Sect. 8.9 of [11] may yield better performances than the SARIMA$(1, 0, 1) \times (0, 1, 1)_{120}$ model in certain individual traffic flow time series, but can also be overfitted with more parameters than necessary in others. The final overall results of the automated SARIMA models are actually slightly worse than the results obtained from the SARIMA$(1, 0, 1) \times (0, 1, 1)_{120}$ model. Therefore only SARIMA$(1, 0, 1) \times (0, 1, 1)_{120}$ results are reported in this work.

Dynamic Regression. If an exogenous explanatory variable is available to help model the response (e.g., using precipitation data to help predict traffic flow in this study), then dynamic regression may be used. In a similar way as described in Sect. 9.1 of [11], the dynamic regression model may be defined as

$$\epsilon_t = \beta x_t + z_t \tag{3}$$

where ϵ_t could be the residual of a SARIMA model described in Eq. 2 (or other time series forecasting models such as exponential smoothing); x_t is the exogenous explanatory variable; and z_t is the residual for this regression model. The dynamic regression model can therefore be viewed as a two-step process of attempting to explain variabilities within a time series by using both a forecasting model and a regression model.

2.2 Exponential Smoothing

Another popular time series forecasting model is exponential smoothing [3, 9]. Since the traffic flow data are inherently seasonal, triple exponential smoothing [29] is required. Two types of seasonality exist for this model, additive and multiplicative. Additive seasonality is more appropriate for a problem like predicting traffic flow as suggested in Sect. 7.5 of [11], and may be defined as.

$$
\begin{aligned}
S_t &= \alpha(Y_t - c_{t-L}) + (1 - \alpha)(S_{t-1} + b_{t-1}) \,, \\
b_t &= \beta(S_t - S_{t-1}) + (1 - \beta)b_{t-1} \,, \\
c_t &= \gamma(Y_t - S_t) + (1 - \gamma)c_{t-L} \,.
\end{aligned}
\tag{4}
$$

where S_t is the smoothed value of Y_t, an observed value of the time series at time t; b_t is the trend factor; c_t is the seasonal factor; L is the seasonal period; and α, β and γ are all smoothing parameters, bounded between 0 and 1, that need to be estimated. The standard practice of parameter estimation is to minimize the one-step ahead within sample forecast sum of squared errors (SSE), which is

$$SSE = \sum_{t=2}^{n} (Y_t - \hat{Y}_{t|t-1})^2 \,, \tag{5}$$

where n represents the number of observations in the data and $\hat{Y}_{t|t-1}$ is the one-step ahead forecast at time t when given the data up to time $t-1$.

In this work, the parameters are optimized by minimizing the 12-step ahead, as opposed to the standard 1-step ahead, within sample forecast SSE. If the parameters are optimized by minimizing the 1-step ahead within sample forecast SSE, then preliminary testings show that forecast performances are only good for one-step ahead and rather bad for all subsequent steps. This could be caused by the optimizer's failure to rely on the seasonal components to minimize the SSE since the most recent lagged value can also be used to effectively minimize SSE for the immediate 1-step ahead. On the other hand, if 12-step ahead within sample forecast SSE needs to be minimized, then the optimizer must rely on the seasonal components of the time series to effectively minimize SSE, and therefore the learned parameters tend to be better suited for both short and relatively long term forecasts.

2.3 Neural Networks

The structure of a neural network consists of multiple layers of artificial neurons. One type of neural networks is feedforward neural networks, in which neurons in a layer may only send signals forward to neurons in the subsequent layer. Other types of neural networks also exist, such as Recurrent Neural Networks (RNN), in which neurons in a layer may also send signals backward. This study uses feedforward neural networks. The process for which a single neuron handles its incoming signals and produces an output signal may be defined as

$$a_{\text{out}} = \sigma(\mathbf{w} \cdot \mathbf{a_{in}} + b) \,, \tag{6}$$

where $\mathbf{a_{in}}$ is the vector of incoming signals; $\mathbf{w}$ is the vector of weights associated with the incoming signals; b represents the bias; σ is the activation function; and a_{out} is the output signal/activation.

A standard choice for the cost function for prediction problems is the Mean Squared Error (MSE) between the final output signals of the network and the observed training outputs, which may be defined as

$$MSE = \frac{1}{n} \sum_{i=1}^{n} \left\| \mathbf{y_i} - \mathbf{a_i} \right\|^2 \,, \tag{7}$$

where n is the number of training instances; $\mathbf{a_i}$ is the vector of final output signals produced by the i-th training/input instance; and $\mathbf{y_i}$ is the i-th observed output vector. The minimization of the cost function can be done using the stochastic gradient descent with backpropagation [23].

The design of an appropriate neural network model for a particular task such as forecasting traffic flow can be extremely flexible or complex, depending on which side of the coin one chooses to look at. With experimentations and trials, a four-layer neural network structure is adopted for forecasting traffic flow in this work. The input layer takes in the day of the week, time of the day, the most recent 24 h of traffic flow data, and the 24 h of traffic flow data in the previous seasonal period (i.e., if forecasting of a Monday's traffic flow is desired, then the previous Monday's traffic flow data were used as inputs). There are two hidden layers, of sizes 40 and 30, and a final output layer of size 24, one for each step ahead forecast. The *tanh* activation function is used in this neural network. Since *tanh* can only output values between −1 and 1, and the magnitude of the gradient is also greatest in the domain between −1 and 1, the time series are normalized using Min-Max Normalization to be within the range of -0.8 and 0.8, in order to leave the neural network some room to output values that are slightly greater and less than the maximum and minimum values in the training set, respectively. In terms of parameter tuning, neural network does not really have a straight forward and intuitive way to choose parameters similar to the Box-Jenkins method used to choose parameters of an ARMA model. At times it can also be difficult to explain why certain parameters tuned a certain way simply yield better performances for a certain dataset. A common, and somewhat expensive, approach is to used an automated grid search. In this study, data from eight randomly selected traffic sensors are used for the parameter tuning purposes. For each of the chosen time series, a random starting day is selected, then 3 months of training data were used to train various neural network models with different parameters and the subsequent 2 months of data were used for testing. The final set of parameters of the four-layer neural network used in this study is as follows: number of training epochs is set to 600, the mini-batch size is 20, and learning rate is set to 0.1.

3   Related Work

Recent work in [18] compared various forecasting techniques for short term traffic flow forecasting. The techniques included ARIMA based models, Support Vector Regression (SVR) based models and feedforward neural networks. A total of nine months of data, from January 2009 to September 2009, for sixteen vehicle detector stations were collected from the California Freeway Performance Measurement System (PeMS).[3] The data were aggregated into 15-minute intervals. The first four months were used to train the models, the next two months for validation and model parameter tuning (for SVR and NN), and the final three months were used for testing. The forecasting accuracies of 15-minute ahead forecasts were used to compare the performances of the models; no forecasts beyond 15 min into the future were produced. The authors concluded that a SARIMA model performed the best overall.

Another recent study in [19] utilized a deep neural network built from stacked autoencoders to predict traffic flow. Data were also collected from a very large number of detectors in the PeMS database for the first three months of 2013. The data were aggregated into 5-minute intervals. The first two months were used for training and the remaining one month was for testing. The proposed deep

---

[3] http://pems.dot.ca.gov/.

neural network was compared against Support Vector Machine (SVM), back-propagation neural network and radial basis function neural network. However, the authors did not make it clear on both the inputs and the parameters used in the aforementioned models that were used to compare with the proposed deep neural network. No statistical models such as SARIMA were included. Forecasts for 15, 30, 45 and 60 min into the future were produced and the authors' proposed deep neural network exhibited superior performances.

Many have studied the impact of weather conditions on traffic. A study done in [13] concluded that traffic flow may be reduced by 14% – 15% during heavy rainfall (>0.25 in/hr) in Toronto. The *Highway Capacity Manual* [20] and a more recent study done in the Twin Cities metropolitan area [21] also reached a similar conclusion. Another study in [25] claimed that light and moderate rainfall (<0.25 in/hr) can reduce freeway traffic flow by 4% to 10%, while heavy rain can reduce freeway traffic flow by 25% to 30% in Hampton Roads, Virginia. Other related studies include [2, 6].

Not as much research has been done that directly incorporates weather data to help with traffic forecasting historically; a couple of such studies have only emerged in the very recent years. There has also been work that used precipitation data to help forecast traffic speed, such as [10, 27], in which neural network and ARIMA based models were used, respectively. A study in [8] may be one of the earliest ones that attempted to use weather information to help forecasting traffic flow. A neural network model was used but the input weather information was encoded in categorical variables of 0 (clear), 1 (rain) and 2 (snow/ice) because detailed information on rainfall were not available to the authors. Another study in [4] included rainfall data as inputs to a neural network model to forecast traffic flow. However, when the model incorporated with rainfall data was compared with the one without, worse performances were obtained. The authors suggested that the counter-intuitive results could be due to the lack of rainy days in the training instances.

A recent study in [7] used a combination of stationary wavelet transform and neural networks to predict traffic flow with the incorporation of rainfall data. Data from two traffic sensors from Dublin, Ireland were used to evaluate the performance of the authors' proposed model and the standard feedforward neural network. The study showed that incorporating rainfall data certainly helps to improve prediction, and the authors' proposed algorithm performed better; though no other models were used for comparisons. A couple more studies using the deep learning approach have emerged in the very recent years. One such study in [16] incorporated weather information such as rain, temperature, humidity, etc., into a deep belief network. Performance comparisons were done with ARIMA and a neural network model of three layers. The authors' proposed deep belief network outperformed ARIMA significantly and did better than the three-layer neural network, though the margins are not as great. Another work in [30] used a combination of recurrent neural network and gated recurrent unit to predict urban traffic flow. The weather data included precipitation, speed and temperature. The authors demonstrated that the incorporation of weather data can improve forecasting accuracies; however, no other models were used for comparison purposes. Only the authors' proposed model, with and without weather data, is included in the performance evaluations. A study in [14] compared multiple models, including ARIMA, backpropagation neural network, deep belief network and long short-term memory neural network for forecasting traffic flow. Rainfall data were incorporated into the aforementioned models. Forecasts are produced for the immediately short terms, 10 and 30 min ahead into the future. The authors concluded that long short-term memory neural network is the top-performing model, and the incorporation of rainfall data generally improves forecasting performance for most models that were tested.

4   Evaluations

This section details the description, selection and pre-preprocessing of the datasets, forecasting procedures and evaluation metrics, and the experimental designs and results.

4.1   Dataset Description and Pre-processing

There are 275 permanent road sensors deployed by the Georgia Department of Transportation. The traffic flow for the two separate directions (north and south, or east and west) on a road, as well as the aggregate traffic flow, are recorded for each sensor. For this study, weekday data from January 2013 to June 2017 are used. The data from weekends are excluded from this study since there are usually much less congestions during weekends and the traffic patterns on weekends are different from those of the weekdays. The practice of removing weekends are very common in the literature, as seen in [15,18,19,26]. In addition, only forecasts from 7:00 am to 7:00 pm are evaluated since traffic flow during night times are usually not congested.

Figure 1 provides an graphical view of traffic flow data of all the Fridays in the year 2013 on a major road in Atlanta, GA. Note there is greater outflowing traffic in the afternoon rush hours as people return home from work. The data from the same traffic sensor but on the opposite direction of the road has a complementary traffic pattern, in which there are more vehicles in the morning rush hours when people attempt to arrive at work on time.

Missing values in the data must be handled since forecasting models expect complete training data. Most commonly, data can be missing for an entire day. This is most likely due to the quality control system that rejects data for an entire day based on rules like "the system will reject any day that does not have data for every hour."[4] For an entire day of missing data, the hourly historical averages of the same weekday from the past four weeks are used for imputations (i.e., if a Tuesday's data are missing, the hourly averages of the last four Tuesdays' data

---

[4] http://www.dot.ga.gov/DriveSmart/Data/Documents/Guides/2017_Georgia_Traffic_Monitoring_Program.pdf.

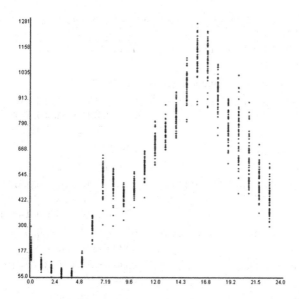

Fig. 1. Friday traffic on US 23 in Atlanta, GA

are used to impute the values). Occasionally, data can be missing for an hour of a day, possibly due to the imperfection of the quality control system. In this scenario, a simple linear interpolation is used to impute the value by computing the average of the data from the hour before and the hour after. Lastly, some sensors simply have too many missing values to be useful. In extreme cases, a sensor may contain no data at all. Therefore any sensor that contains more than one year of missing data are excluded from this study. The number of remaining usable traffic sensors is 157.

The usable sensors are further filtered to include mostly urban areas and busy freeways where congestions are most likely to occur on a regular basis. Other sensors that scatter across the state of Georgia, including more rural areas and freeways that are not very busy, are excluded from this study. In particular, traffic sensors from 15 counties, as summarized in Table 1, are included in this study. The total number of sensors from the selected counties is 74, and the overall percentage of missing values is close to 8.5%.

The precipitation data come from the Automated Surface Observing System (ASOS)[5], a joint program maintained by the National Weather Service (NWS) and the Federal Aviation Administration (FAA). The ASOS sensors are typically placed in airports or air bases. Data are downloaded in hourly resolution through a convenient web interface[6] provided by the Department of Agronomy of Iowa State University. A total of 57 ASOS stations contain records from January 2013 to June 2017, but unfortunately more than half of them contain all 0's or very little data. The number of usable ASOS sensors is 22.

---

[5] https://www.weather.gov/asos/.

[6] https://mesonet.agron.iastate.edu/request/download.phtml?network=GA_ASOS.

Table 1. Summary of Traffic Sensors

County	Nearby City, Freeway or Point of Interest	Traffic Station ID
Banks	Commerce, I-85, Tanger Outlets	011-0103
Bibb	Macon, I-16, I-75,	021-0116 021-0132 021-0158 021-0258 021-0267 021-0334 021-0349 021-0372 021-0376 021-0541 021-0587
Bryan	Savannah, I-16	029-0103
Camden	I-95	039-0145 039-0218
Chatham	Savannah, I-16, I-516, I-95	051-0107 051-0109 051-0132 051-0137 051-0138 051-0318 051-0334 051-0383 051-0443 051-0509 051-0649
Clarke	Athens, University of Georgia (UGA)	059-0014 059-0087 059-0118 059-0367 059-0611 059-0613
Clayton	Atlanta, I-285	063-0383 063-1023 063-1032 063-1085 063-1172 063-1201
Cobb	Atlanta	067-2334 067-2623
Fulton	Atlanta, I-75, I-85	121-0178 121-0190 121-5110 121-5114 121-5225 121-5374 121-5463 121-5468 121-5486 121-5524 121-5633 121-5969 121-6370
Glynn	Brunswick, St. Simon's Island	127-0105 127-0107 127-0236 127-0289 127-0456
Gwinnett	Atlanta, I-85, Mall of Georgia	135-0298 135-0305 135-0563
Houston	South of Macon	153-0143 153-0189 153-0332 153-0365
Muscogee	Columbus, I-185	215-0165 215-0336
Oconee	Watkinsville, UGA	219-0203
Richmond	Augusta, I-20, I-520	245-0214 245-0218 245-0223 245-0233 245-0303 245-0947

Each traffic sensor is then paired with its closet ASOS sensor based on GPS coordinates. Roughly 44% of the traffic sensors are paired with ASOS sensors that are located within 5 miles; 76% of the traffic sensors can find ASOS sensors within 10 miles; 86% of the traffic sensors can pair with ASOS sensors within 15 miles; and 90% of the traffic sensors may be paired with ASOS sensors within 20 miles. For the traffic sensors located in two coastal counties in southeastern Georgia, Camden (2 sensors) and Glynn (5 sensors) counties, the closest ASOS sensor is about 75 and 50 miles away, respectively. In the end, 14 out of 22 ASOS sensors are used for pairing with traffic sensors. The 8 remaining ASOS sensors are not located close enough to at least one of the traffic sensors included in this work.

The percentage of missing values in the 14 ASOS sensors is about 1.7%. The missing values are imputed by generating random values from Gaussian distributions, for which the means and variances are computed from the most recent 5 observations.

4.2   Forecast Validation and Performance Metrics

Rolling forecast validation with a fixed window size of w, which is the size of the training set, is used to test the models. The forecasting horizon (h) is the number of steps ahead into the future to produce forecasts. Each forecasting model is trained from w observations, and then tested in the subsequent 8 weeks, which serves as a testing set. Since it may not be feasible to forecast all of 8 weeks at once, only h-step ahead forecasts are produced each time. The forecasting models will continually forecast h steps into the future, each time using the most recent hourly data as inputs, until the end of the testing set has been reached. No forecasts are produced outside the 7:00am to 7:00pm range and the imputed values are excluded from performance comparisons as well. After all forecasts have been made in the testing set, the window of training data would then slide over by 8 weeks, including 8 weeks of new observations and dropping the oldest 8 weeks of data, and then the training and forecasting processes are repeated.

Forecasting accuracies are measured using the Mean Absolute Percentage Error (MAPE) metric, which may be defined as

$$MAPE = \frac{1}{n} \sum_{i=1}^{n} \left| \frac{y_i - \hat{y}_i}{y_i} \right|, \tag{8}$$

where n is the total number of forecast values, $\hat{y}_i$ is the i-th forecast value, and y_i is the corresponding observed value.

4.3   Experimental Setups and Empirical Results

The first experiment focuses on evaluating the forecasting performances of various models for both short and relatively long terms. The sliding window w is set to be 12 weeks of hourly traffic flow data, and the forecasting horizon h is set to 24, meaning 24 forecasts are produced ranging from 1-hour ahead to 24-h ahead into the future.

A baseline, weekly historical averages computed from instances in the training set, is also included for comparing with the aforementioned models. Each of the 74 traffic sensors contains two separate time series, one for each direction of the road, yielding a total of 148 univariate traffic flow time series. Roughly over 1.7 million forecast values are produced per model (excluding baseline) per step for all 148 time series. The total number of forecast values for all models, all 24 steps, and all 148 time series is close to 130 million. Since the time series contain different numbers of forecast values due to different numbers of missing values, the final results are aggregated by computing weighted averages across multiple time series and summarized in Fig. 2. All testings are done on a 48-core AMD Opteron Machine from the Sapelo cluster of Georgia Advanced Computing Resource Center[7] to facilitate parallel processing.

---

[7] https://gacrc.uga.edu/.

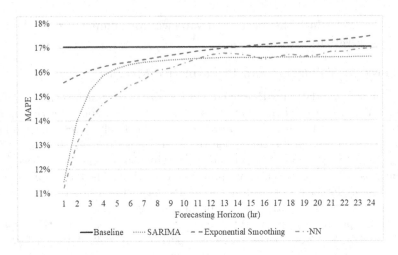

Fig. 2. Performance comparison of forecasting models

It comes as no surprise that in the immediate short terms, the forecasting models tend to perform well. The SARIMA model produces reasonably accurate forecasts in the immediate short terms and experiences sharp declines in forecasting accuracies up to about step 5. Then the performance drops gradually for all the remaining steps, yet always yielding better results than the baseline. The exponential smoothing model yields the lowest performance overall, and experiences a slow and stead decline in performances. At around step 20, the exponential smoothing model is no longer more effective than the baseline. The performance of the neural network is generally the best, leading in terms of forecasting accuracies up to about 12 steps. Then the neural network exhibits similar performances with SARIMA, and then starts to perform slightly worse at around step 20, yet always remain under the baseline.

The second experiment focuses on using precipitation data to aid in traffic flow forecasting. The sliding window w is expanded to 48 weeks of hourly traffic flow data in order to include more training instances with rainfall. The forecasting horizon h is also reduced to 1 since at any given time point, it's difficult to make reliable predictions of precipitation in the long term without great expertise in the field of weather forecasting and possibly additional data such as satellite images. The dynamic regression models are used to regress residuals from SARIMA and exponential smoothing models on rainfall data in order to help explain additional variabilities. Only training instances that experience at least a moderate amount of rain (>0.1 in/hr) are considered for the regression.

As for neural network, an additional input neuron representing rainfall is added to the input layer. An additional neuron is also added to each of the two hidden layers, and the number of neurons in the output layer has been reduced to one. It is reasonable to assume that there is a great difference in traffic patterns during hours with no rain at all and hours with some rain. In other words, it would be helpful for the neural network to recognize that there is a greater gap between 0 in. of rain and 0.1 in. of rain than 0.1 in. of rain and 0.2 in. of rain. To simulate this effect, all values of 0's in the ASOS precipitation datasets were replaced with −1.5 before the data are normalized for training of a neural network.

To forecast, the models first require a prediction of the rainfall 1 h later in order to produce 1-step ahead traffic flow forecasts. The future rainfall is estimated by examining the most recent 3 h and average the values that are greater than 0 in/hr. Forecasts are only produced for the instances in the testing sets with at least moderate rainfall (>0.1 in/hr). Overall, close to 80 thousand forecasts are produced for all models across all time series. The results are summarized in Table 2.

Table 2. Short term forecasting in rainy weather

	Baseline	Baseline (12 weeks)	SARIMA	Exponential Smoothing	Neural Network
Traffic Flow Data Only	42.43%	41.62%	20.11%	34.05%	15.12%
With Rainfall Data	38.03%	37.15%	19.79%	33.41%	14.95%

An additional baseline representing the weekly historical averages in the most recent 12 weeks before the testing set is also included to draw comparisons with the previous experiment. The first baseline is the weekly historical averages computed from the entire training set, which has been expanded to 48 weeks. The manner for which the two baselines take account of the rainfall data is to reduce the historical averages by 15% during heavy rain (>0.25 in/hr) or 10% during moderate rain (>0.1 in/hr), as suggested by a couple of studies discussed in Sect. 3.

From the numbers in Table 2, it is immediately obvious that the two baselines perform very poorly during rainy days. The performances of models are all improved with the inclusion of rainfall data. Neural network remains as the top performer. It is also interesting to note that for 1-step ahead forecast, there is greater gap between the performances of SARIMA and neural network in this experiment than those of the previous experiment. This could be due to the ability of a neural network to better handle complex traffic situations and sudden changes that are potentially due to rainfall or other factors since a neural network can make use of a very large number of parameters.

5   Conclusion and Future Work

In this work we evaluated several commonly used forecasting models for some currently under-researched items such as both short and relatively long term traffic flow forecasting and the incorporation of precipitation data into forecasting models to make better forecasts on rainy days. The neural network model is the top performer overall in both experiments.

In terms of future work, there are several directions that can be pursued in order to further improve upon this work. Bigger datasets with higher resolutions are always helpful. The hourly resolution may not always be sufficient to capture the dynamic nature of traffic patterns in major urban areas and freeways. The Caltrans Performance Measurement System (PeMS)[8] from the state of California would be a great source of such data. More types of weather data besides precipitation can also be used, similar to a few studies mentioned in the Related Work section. Data on special events, major holidays and traffic accidents can all be incorporated into forecasting models as well to handle the irregular patterns. Spatial dependencies should also be exploited, as traffic flowing into a certain direction at a particular location should be used to help forecast traffic further done the road. Additional forecasting models, such as Recurrent Neural Networks and in particular, Long Short-Term Memory Neural Networks, should be included in order to create a more comprehensive evaluation of forecasting techniques for traffic flow.

References

1. Box, G.E., Jenkins, G.M.: Time Series Analysis Forecasting and Control. Technical report, DTIC Document (1970)
2. Brilon, W., Ponzlet, M.: Variability of speed-flow relationships on German autobahns. Transportation research record. J. Transp. Res. Board **1555**, 91–98 (1996)
3. Brown, R.: Exponential Smoothing for Predicting Demand. Little, Boston (1956). https://books.google.com/books?id=Eo_rMgEACAAJ
4. Butler, S., Ringwood, J., Fay, D.: Use of weather inputs in traffic volume forecasting. In: ISSC 2007 Conference (2007). Submitted
5. Castro-Neto, M., Jeong, Y.S., Jeong, M.K., Han, L.D.: Online-SVR for short-term traffic flow prediction under typical and atypical traffic conditions. Expert Syst. Appl. **36**(3), 6164–6173 (2009)
6. Chung, Y.: Assessment of non-recurrent congestion caused by precipitation using archived weather and traffic flow data. Transp. Policy **19**(1), 167–173 (2012)
7. Dunne, S., Ghosh, B.: Weather adaptive traffic prediction using neurowavelet models. IEEE Trans. Intell. Transp. Syst. **14**(1), 370–379 (2013)
8. Florio, L., Mussone, L.: Neural-network models for classification and forecasting of freeway traffic flow stability. Control Eng. Pract. **4**(2), 153–164 (1996)
9. Holt Charles, C.: Forecasting trends and seasonal by exponentially weighted averages. Int. J. Forecast. **20**(1), 5–10 (1957)
10. Huang, S.H., Ran, B.: An Application of Neural Network on Traffic Speed Prediction Under Adverse Weather Condition. Ph.D. thesis, University of Wisconsin-Madison (2003)
11. Hyndman, R.J., Athanasopoulos, G.: Forecasting: principles and practice. OTexts (2014)
12. Hyndman, R.J., Khandakar, Y., et al.: Automatic time series forecasting: The forecast package for R. No. 6/07, Monash University, Department of Econometrics and Business Statistics (2007)

---

[8] http://pems.dot.ca.gov/.

13. Ibrahim, A.T., Hall, F.L.: Effect of Adverse Weather Conditions on Speed-Flow-Occupancy Relationships. No. 1457 (1994)
14. Jia, Y., Wu, J., Xu, M.: Traffic flow prediction with rainfall impact using a deep learning method. J. Adv. Transp. 2017 (2017)
15. Kamarianakis, Y., Prastacos, P.: Space-time modeling of traffic flow. Comput. Geosci. **31**(2), 119–133 (2005)
16. Koesdwiady, A., Soua, R., Karray, F.: Improving traffic flow prediction with weather information in connected cars: a deep learning approach. IEEE Trans. Veh. Technol. **65**(12), 9508–9517 (2016)
17. Lin, L., Wang, Q., Sadek, A.: Short-Term forecasting of traffic volume: evaluating models based on multiple data sets and data diagnosis measures. Transportation research record. J. Transp. Res. Board **2392**, 40–47 (2013)
18. Lippi, M., Bertini, M., Frasconi, P.: Short-Term traffic flow forecasting: an experimental comparison of time-series analysis and supervised learning. IEEE Trans. Intell. Transp. Syst. **14**(2), 871–882 (2013)
19. Lv, Y., Duan, Y., Kang, W., Li, Z., Wang, F.Y.: Traffic flow prediction with big data: a deep learning approach. IEEE Trans. Intell. Transp. Syst. **16**(2), 865–873 (2015)
20. Manual, H.C.: Highway Capacity Manual. Washington, DC p. 11 (2000)
21. Maze, T., Agarwai, M., Burchett, G.: Whether weather matters to traffic demand, traffic safety, and traffic operations and flow. Transportation research record. J. Transp. Res. Board **1948**, 170–176 (2006)
22. Prevedouros, P.D., Chang, K.: Potential effects of wet conditions on signalized intersection LOS. J. Transp. Eng. **131**(12), 898–903 (2005)
23. Rumelhart, D.E., Hinton, G.E., Williams, R.J., et al.: Learning representations by back-propagating errors. Cogn. Model. **5**(3), 1 (1988)
24. Shekhar, S., Williams, B.: Adaptive seasonal time series models for forecasting short-term traffic flow. Transp. Res. Record. J. Transp. Res. Board **2024**, 116–125 (2008)
25. Smith, B.L., Byrne, K.G., Copperman, R.B., Hennessy, S.M., Goodall, N.J.: An investigation into the impact of rainfall on freeway traffic flow. In: 83rd Annual Meeting of the Transportation Research Board, Washington DC. Citeseer (2004)
26. Sun, H., Liu, H.X., Xiao, H., He, R.R., Ran, B.: Short term traffic forecasting using the local linear regression model. In: 82nd Annual Meeting of the Transportation Research Board, Washington, DC (2003)
27. Tsirigotis, L., Vlahogianni, E.I., Karlaftis, M.G.: Does information on weather affect the performance of short-term traffic forecasting models? Int. J. Intell. Transp. Syst. Res. **10**(1), 1–10 (2012)
28. Williams, B.M., Hoel, L.A.: Modeling and forecasting vehicular traffic flow as a seasonal ARIMA process: theoretical basis and empirical results. J. Transp. Eng. **129**(6), 664–672 (2003)
29. Winters, P.R.: Forecasting sales by exponentially weighted moving averages. Manag. Sci. **6**(3), 324–342 (1960)
30. Zhang, D., Kabuka, M.R.: Combining weather condition data to predict traffic flow: a GRU based deep learning approach. IET Intelligent Transport Systems (2018)

Approximate Query Matching
for Graph-Based Holistic Image Retrieval

Abhijit Suprem[1(✉)], Duen Horng Chau[2], and Calton Pu[1]

[1] School of Computer Science, Georgia Institute of Technology, Atlanta, USA
asuprem@gatech.edu, calton@cc.gatech.edu
[2] School of Computational Science and Engineering,
Georgia Institute of Technology, Atlanta, USA
polo@gatech.edu

Abstract. Image retrieval has transitioned from retrieving images with single object descriptions to retrieving images by using complex natural language to describe desired image content. We present work on holistic image search to perform exact and approximate image retrieval that returns images from a database that most closely match the user's description. Our approach can handle simple queries for single objects (ex: cake) to more complex descriptions of multiple objects and prepositional relations between objects (ex: girl eating cake with a fork on a plate) in graph notation. In addition, our approach can generalize to retrieve queries that are semantically similar in case specific results are not found. We use the scene graph, developed in the Visual Genome dataset as a formalization of image content stored as a graph with nodes for objects and edges for relations describing objects in an image. We combine this with approximate search techniques for large-scale graphs and a semantic scoring algorithm developed by us to holistically retrieve images based on given search criteria. We also present a method to store scene graphs and metadata in graph databases using Neo4 J.

Keywords: Image retrieval · Graph search · Approximate search
Scene graphs

1 Introduction

Given the explosion of image-based content shared digitally, there has been similar research focus on developing object recognition, image recognition and variants, and image segmentation tools and systems for understanding, identifying, storing, and querying images. The major research thrusts include automatic image captioning, image annotation (including relationship detection), and content-based image retrieval. Automatic image captioning attempts to generate natural language captions for image images that try to represent the holistic meaning of the image. As the saying goes, 'A picture is worth a thousand words', and finding the right words is a non-trivial task. Image annotation focuses on object detection combined with relationship detection and grounding to identify canonical relationships within an image. Following [1], these are of the form <subject - predicate - object>. Finally, content-based image

© Springer International Publishing AG, part of Springer Nature 2018
F. Y. L. Chin et al. (Eds.): BIGDATA 2018, LNCS 10968, pp. 72–84, 2018.
https://doi.org/10.1007/978-3-319-94301-5_6

retrieval is a wide net; we can consider some recent work from [2], who focus on using automatic image annotation via scene graph grounding to retrieve images similar to a query scene graph.

1.1 Motivation

The motivation for this work comes from the need for holistic image retrieval systems. As [3] notes, text-based queries for image retrieval encode high levels of reasoning and abstraction about an image that is difficult to represent textually. In addition, text-based queries necessarily represent a semantic gap – a disconnect between human semantics and visual features. [2] demonstrates some drawbacks of current text-based image retrieval systems, namely the inability of such systems to abstract queries beyond index-based searches by recognizing the relationships between terms. We show this in Fig. 2, where common image search utilities return results that may not match the given query. We note in the first search engine's results that the second through fifth results feature a woman eating a cake, while in the second, none of the first three feature a girl eating a cake. Our results, however, return images that contain both features of the query, where possible, at least one (image 4 in our results).

1.2 Overview

An overview of the process is shown in Fig. 1. Given the query, we extract the canonical forms - what we deem to be the smallest query unit. In the Visual Genome database, this is a subgraph of the form `<subject - predicate - object>`, where each of the *subject*, *predicate*, and *object* are nodes with edges between them labeled with the type of relation (subject-to-predicate or predicate-to-object). We the generate approximates of the canonical forms to broaden our search. In this case, approximates of `<girl - eating - cake>` include `<girl.n.01 - eat.v.01 - patty.n.01>` and `<girl.n.01 - eat.v.01 - cake.n.03>`. Note that these approximates are represented as WordNet synsets as they are sued to search the Visual Genome database, whose scene graphs also represent nodes as WordNet synsets. We retrieve images and ground them to the query – a process during which we identify which portions of the query graph the image satisfies and to what degree – using phrase similarities derived from high dimensional word embeddings (we use LexVec [4] for its state-of-the-art performance on Word Similarity measurements), Finally, we score each image on its grounding and rank the results.

Our work partially bridges the semantic gap by leveraging the idea of a scene graph as presented in the Visual Genome dataset [1]. A key contribution of [1] is the formalization of a scene graph – each image in the Visual Genome Dataset contains a human-generated graph of nodes denoting subjects, objects, and predicates, with the edges denoting the relations between them. The graph annotates the relational content in the image. The dataset also contains region-level and image-level natural language captions, along with WordNet synsets for objects and relationships. There is a preponderance of research surrounding the scene-graph formalization – Relationship Detection [5], Dense Caption Generation [6], and Scene Graph Generation [7]. We

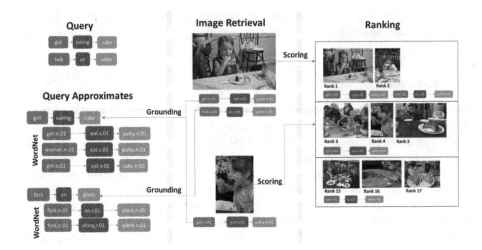

Fig. 1. The user provides a query with two independent requests: girl - eating - cake and fork - on - plate. We generate queries that approximately match the given query using WordNet synsets (girl.n.01 - eat.v.01 - patty.n.01, fork.n.01 - along. r.01 - plate.n.01). We then retrieve images that contain these subgraphs in their scene graphs and ground the subgraphs to our top-level natural language queries. We then map the retrieved synsets and the query nodes to the same word embedding space, and measure similarities to determine holistic image similarity. the image similarity scores are then used to rank the images.

propose leveraging existing capabilities showcased in the above works for an efficient and scalable image retrieval pipeline.

Our contributions include:

Query Approximation and Ranking. We present a model for performing approximate search on a scene graph database. Given a query of the form <subject - predicate - object> (e.g. <subject girl - eating - cake>, we generate approximate queries searching for terms similar to the query terms and evaluating phrases for their similarities to the query. Using this, we can generate, for the given example, query approximates that include woman - eating - cake, person - eating - cake, girl - eating - patty, and woman - eating - patty. In addition, we present an algorithm for grounding and evaluating retrieved scene graphs to the query graph for holistic image ranking.

Aggregate Graph Representations. We present several graph database representations that are used in generating plausible query approximates by reducing the search space to the current semantic context. Specifically, we show how to create three aggregate graph representations that each index a different query node type: (i) a Subject Aggregate Graph (SAG) for obtaining subject approximates, (ii) an Object Aggregate Graph (OAG) for obtaining object approximates, and (iii) a Predicate Aggregate Graph (PAG) for obtaining predicate approximates, all in the current semantic context.

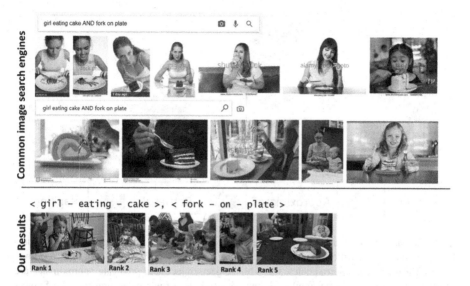

Fig. 2. We show the results from common search engines for the same query. Our results are also presented. We note in the first set of results that the first two images do not include cake - in fact, the woman is eating peas and peppers. A woman eating cake appears third. In the second set of results, a profile view of a woman eating cake appears third in the results. Our results show the first two results with a woman (approximated as girl, as our far smaller image database did not contain any images that exactly matched the query) eating cake, with a fork on the plate.

1.3   The Image Retrieval Pipeline

Figures 1 and 3 summarize our approach. We now focus on the pipeline in Fig. 3. Given a natural language graph query (e.g. a small scene graph as considered by [2]), we reduce the query to its canonical forms (Sect. 4.1); we define a canonical form as the basic unit of query consisting of a subject, predicate, and object of the form <subject - predicate - object>. We then generate approximate queries using WordNet synsets; we retrieve the set of candidate images that contain these approximates and ground the images' scene graphs to the query. We then use word embeddings and project each canonical form to the vector space and score each image's scene graph to the query graph, penalizing images with missing or inexact results.

Three sample queries of varying complexity are presented in Fig. 4. Our approach handles both simple and complex queries. Our grounding ensures images that most closely match the entire query are ranked higher than images that offer a partial match. In addition, we bridge the 'semantic gap' by searching for approximates. This allows us to return images that closely match the provided query in situations where an exact match may not be possible; an example is shown in Fig. 5: there is no image that exactly matches the query; however, our approach approximates woman into girl, and finds images that contain most of the query: a girl eating cake, with the cake on a plate,

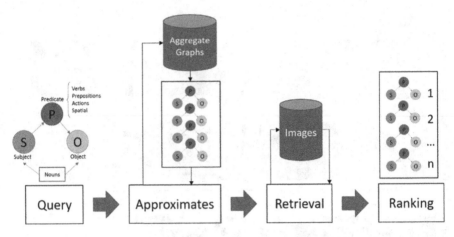

Fig. 3. The user provides a natural language graph query. We convert this to the canonical form (the basic query unit of the form <subject - predicate - object>) for each top-level query and generate the approximates. Plausible subject, object, and predicate approximates are generated using the aggregate graphs we have devised. These aggregate graphs – the Subject Aggregate Graph, the Object Aggregate Graph, and the Predicate Aggregate Graph – allow us to reduce the search space of subjects, objects, and predicates, respectively. Using the node approximates, we generate query approximates and retrieve the set of candidate images using an inverted index of the canonical queries from the Visual Genome dataset. We then ground the images to determine and rank how well the image represents our natural language query.

and a fork on the plate as well. While the highest ranked image in the figure does contain cake with frosting, this is a happy coincidence; the scene graph itself does not contain the annotation, so we say the image mostly matches the query. We also show the second through fifth ranked images and their matches scene subgraphs.

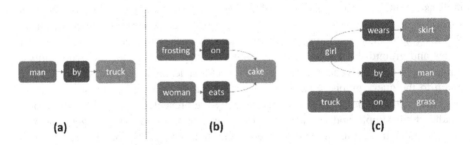

Fig. 4. (a) A simple query. This is also the canonical form described in Sect. 4.1. (b) A more complex query. The same **noun** (*cake*) is the object of two separate subjects (*woman* and *frosting*), each connected by a different predicate. (c) Two independent top-level queries for the same image, i.e. they do not share any nodes between them. This consists of a 'simple' top-level query and a 'complex' top-level query, as per the informal terms adopted in (a) and (b).

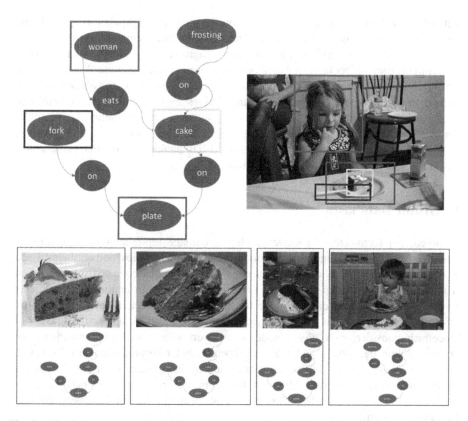

Fig. 5. We show our system matching a complex query and returning images that most closely match all facets of the query.

2 Related Work

Content-Based Image Retrieval. Content-based image retrieval involves retrieving and ranking images given either a text or image query. The former can be used to retrieve images that incorporate some of the query requirements, while the latter can be used to retrieve images relevantly similar to the query – here evaluation is context-specific, e.g. whether images have similar objects, similar colors, similar poses, or whether the images are exactly the same. Some implementations include Google Reverse Image Search and TinEye. Our focus is on the former approach, as we feel text-based queries allow users more freedom in specifying images to retrieve. It is not the case that a user always has a sample of the image she would like, or can sketch a faithful representation in, e.g., sketch-based image retrieval. [2] 's key contribution is as follows: the authors develop a framework for generating accurate groundings of a query scene graph (either complete or partial) and use grounding likelihood to rank images for retrieval and display.

Semantic Similarity. An integral aspect of our image retrieval pipeline is approximate query generation. We work with canonical relationship phrases of the form `<subject - predicate - object>` to generate query approximates. As the Visual Genome dataset maps each object to available WordNet synsets, we incorporate WordNet based semantic similarity measures. [8] proposes a domain-specific corpus-based training method to identify correct word sense and derives more accurate cosine-similarity measures between source and target words. [9] shows a simple baseline for WordNet synset similarity using vector embedding cosine similarity averages. [10] shows a sentence-based similarity measure that uses a TF-IDF analogue to compute similarities between a source word and its synset lemmas.

3   Graph Databases

As noted, [1] formalizes the scene graph – an image representation using human annotations on bounded regions that grounds objects, relationships, and attributes. Each object (usually a noun) is considered a node with a directed edge towards a relation (a predicate with a part-of-speech tag of verb, preposition, or action), The predicate may or may not have a directed edge towards another object instance. In addition, objects also have attributes (usually adjectives, but may also include actions). We will henceforth consider each *object* node as a **noun** with two forms: `subject` or `object`. Note that *object* and `object` operate under separate domains; however, notation confusion suggests these terms as an appropriate choice. A canonical form is a set of three nodes and two edges of the form `<subject → predicate → object>`: this triplet indicates a base query that we use as the smallest query unit (i.e. the canonical form).

3.1   Full Scene Graphs

We generate the full scene graph for each image using the `scene_graphs` from the Visual Genome dataset. The full scene graph is stored in a Neo4J database. Each full scene graph consists of at least one subgraph with at least one canonical triplet of the form `<subject - predicate - object>`. We note that there may be multiple independent subgraphs corresponding to different regions in the image.

3.2   Object Aggregate Graph

Our query approximation requires us to generate the candidate subjects, objects, and predicates for each top-level query. We develop a novel aggregate characterization of the aggregate graph that is introduced in [1] – we maintain a unique index of `<subject → predicate>` pairs, and for each pair, we maintain a unique list of `objects` that are associated with that `subject → predicate` pair. `Objects` are not unique across each subgraph in the aggregate graph: `apple.n.01` may appear in multiple `<subject → predicate>` subgraphs; however, within a subgraph, each WordNet synset occurs once.

With the **Object Aggregate Graph**, we can, given a `subject` → `predicate` pair as well as several candidate `objects`, reduce the set of candidate objects to plausible candidates for that `subject` → `predicate`. Given candidate objects C_O and child objects K_O in the **Object Aggregate Graph**, we return $C_O \cup K_O$. Figure 6 shows a subgraph within our **Object Aggregate Graph**.

Fig. 6. This subgraph from the Object Aggregate Graph shows all possible objects that a Fire Engine (`fire_engine.n.01`) can be next to (`along.r.01`). These include streets (`street.n.01`), sidewalks (`sidewalk.n.01`), and highways (`highway.n.01`). Given a set of approximate synsets for *street*, as well as the `fire_engine.n.01` → `along.r.01` pair, we can identify the appropriate set of plausible objects using the process from Sect. 3.2.

3.3   Subject Aggregate Graph

To generate candidate subjects, we devise a **Subject Aggregate Graph**: we maintain a set of unique `predicate` → `object` pairs, and for each pair, we keep a unique set of `subjects` that are associated with that pair. So, given a `predicate` → `object` pair, and candidate subjects C_S, we reduce it to the plausible set of subjects by returning $C_S \cup K_S$ (where K_S are the parent objects in each subgraph in the Subject Aggregate Graph). We show this in Fig. 7.

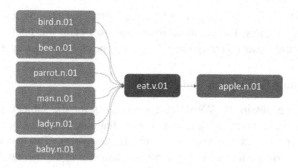

Fig. 7. This subgraph from the Subject Aggregate Graph shows all possible objects that can eat (`eat.v.01`) an apple (`apple.n.01`). These include woman (`woman.n.01`), child (`child.n.01`), and girl (`girl.n.01`).

3.4   Predicate Aggregate Graph

We adapt [1] 's aggregate graph here; however, we include all **nouns** and **predicates** from the scene graphs, instead of the top-k **nouns** and **predicates**. Given a set of candidate `subjects` and candidate `objects`, we can obtain the set of plausible predicates, and return the union of the candidate and plausible predicates. We show a subgraph in Fig. 8.

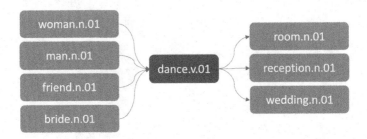

Fig. 8. This partial predicate aggregate graph shows all possible nodes that are subjects and objects for the predicate *dance* (`dance.v.01`). This allows us to determine plausible predicates given a subject-object pair.

4   Query Matching and Image Retrieval

4.1   Canonical Form

The canonical form of a query is a triplet (Fig. 4a) of the form `<subject →
predicate → object>`. Given a graph query, we extract canonical forms from the query and operate on each independently, within scope of the query subgraph. Once again, we refer to Fig. 4c – this query consists of two independent subgraphs: one about a `girl` by a `man`, with the `girl` wearing a `skirt` and one about a `truck` on `grass`. We split the first query into `<girl → wears → skirt>` and `<girl → by → man>`. Similarly, we extract from `truck on grass` the triplet `<truck → on → grass>`.

We perform this triplet extraction under an independence assumption – that we can extract individual triplets, the combine them later to obtain the final rankings. This allows us to operate independently on each triplet and its set of query approximates.

4.2   Approximate Generation and Retrieval

Given a triplet, we obtain the WordNet approximates for the subject, predicate, and object. There are three levels of approximates – obtaining the sister synsets, the child synsets (hyponymy), and the parent synsets (hypernymy). The sister synsets are obtained by performing a lookup in the WordNet database of our natural language node label (i.e. 'girl' or 'skirt' or 'wears'). there are four scopes available for synset lookup: we can take (i) just the sister synsets, (ii) the sister and child synsets, (iii) the sister and parent synsets, (iv) or all three hierarchies: sister, child, and parent synsets. We limit our choice for candidate subjects and objects to the sister synsets and for candidate predicates to sister and child synsets. This is to speed up computation time on our local machines; on parallelized clusters, such a limitation is not necessary, and we can use the complete closure of a synset: sisters, children, and parents.

Given the set of approximates for the subject and object, we obtain plausible predicates from the aggregate graph. The predicates are ranked to the provided predicate sister synsets: we measure the Wu & Palmer (WUP) Similarity between the plausible predicates and set of sister predicates, and taker the average similarity score. Here, we take the top 2/3 predicates as our set of predicate approximates. We choose WUP similarity as it weights synset edges by distance in the hierarchy, i.e. semantically similar predicates are ranked higher than synonymy predicates without semantic relation. This allows us to narrow results to more appropriate relations by context.

4.3 Synset Embeddings

For ranking, we prefer to use a Euclidean metric for measuring triplet distance between the query and approximates. However, the WordNet hierarchy does not operate on such a metric. We use a model similar to [9] – we determine the embeddings for each synset in our database following the baseline method for [9], but instead of the sum, we take the average of the embeddings sum to obtain the synset centroid in the embedding space with

$$v_s = \frac{1}{|L_S|} \sum_{l \in L_S} v_l \qquad (1)$$

where v_S is the embeddings vector for a synset S, L_S is the set of lemmas for the synset S, and V_l is the embedding for each lemma $l \in L_S$. We use LexVec [4] embeddings as the lookup table for v_l as LexVec has shown state-of-the-art performance in word similarity and analogy. We deal with compound words by averaging the embeddings for the compound word itself and the embeddings sum of its component words. If the compound word does not exist, we take only the embeddings sum of component words: given a compound word $w = w_1, w_2, \cdots w_n$ (i.e. for $w = $ christmas tree, $w_1 = $ christmas and $w_2 = $ tree), we take $v'_l = \frac{1}{2}\left(v_l + \sum_{l=l_1,l_2,\cdots,l_n} v_{l_i}\right)$.

4.4 Image Ranking

Approximate-Triplet Image Retrieval. After triplet retrieval, we have, for each canonical triplet T_i, a set of approximates $A^{T_i} = a_1, a_2, \cdots, a_k$. Each of these approximate is a triplet that is semantically similar to the query triplet. For each approximate a_i, we have a set of images $I_1, I_2, \cdots, I_n$ that contain the approximate triplet within them. We generate an inverted index of images for each image, we collect all approximates a_k for each triplet T_i (Fig. 9). This allows us to work on an image-by-image basis by scoring each image based on its holistic similarity to the query triplet.

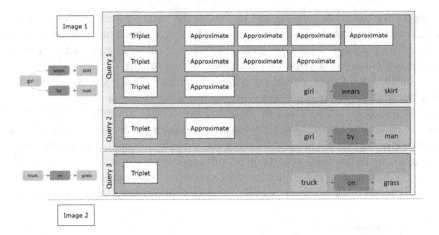

Fig. 9. We store all approximates for each triplet for each image. It is possible that images may not contain a relation, in which case the index will not contain any approximate for a given query (truck on grass).

Inverted Index. Each approximate is a triplet of synsets, and we obtain the synset embeddings using the method in Eq. 1. We also obtain the embeddings for the canonical representations from the embeddings lookup (LexVec, in this case). At this point, we note that each approximate triplet is independent of every other approximate triplet. We further note that each approximate $a_i \in I_n$, there may be multiple triplets that match the approximate. Consider an approximate triplet of the form man on grass: an image may contain multiple instances of this approximate, and each is stored in the inverted index. Each of these approximate primitives ap_j contains a grounding of its nodes to nodes in the parent query approximate a_k, which itself contains a grounding to its parent triplet T_i. As such, for each $ap_{i \in [1,j]}$, we know what it grounds to in T_i. We further note that a single image triplet may appear in multiple primitives ap_j as the parent queries could have similar triplets in multiple subgraphs.

Subgraph Scoring. We first 'collapse' the obtained triplets to their subgraphs s_a – we generate the image subgraphs that contain all approximate primitives within a single top-level query triplet $T_i = s_1, s_2, \cdots s_a$. As such, we reduce all redundant copies of each unique node within the inverted index and connect the independent triplets wherever they share subjects or predicates. This is necessary for the query matching we will perform to rank each image approximate to the provided query. We want a notion of subgraph isomorphism to match out collapsed primitives to our top-level query T_i. From Fig. 9, we note that there may be some T_i that do not contain approximates. For each s_a, we obtain the cosine similarity between its synset embedding and its ground embedding, which is the embedding of the node in the top-level query T_i. The cosine similarity is in $[0, 1]$, where 1 indicates the highest similarity. For each node, we instead store $n_{score} = 1 - \text{cosine\_sim}(node, query)$. For each missing node in s_a, we add a null node with a distance of 1, representing a missing node.

Subgraph Ranking. With this representation, we can now formulate this as a minimization problem: we wish to select the subgraph with the smallest score. Since each subgraph $s_a \in T_i$ may contain a combination of approximates, it is intractable to calculate the global minimum selection of subgraphs. We instead model this as an analogue of vertex cover. By construction, each subgraph contains a unique set of approximates for the top-level query. As such, larger subgraphs are inherently more 'isomorphic' to each T_i simple because they contain more grounded nodes.

We then sort our subgraphs by size, and within each set of subgraphs with the same size, we pick the subgraph s_m with the smallest score. This subgraph s_m has a subset of the top-level approximates in T_i. We remove these approximates from consideration, and again pick, from the remaining, the largest subgraphs. We repeat this until all top-level queries are satisfied for each top-level subgraph. Each of the selected subgraphs S_m's scores $S_i = score(s_1, s_2, \cdots, s_m)$ are summed and normalized by the number of nodes to find the score on I_n for query T_i. We repeat this for each query in I_n to obtain scores $S = [S_1, S_2, \cdots, S_n]$ for each query T_i.

Image Scoring and Ranking. We note that for each of the scores $S = [S_1, S_2, \cdots, S_n]$ is in $[0, 1]$, with a lower score corresponding to a more similar match for each top-level query. It is straightforward then to represent these scores as an i-dimensional score vector $[S_1, S_2, \cdots, S_i]$ and measure its Euclidean distance from the origin. This gives us a score for image I_n. We then sort the image score to obtain the image ranking under query, where the smallest score is the most relevant image.

5 Conclusions and Future Work

We have presented work on image retrieval using graph based approximate querying and ranking. Representative examples are shown in Figs. 1 and 5, as well as a comparison in Fig. 2. We note from Fig. 2 that our system returns relevant results at higher ranks than two leading Search Engines. We also note from Fig. 5 that even for more complex queries, out system can return results that closely match the provided query. Top ranked results match as much of the query as possible *with holistic meaning* – we reduce the 'semantic gap' by considering relations between objects in the image in lieu of using a document-based that eschews a focus on image relational content.

As such, the scene graph of an image is a key factor of our work. With regard to this, future work is two-fold:

- **Accurate scene graph generation:** There is some work on scene graph generation in [7]. However, the authors note that the performance is subpar. Better state-of-the-art performance in automated scene graph generation from unannotated images would allow creation of image databases at scale. Our system can then be implemented on top of such a database for approximate image retrieval.
- **Query graph generation:** We provide an informal comparison of our results to current search engine results in Fig. 2. However, this is not a robust comparison as search engines accept natural language input while we provide input directly as graph queries. This is due to a major input limitation in the conversion of natural

language to scene graphs. Future work would focus on graph query generation from natural language that more closely matches desired human queries, using, e.g. dependency parsing or similar methods.

References

1. Krishna, R., Zhu, Y., Groth, O., Johnson, J., Hata, K., Kravitz, J., Chen, S., Kalantidis, Y., Li, L.-J., Shamma, D.A.: Visual genome: connecting language and vision using crowdsourced dense image annotations. Int. J. Comput. Vis. **123**(1), 32–73 (2017)
2. Johnson, J., Krishna, R., Stark, M., Li, L.-J., Shamma, D., Bernstein, M., Fei-Fei, L.: Image retrieval using scene graphs. In: Proceedings of the IEEE Conference on Computer Vision and Pattern Recognition, pp. 3668–3678 (2015)
3. Liu, Y., Zhang, D., Lu, G., Ma, W.-Y.: A survey of content-based image retrieval with high-level semantics. Pattern Recogn. **40**(1), 262–282 (2007)
4. Salle, A., Idiart, M., Villavicencio, A.: Enhancing the lexvec distributed word representation model using positional contexts and external memory. *CoRR* (2016)
5. Lu, C., Krishna, R., Bernstein, M., Fei-Fei, L.: Visual relationship detection with language priors. In: Leibe, B., Matas, J., Sebe, N., Welling, Max (eds.) ECCV 2016. LNCS, vol. 9905, pp. 852–869. Springer, Cham (2016). https://doi.org/10.1007/978-3-319-46448-0_51
6. Johnson, J., Karpathy, A., Fei-Fei, L.: Densecap: Fully convolutional localization networks for dense captioning. In: Proceedings of the IEEE Conference on Computer Vision and Pattern Recognition, pp. 4565–4574 (2016)
7. Xu, D., Zhu, Y., Choy, C.B., Fei-Fei, L.: Scene graph generation by iterative message passing. In: Computer Vision and Pattern Recognition (CVPR) (2017)
8. Patwardhan, S., Pedersen, T.: Using WordNet-based context vectors to estimate the semantic relatedness of concepts. In: Proceedings of the EACL 2006 Workshop: Making Sense of Sense-Bringing Computational Linguistics and Psycholinguistics Together, vol. 1501, pp. 1–8 (2006)
9. Khodak, M., Risteski, A., Fellbaum, C., Arora, S.: Automated WordNet Construction Using Word Embeddings. In: SENSE 2017 (2017)
10. Arora, S., Liang, Y., Ma, T.: A simple but tough-to-beat baseline for sentence embeddings. In: International Conference on Learning Representations (2017)

Research Track: BigData Analysis

PAGE: Answering Graph Pattern Queries via Knowledge Graph Embedding

Sanghyun Hong[1], Noseong Park[2(✉)], Tanmoy Chakraborty[3],
Hyunjoong Kang[4], and Soonhyun Kwon[4]

[1] University of Maryland, College Park, MD, USA
shhong@cs.umd.edu
[2] University of North Carolina, Charlotte, NC, USA
npark2@uncc.edu
[3] Indraprastha Institute of Information Technology Delhi, Delhi, India
tanmoy@iiitd.ac.in
[4] Electronics and Telecommunications Research Institute, Daejeon, South Korea
{hjkang,kwonshzzang}@etri.re.kr

Abstract. Answering graph pattern queries have been highly dependent on a technique—i.e., subgraph matching, however, this approach is ineffective when knowledge graphs include incorrect or incomplete information. In this paper, we present a method called PAGE that answers graph pattern queries via knowledge graph embedding methods. PAGE computes the energy (or uncertainty) of candidate answers with the learned embeddings and chooses the lower-energy candidates as answers. Our method has the two advantages: (1) PAGE is able to find latent answers hard to be found via subgraph matching and (2) presents a robust metric that enables us to compute the plausibility of an answer. In evaluations with two popular knowledge graphs, Freebase and NELL, PAGE demonstrated the performance increase by up to 28% compared to baseline KGE methods.

Keywords: Graph databases · Graph query answering
Knowledge graph embedding

1 Introduction

Graphs/networks are widely used in various fields, e.g., knowledge graphs (KGs) in the Semantic Web, social networks in Social Analytics, protein-protein interaction (PPI) networks in Bioinformatics, etc. As their applications are diverse, many different graph mining paradigms have been proposed: the Semantic Web has its own knowledge graph query language called SPARQL [14], and Neo4j [10], the market-leading graph database management system, also has a graph query language called Cypher. Unfortunately, that progress has been made at search *subgraph patterns* from underlying graphs via subgraph isomorphism, often hard to find answers when the graphs are incomplete or carry incorrect information [11].

Graph embedding methods have come into the light nowadays because of their promising performance in various tasks such as community detection [7],

© Springer International Publishing AG, part of Springer Nature 2018
F. Y. L. Chin et al. (Eds.): BIGDATA 2018, LNCS 10968, pp. 87–99, 2018.
https://doi.org/10.1007/978-3-319-94301-5_7

link prediction in the social network [12,17], and query answering on knowledge graphs [2]. Those methods learn *latent vector representations (or embeddings)* of vertices and relations[1]. Prior works have reported that using embeddings can provide a way to answer factoid queries[2] even with incorrect and incomplete information [2–4]. However, KGE methods have only considered simple queries consisting of a single edge or multiple unidirectional edges—i.e., it has not been explored whether we can use them to answer general graph queries.

In this paper, we introduce PAGE (**P**attern query **A**nswering through knowledge **G**raph **E**mbedding) that delivers a new paradigm of querying KGs. To the best of our knowledge, we are the first effort to combine them and explore the potential of KGE methods in answering graph queries. Advantages of the proposed approach are twofold:

1. *Our method can discover latent patterns which remain hidden in the incomplete or incorrect KGs.* Rather than relying on the subgraph matching, PAGE chooses candidate answers for a graph query based on the energy computed with embeddings, which enables our method to return complete answers.
2. *We present a robust metric that can compute the plausibility of an answer.* This metric aids in post-processing after querying KGs. There can be numerous subgraph patterns matched to a graph query; processing all of them is computationally too expensive. In PAGE, we only identify highly plausible subgraph patterns and provide them as candidate answers.

In evaluations, we conduct two experiments: (1) factoid query answering and (2) graph query answering with two popular KGs, Freebase and NELL. Our result demonstrates that PAGE outperforms baseline KGE methods by at most 28% in terms of standard metrics such as mean rank and Hits@10/100/1000. The evaluation results show the potential of using KGE methods for answering graph queries in KGs even though the KGs carry incomplete information.

2   Background

In this section, we introduce the basic concepts of graph pattern query answering and KGE. Let $\mathcal{G} = (\mathcal{V}_\mathcal{G}, \mathcal{E}_\mathcal{G})$ be a KG and L be a set of relations. $\mathcal{V}_\mathcal{G}$ is a set of vertices and $\mathcal{E}_\mathcal{G}$ is a set of edges labeled by one of the relations in L. A relation in L means a certain type of relationship between vertices.

2.1   Graph Query Answering

Given a KG $\mathcal{G}$ and a graph query $\mathcal{Q} = (\mathcal{V}_\mathcal{Q}, \mathcal{E}_\mathcal{Q})$, the task of conventional graph query answering is to find all subgraph patterns of $\mathcal{G}$ matched to $\mathcal{Q}$ via subgraph isomorphism [16].

---

[1] A relation is an edge label. A KG is an edge-labeled graph.

[2] Factoid queries are the simplest type of graph queries such as "who visited Canada?", denoted as $?x \xrightarrow{visited} Canada$.

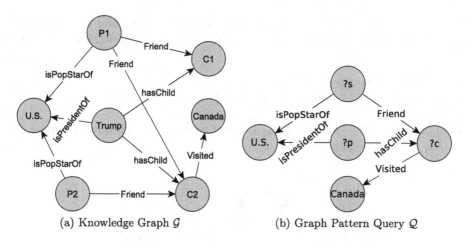

(a) Knowledge Graph $\mathcal{G}$ (b) Graph Pattern Query $\mathcal{Q}$

Fig. 1. An example KG (a) and graph pattern query (b). (Note that $?p$, $?c$, and $?s$ are variables in the query (b), and the answers are $Trump$, $C2$, and $P2$ of $\mathcal{G}$.)

Example 1 (Graph Pattern Query with Projection). In Fig. 1, the graph query searches for a child $?c$ of the president $?p$ of the United States, who had visited Canada before and is a friend of a pop star $?s$. The answer is C2 because $\{?c = C2, ?p = \text{Trump}, ?s = P1\}$ is a valid subgraph that matches the query.

In query languages, a graph pattern query is expressed as a series of path queries. We write the example query in Neo4j's Cypher query language as follows:

Query 1.1: Cypher representation of the query in Figure 1 (b).
```
MATCH (US)<-[:isPresidentOf]-(?p)-[:hasChild]->(?c)
MATCH (US)<-[:isPopStarOf]-(?s)-[:Friend]->(?c)
MATCH (?c)-[:Visited]->(CANADA)
RETURN ?c
```

Note that each path query is projected to a vertex matched to $?c$. As shown in the above expression, *the problem of answering a graph query can be decomposed into: answering each path query, coming up with a set of candidate answers, and choosing the common answer among candidates.* This observation sheds light on how we can answer the general form of graph queries.

2.2 Knowledge Graph Embedding

KGE methods map vertices and relations into a d-dimensional continuous vector space. TransE, SE, and SME [2–4] are the most popular and pioneering works; those methods answer a factoid query by using the concept of energy. Given a semantic triple $t = (\mathbf{v}, \mathbf{r}, \mathbf{u})$ in a KG, the energy of the triple indicates the uncertainty (or error) such that a high energy level means a high uncertainty of the triple.

The v and u stand for vertices, r is a relation between them, and we use a bold character to denote an embedding vector. The energy can be defined in various ways:

1. In TransE, $e(v, r, u) = \|\mathbf{v} + \mathbf{r} - \mathbf{u}\|_{1 \text{ or } 2}{}^{3}$.
2. In SME, $e(v, r, u) = g_v(\mathbf{r}, \mathbf{v})^{\mathrm{T}} \cdot g_u(\mathbf{r}, \mathbf{u})$, where g_v and g_u are linear or bilinear functions.
3. In SE, $e(v, r, u) = \|\mathbf{r_l}\mathbf{v} - \mathbf{r_r}\mathbf{u}\|_{1 \text{ or } 2}$, where $\mathbf{r_l}$ and $\mathbf{r_r}$ are left and right projection matrices[4] representing r.

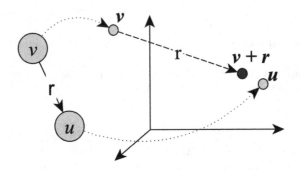

Fig. 2. An example of TransE embedding: in TransE, a triple v, r, u (on the left side) are mapped into the 3-dimensional space (on the right side). $\mathbf{v} + \mathbf{r}$, represented by the blue vertex, is the prediction of v's neighbor connected through r that minimizes the energy (error) $\|\mathbf{v} + \mathbf{r} - \mathbf{u}\|$, which refers to the distance between the blue vertex and $\mathbf{u}$.

In training, KGE methods learn the vector representations (or embeddings) of vertices and relations by minimizing the following loss function:

$$\mathcal{L} = \sum_{t^+ \in \mathcal{E}} \sum_{t^- \in \mathcal{N}(t^+)} \max\left(0, \gamma + e(t^+) - e(t^-)\right), \qquad (1)$$

where $\mathcal{E}$ is a set of triples, $\mathcal{N}(t^+)$ is a set of negative triples t^- derived from true triples $t^+ \in \mathcal{E}$, γ is a margin, and $e(\cdot)$ is the energy of a triple. For instance, $e(\mathbf{v}, \mathbf{r}, \mathbf{u}) = \|\mathbf{v} + \mathbf{r} - \mathbf{u}\|$ in TransE such that a true triple that exists in the KGs makes the $e(\cdot)$ zero—i.e., $\mathbf{v} + \mathbf{r} = \mathbf{u}$ (see Fig. 2). Thus, training with this loss function decreases the energy of the true triples while increasing the energy of false triples such that they differ by at least γ.

With the learned embeddings, existing works answer two types of queries: (1) factoid queries $u \xrightarrow{r} ?x$ and (2) unidirectional path queries $u \xrightarrow{r_1, r_2, \cdots} ?x^5$ [6] by

---

[3] $\| \cdot \|_1$ (resp. $\| \cdot \|_2$) refers to the ℓ_1-norm (resp. ℓ_2-norm) of a vector.
[4] After vectorization, a matrix can still be represented by a vector.
[5] Note that r_i means an intermediate relation in the path between u and $?x$, and all the relations have the same direction.

finding the top-k answers that have the smallest energy among all the candidate answers for $?x$.

Finding a set of candidate answers relying on the energy provides two advantages: (1) this enables us to find the latent answers, and (2) the methods works with KGs that include incomplete information. In this work, we *further extend the KGE methods to answer the general form of a graph pattern query*, which enables to answer the query without subgraph isomorphism (or subgraph matching).

3 Graph Query Answering via KGE

Using existing KGE methods to answer graph pattern queries has a major problem—i.e., those methods are limited to answer factoid or unidirectional path queries (see Sect. 2.2). However, general graph queries involve *bi-directional path queries* as shown in Query 1.1. In addition, it is well-known that considering multiple-hop paths makes KGE methods vulnerable to accumulated errors because an error in an edge can be amplified after multiple hops [6]. This motivates us to present a new query model and a training method. In this section, we introduce PAGE, a novel method that enables us to answer graph queries on incomplete KGs via knowledge graph embeddings without relying on subgraph isomorphism.

3.1 PAGE Energy Definition and Query Model

Dropping Subgraph Isomorphism. Subgraph isomorphism (or subgraph matching) has traditionally been considered as the key to answer graph pattern queries. However, the quality of the answers drastically decreases when the underlying KG is not complete or contains incorrect knowledge. This is the well-known problem since constructing a KG from web pages or documents is challenging, and the KG usually carries incorrect knowledge representation [15]. To overcome this issue, we drop the subgraph isomorphism in the proposed PAGE query model. Instead, we utilize KGE methods that provide high accuracy in answering factoid or unidirectional queries and is able to rectify incorrect knowledge [15].

The Energy of a Bidirectional Path Query. In our model, we consider a graph query as a set of bidirectional path queries (see Sect. 2.1). To answer a graph query, we first need to answer each bidirectional path query via KGE methods. However, most KGE methods have been proposed without considering bidirectional path queries—i.e., the operators used to compute the energy are not invertible. Thus, we define the regular and inverse energy operations as follows:

Definition 1 (Regular Operation). *Given a query $v \xrightarrow{r} ?x$, the regular operation is to find x such that $energy(\mathbf{v}, \mathbf{r}, \mathbf{x}) = 0$, e.g., $\mathbf{x} = \mathbf{v} + \mathbf{r}$ in* TransE. *This answers a query $v \xrightarrow{r} ?x$.*

Definition 2 (Inverse Operation). *Given a query $?x \xrightarrow{r} u$, the inverse operation is to find x such that $energy(\mathbf{x}, \mathbf{r}, \mathbf{u}) = 0$, e.g., $\mathbf{x} = \mathbf{u} - \mathbf{r}$ in* TransE. *This answers a query $?x \xrightarrow{r} u$.*

For instance, suppose that we compute the energy of a bidirectional 2-hop path query $?c \xleftarrow{r_1} Trump \xrightarrow{r_2} US$ in TransE. The inverse operation of the energy is derived in a straightforward manner (see Sect. 3.3). Thus, the energy is computed as $e(\mathbf{e}) = |\mathbf{u} - \mathbf{r2} + \mathbf{r1}|$, where $\mathbf{u}$ is US, and the answers can be C1/C2 close to the vector $\mathbf{e}$. With the operations, we define the energy of a bidirectional path.

Definition 3 (Energy of Bidirectional Path). *Given an h-hop bidirectional path p in a KG, whose left-end is a vertex $u \in V$ and right-end is a vertex $v \in V$ with a series of intermediate relations $r_1, \cdots, r_h$, let $\mathbf{x}$ be a vector calculated after a series of regular and inverse operations starting from $\mathbf{u}$ up to r_{h-1}. The energy of the bidirectional path is then defined in* TransE *as:*

$$energy(p) = \begin{cases} ||\mathbf{x} + \mathbf{r_h} - \mathbf{v}||, & \text{if the last edge is } \xrightarrow{r_h} v \\ ||\mathbf{v} + \mathbf{r_h} - \mathbf{x}||, & \text{if the last edge is } \xleftarrow{r_h} v \end{cases} ,$$

Note that this bidirectional path energy definition is independent from underlying triple energy definitions. We also test a couple of different triple energy definitions in our experiments. Now we can define the energy of a graph query by using the sum of all the energy of bidirectional paths in the query.

Definition 4 (Energy of a Graph Query). *Let Q be a pattern query and q be an answer candidate to Q, the energy of the graph query, denoted as $e(q)$, is defined as:*

$$energy(q) = \sum_{p \in paths(q)} energy(p), \tag{2}$$

where paths(q) returns bidirectional paths of q that are matched to bidirectional path queries of Q.

Therefore, answering a graph query Q is to find a bidirectional paths p in a KG such that $energy(p)$ is minimized.

3.2   Improve the Training Step of KGE Methods

To leverage the new energy definition that supports the regular and inverse operations in PAGE, we need to improve the training process of KGE methods. We first sample spanning trees from KGs and decompose each tree into a set of bidirectional paths between two terminal vertices (or degree 1 vertices) in the tree. We then create a set of false paths by altering one terminal vertex of a true path and use both true and false path sets as our training data. This improvement enables KGE methods to learn embeddings of vertices and relations used for answering graph pattern queries.

Sampling Spanning Trees. We sample spanning trees from the training sets in Sect. 4.1 by performing the following procedure.

1. Randomly choose a vertex from a KG $\mathcal{G}$.
2. Perform the Join 3(b) of the FFSM [8] e times so that a spanning tree with e edges will be sampled[6].
3. Repeat 1 and 2 until all vertices and edges of the graph $\mathcal{G}$ are covered by at least c different sampled spanning trees.

To ensure a set of comprehensive samples, we utilize the sampling procedure with multiple e values—i.e., we use e up to 4 in our experiments. This method also allows frequently appearing edges in $\mathcal{G}$ to be more sampled than others, which is fair because those edges are more likely to be the part of answers for a graph query. We derive false spanning trees from a true spanning tree by applying the Join 3(b) of the FFSM e times with the edges not in $\mathcal{G}$.

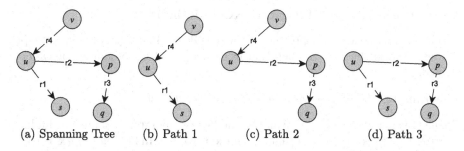

(a) Spanning Tree   (b) Path 1   (c) Path 2   (d) Path 3

Fig. 3. An example spanning tree (a), and its three terminal-to-terminal decompositions (b), (c), and (d).

Decomposing Spanning Trees into Bidirectional Paths. Since our method considers each bidirectional paths p from a spanning tree t as a training case, we compute the energy of a path by decomposing the sampled spanning trees. For instance, in Fig. 3, we can extract three terminal-to-terminal bidirectional paths from a sampled spanning tree. We utilize the concept introduced in Definition 3 to compute the energy of each case.

Margin-Based (Hinge) Loss Function. PAGE does not only minimize the total energy of training paths decomposed from sampled spanning trees, but also tries to obtain a reasonable energy margin between a training path and false paths derived from the training path. We utilize the same loss function as described in Eq. (1) with our own energy definition. Given a training path t^+ and a false path t^- created by randomly modifying one terminal vertex of t^+, t^+'s energy is required to be smaller than that of t^- by a margin of γ. If this is the case for all true and false paths, then the loss function becomes 0, which means a perfect embedding.

---

[6] The Join 3(b) operation simply appends one random vertex to the terminal position of the current tree such that the extended tree can be also a valid subtree in $\mathcal{G}$.

Input: KG $\mathcal{G} = (\mathcal{V}_\mathcal{G}, \mathcal{E}_\mathcal{G})$, Max iteration max_iter, Learning rate δ

Output: Embedding matrix M for vertices and relations

---

1 M $\leftarrow d \times n$ matrix with random initialization
2 $\mathcal{T} \leftarrow$ sampled true spanning trees
3 **while** $iter \leq max\_iter$ **do**
4 $\quad$ $\mathcal{T} \leftarrow random\_permute(\mathcal{T})$
5 $\quad$ **foreach** $\mathcal{T}_{batch} \in minibatch\_split(\mathcal{T})$ **do**
6 $\quad\quad$ $\mathcal{N}(\mathcal{T}_{batch}) \leftarrow$ sampled false spanning trees from each $t^+ \in \mathcal{T}_{batch}$
7 $\quad\quad$ M $=$ M $- \delta \times \nabla_\mathsf{M}\mathcal{L}(\mathcal{T}_{batch}, \mathcal{N}(\mathcal{T}_{batch}))$
8 $\quad\quad$ M $= \frac{\mathsf{M}}{||M||_F}$
9 **return** M

Algorithm 1: The embedding algorithm used in `PAGE`

3.3   Infeasible Cases of Existing KGE Methods

In this section, we formally prove that some KGE methods cannot answer graph queries because the inverse operation in each method is not unique or cannot defined.

Theorem 1. *The inverse operator of* `SME` *is not unique.*

Proof. Given a query $?x \xrightarrow{r} u$, $\mathbf{r}$ and $\mathbf{u}$, the inverse operator is defined as finding a vertex x mapped to $?x$ such that $energy(\mathbf{x}, \mathbf{r}, \mathbf{u}) = 0$. In SME, $energy(\mathbf{x}, \mathbf{r}, \mathbf{u}) = g_v(\mathbf{r}, \mathbf{x})^\mathrm{T} \cdot g_u(\mathbf{r}, \mathbf{u})$. Let $X = g_v(\mathbf{r}, \mathbf{x})$ and $Y = g_u(\mathbf{r}, \mathbf{u})$; thus, $energy(\mathbf{x}, \mathbf{r}, \mathbf{u}) = X^\mathrm{T} \cdot Y$. Note that X and Y are both $d \times 1$ column vectors. When $X^\mathrm{T} \cdot Y = 0$, $w \cdot X$, where w is any scalar coefficient, an energy of 0 also arises, which implies that any $\mathbf{x}'$ can be a solution as long as $g_v(\mathbf{r}, \mathbf{x}') = w \cdot X$. There are so many such w that $w \cdot \mathbf{x}'$ can be a solution of the inverse operator.

Theorem 2. *In some variations of* `TransE`, *the inverse operator cannot be defined or is not computationally desired.*

Proof. Due to space constraints, we sketch a proof. The key idea is: (i) to prove the existence (or uniqueness) of the inverse of a generative model (`TransG`) and (ii) to discuss the inverse matrix computation time during the loss minimization (`TransH`, `TransD` and so on). For instance, `TransG` learns multiple vector representations for a relation r. Thus, $energy(v, r, u)$ is a weighted sum of several different energy levels. Each vector representation leads to a different energy level, which can be simply written as $\sum w_i \cdot energy_i(v, r, u)$, where $energy_i(v, r, u)$ is the energy level defined by the i_{th} vector representation of r. Given a query edge $?x \xrightarrow{r} u$, there are many such candidates of $?x$ that the weighted sum equals zero. Thus, the inverse operator solution of `TransG` also cannot be uniquely defined.

3.4   Embedding Algorithm

Let M be a $d \times n$ embedding matrix, where $n = |\mathcal{V}_G| + |\mathsf{L}|$—i.e., each column of M is an embedding of a vertex or a relation. We use the *projected stochastic gradient descent (SGD)* method described in [7] to compute the M that minimizes the loss function in Eq. 1. In Algorithm 1, we first randomly initialize M (line 1) and sample spanning trees from $\mathcal{G}$ (line 2). In each iteration, we randomly permute sampled spanning trees $\mathcal{T}$ (line 4) and update M w.r.t. the gradient of the loss term (line 7). The SGD computation is efficiently done by various deep learning platforms with the support of GPUs[7]. At the end of each iteration (line 8), we project M onto the unit sphere to prohibit M from being extremely large during iterations.

4   Evaluation

We evaluate PAGE in two tasks: (1) factoid query answering and (2) graph query answering. We expect that PAGE based on KGE methods outperform in the aforementioned two tasks than the baseline KGE methods.

Baselines. We use TransE and SE as our baseline methods because they support both regular and inverse operations. Other KGE methods such as SME and some variations of TransE are excluded because they cannot define unique inverse operations from their energy definitions (see Sect. 3.3). In our experiments, we denote the TransE improved by the proposed training process as PAGE-TransE and the improved SE as PAGE-SE.

Experimental Setup. We implement PAGE in Python 2.7 with Theano deep learning library[8]. In evaluations, we run PAGE on Amazon EC2 instances of type g2.2xlarge equipped with an Intel Xeon E5-2670 processor that has eight processor cores, 15 GB of RAM, and a Nvidia Tesla GPU with 4 GB of video memory and 1,536 CUDA cores.

4.1   Databases and Evaluation Metrics

In this subsection, we discuss our databases and metrics.

Databases. We conducted experiments on datasets from two popular KGs: *FB15K* [3] is a subset of Freebase, and *Nell186* [5] is a subset of NELL containing the most frequent 186 relations. In both KGs, we have well-defined training graphs and factoid testing/validating queries. We sample spanning trees from training graphs and also create random graph pattern queries as follows.

1. Merge training and test sets into one KG.
2. Randomly select a vertex v from the test set.
3. Create z paths by iterating the following steps z times.
   (a) Choose a length in between 2 and 4.

---

[7] We used the Theano [1], one of the most popular deep learning platform.

[8] Theano is one of the most popular deep learning platforms. Optimizing the loss function with the SGD method can be performed efficiently with the support of GPUs.

(b) Randomly select a path of the chosen length starting from v. This path should have at least one edge in the original test set.
4. Convert v and all intermediate vertices of the sampled paths into variables and create a graph query Q.
5. The correct answer to the query Q is v, i.e., we are interested only in finding an entity mapped to the variable converted from v.

The statistics of our datasets are summarized in Table 1.

Table 1. Statistics of the *FB15K* and *Nell186* databases

Database	Vertices	Relations	Training edges	Testing queries	Validating queries
FB15K	14,951	1,345	483,142	50,000	59,071
Nell186	14,463	186	31,134	5,000	5,000

Metrics. We use the same evaluation metrics as in previous studies: (1) the average rank of the correct answers among the entities sorted in ascending order of energy (mean rank), and (2) the proportion of correct answers ranked in the top 10/100/1000 (Hits@10/100/1000).

4.2   Factoid Query Answering

Table 2 summarizes the results of the factoid query answering task. The best performances are shown in TransE cases, and SE shows the worse performance than TransE for all the datasets across all the metrics. Thus, our discussion focuses on the TransE and PAGE-TransE cases. In terms of the mean rank, PAGE-TransE demonstrates at most 13% better performance than the baseline TransE. The performance of TransE and PAGE-TransE in terms of Hits@10/100 is similar in *FB15K* whereas TransE performs slightly better in *Nell186*.

Table 2. Mean ranks and Hits@10/100/100 for the factoid query task. (The best values are marked in bold font.)

Database	Metric	Type	TransE	PAGE-TransE	SE	PAGE-SE
FB15K	Mean rank	Micro	181.76	**178.98**	408.69	375.48
		Macro	109.02	**106.09**	412.04	364.24
	Hits@10/100	Micro	43.4%/**76.6%**	**44.2%**/76.2%	21.9%/59.2%	22.3%/59.3%
		Macro	49.2%/**81.3%**	**49.4%**/80.9%	27.6%/62.8%	28.9%/62.8%
Nell186	Mean rank	Micro	885.54	**784.02**	3412.0	3752.5
		Macro	885.54	**784.02**	4492.2	4736.8
	Hits@10/100	Micro	**41.5%/74.6%**	38.6%/72.4%	10.1%/15.75%	9.2%/14.5%
		Macro	**41.5%/74.6%**	38.6%/72.4%	3.3%/8.0%	3.0%/7.1%

PAGE that involves proposed training process did not show the best performance in all the factoid query answering tasks. However, PAGE demonstrates similar accuracy in terms of Hit@100 and is better in graph query answering (see Sect. 4.3. In more than 70% to 80% of the testing queries, correct answers are part of the top-100 candidates, which means our approach is generally applicable.

4.3   Graph Query Answering

In Table 3, we summarize the results of the graph query answering task. Since the graph query answering is a more difficult task than the factoid query answering, we use Hits@100/1000 instead of Hits@10. Similar to the factoid query answering results, SE exhibits worse performance than TransE, thus, our comparison focuses on the TransE cases. As expected, PAGE-TransE significantly outperforms TransE in all cases, which implies that considering of terminal-to-terminal bidirectional paths in the training process enables answering graph queries. In detail, PAGE-TransE demonstrates 9% to 28% enhancements for Hits@100 (19% to 24.3% in *FB15K* and 60.2% to 65.4% in *Nell186*) compared to the original TransE.

Table 3. Mean ranks and Hits@10/100/100 for the graph task. (The best values are marked in bold font.)

Database	Metric	Type	TransE	PAGE-TransE	SE	PAGE-SE
FB15K	Mean rank	Micro	1150.5	**1088**	7493.5	7514.0
		Macro	2509.9	**2362.8**	7571.4	7933.7
	Hits@100/1000	Micro	18.3%/56.7%	**25%/60%**	1.7%/5.0%	1.7%/10%
		Macro	19%/58.7%	**24.3%/61.7%**	2.0%/6.0%	2.0%/7.7%
Nell186	Mean rank	Micro	38.5	**38**	4803.6	4960.1
		Macro	769.4	**491.1**	5240.3	5431.8
	Hits@100/1000	Micro	64.8%/89.4%	**66.5%/94.6%**	15.75%/31.4%	14.5%/28.9%
		Macro	60.2%/80.7%	**65.4%/87.2%**	7.9%/23.4%	7.1%/21.3%

5   Discussion

Complexity Issue of the PAGE Query Model. Dropping the subgraph isomorphism enables to find latent answers since any vertex can be a candidate answer of a variable. However, considering entire vertices as candidate answers is not computationally preferred. For instance, in the mixed-directional path query $?c \xleftarrow{r_1} ?y \xrightarrow{r_2} ?z \xleftarrow{r_3} US$, the number of candidates can be exponentially increased once we decided to search candidates for the intermediate variables $?y$ and $?z$. Instead, our method excludes intermediate variables in a mixed-directional query path from candidates and computes the energy between $?c$ and US by considering only the relations r_1, r_2, and r_3 (and their directions). This approach decreases computations and enables a lightweight query processing time complexity—i.e., k^n rather than k^m, where k is the number of candidates for a variable, m is the number of all variables, and $n \ll m$ is the number of non-intermediate variables.

Qualitative Comparison with Approximated Graph Query Model. Many approximated graph query answering models have been proposed [9,13]. These models partially ignore the subgraph isomorphism by allowing missing edges in a KG or considering only semantically similar edges. However, there is a case in which the answers from such models cannot be one of the top-rank answers whereas our model ranks any low-energy candidate highly. For instance, in the worst case, suppose that the query is "Who is the athlete who won the U.S. Open against Roger Federer and is a teammate of Andy Murray?". The correct answer is Andy Roddick, however, the following two triples are not contained in the training set of *Nell186* [5]:

$$Andy\ Roddick \xrightarrow{won} US\ Open$$

$$Andy\ Roddick \xrightarrow{isTeammateOf} Rodger\ Federer$$

No existing approximate query model can answer this query correctly because all query edges are not matched for Andy Roddick, however, our PAGE model had listed Andy Roddick as one of the top-20 candidates (more precisely, the 18th candidate in terms of energy) among all the vertices.

6   Conclusion

This paper is the first work that tackles the problem of subgraph matching by utilizing KGE methods. We propose PAGE, a novel query model that enables to answer general graph queries on incorrect or incomplete KGs, which provides a new paradigm of querying KGs. Our work has two contributions to data mining and KGE research: (1) we demonstrated that a graph query (or a complex form of a query) can be answered through KGE methods by decomposing the query into multiple mixed-directional path queries, and (2) we achieved the same performance in simple query answering task and the better performance in graph query answering task with two popular KGs. In evaluations, the performance enhancement is at most 28% compared to the baseline KGE methods.

References

1. Bergstra, J., Breuleux, O., Bastien, F., Lamblin, P., Pascanu, R., Desjardins, G., Turian, J., Warde-Farley, D., Bengio, Y.: Theano: a CPU and GPU math compiler in python. In: Proceedings of 9th Python in Science Conference, pp. 1–7 (2010)
2. Bordes, A., Glorot, X., Weston, J., Bengio, Y.: A semantic matching energy function for learning with multi-relational data - application to word-sense disambiguation. Mach. Learn. **94**(2), 233–259 (2014)
3. Bordes, A., Usunier, N., García-Durán, A., Weston, J., Yakhnenko, O.: Translating embeddings for modeling multi-relational data. In: NIPS, pp. 2787–2795 (2013)

4. Bordes, A., Weston, J., Collobert, R., Bengio, Y.: Learning structured embeddings of knowledge bases. In: AAAI. AAAI Press, San Francisco (2011)
5. Guo, S., Wang, Q., Wang, B., Wang, L., Guo, L.: Semantically smooth knowledge graph embedding. In: ACL, pp. 84–94. The Association for Computer Linguistics (2015)
6. Guu, K., Miller, J., Liang, P.: Traversing knowledge graphs in vector space. In: Empirical Methods in Natural Language Processing (EMNLP) (2015)
7. Hong, S., Chakraborty, T., Ahn, S., Husari, G., Park, N.: SENA: preserving social structure for network embedding. In: Proceedings of the 28th ACM Conference on Hypertext and Social Media, pp. 235–244. ACM (2017)
8. Huan, J., Wang, W., Prins, J.: Efficient mining of frequent subgraphs in the presence of isomorphism. In: Proceedings of Third IEEE International Conference on Data Mining 2003, pp. 549–552, November 2003. https://doi.org/10.1109/ICDM.2003.1250974
9. Khan, A., Wu, Y., Aggarwal, C.C., Yan, X.: NeMa: fast graph search with label similarity. In: Proceedings of the 39th International Conference on Very Large Data Bases, pp. 181–192 (2013)
10. Neo4j: The world's leading graph database (2017)
11. Paulheim, H.: Knowledge graph refinement: a survey of approaches and evaluation methods. In: Semantic Web, pp. 1–20 (2016). (Preprint)
12. Perozzi, B., Al-Rfou', R., Skiena, S.: DeepWalk: online learning of social representations. In: KDD, pp. 701–710. ACM (2014)
13. Pienta, R., Tamersoy, A., Tong, H., Chau, D.H.: MAGE: matching approximate patterns in richly-attributed graphs. In: BigData Conference, pp. 585–590 (2014)
14. Prud'hommeaux, E., Seaborne, A.: SPARQL Query Language for RDF. W3C Recommendation (2008)
15. Ren, X., Wu, Z., He, W., Qu, M., Voss, C.R., Ji, H., Abdelzaher, T.F., Han, J.: CoType: joint extraction of typed entities and relations with knowledge bases. In: Proceedings of the 26th International Conference on World Wide Web, Geneva, Switzerland, pp. 1015–1024 (2017). https://doi.org/10.1145/3038912.3052708
16. Ullmann, J.R.: An algorithm for subgraph isomorphism. J. ACM **23**(1), 31–42 (1976). https://doi.org/10.1145/321921.321925
17. Wang, D., Cui, P., Zhu, W.: Structural deep network embedding. In: Proceedings of the 22nd ACM SIGKDD International Conference on Knowledge Discovery and Data Mining, pp. 1225–1234. ACM, New York (2016). https://doi.org/10.1145/2939672.2939753

Distributed Big Data Ingestion at Scale for Extremely Large Community of Users

Venkat Tipparam[(✉)], Belinda Liu, Yifei Chen, Zoe Lang, Gang Ye, Diana Li, Hong-Yen Nguyen, CP Lai, and Steve Chan

eBay Inc., 2145 E Hamilton Ave, San Jose, CA 95125, USA
{vtipparam,beliu,yifchen,zlang,gye,diali,ynguyen,
chulai,stevechan}@ebay.com

Abstract. To make big data analytics available to mass online users, in the range of tens of millions, a different architecture other than those in the market has been designed and implemented which employs distributed blob store, custom compression, and custom query algorithm, including filtering, joins and group by. The system has been in operation at eBay for years and is described in [1]. However, large scale ingestion of data to a distributed blob store presents a unique challenge. This paper outlines an approach to solve the problem and uses an example of ingesting one trillion real time impressions per day, or 11+ millions per second, to illustrate how the proposed approach work. As discussed in the paper, the approach manages to consume 1 trillion real time impressions per day and is capable of making the data available to 100 million online users for analytics in just a few minutes. The incoming stream is partitioned first and then combined for ingestion. The ingestion is also divided into two stages, while data are available for query immediately after the first stage. Techniques are discussed to distribute volume of the data among system components to bring down the load on each component to a reasonable level.

Keywords: Large scale ingestion · Distributed blob · Compression

1 Introduction

In online eCommerce world, a site will have suppliers with services or goods that meet the needs of consumers, e.g., shops selling their wares, property owners open up for short term rental, advertisers competing for impressions and clicks on social media sites. Enterprises operating the sites always have had powerful analytical platforms for in-house business analysts to make sense of the impressions, clicks, sales, rents, rides, etc. However, suppliers on the site, big or small, are business owners in their own rights and they need to have analytical capability of their own to have deeper insights and to optimize their businesses. For the rest of the paper, we use the term user to refer to a supplier.

While a site has huge amount of data, each supplier or user on the site is usually a small fraction of the whole and one user cannot access the data of another of user. There can be up to a hundred millions of users but each user usually has no more than tens of millions of records or rows of data. So, an analytical platform for these users

© Springer International Publishing AG, part of Springer Nature 2018
F. Y. L. Chin et al. (Eds.): BIGDATA 2018, LNCS 10968, pp. 100–109, 2018.
https://doi.org/10.1007/978-3-319-94301-5_8

will have to provide maximum concurrency for the large community but provide access to only a fraction of the whole data set to each user.

This leads to an architecture that compress each user's data into a finite set of blobs. Each blob will contain a user's data organized as rows of certain data type for a given month. Each of the blobs can be accessed via a composite key of user-type-month. A key-value store, where a value is a persisted blob, or a distributed blob store, can deliver a blob in <50 ms to a query node and can support large number of concurrent reads. A thread on a query node, using the compression and query algorithm described in [1], can join, filter and aggregate millions of rows contained in a blob in a second.

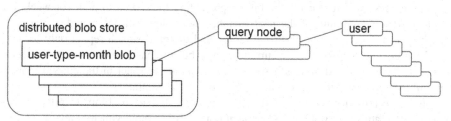

This architecture decouples compute from store and allows any compute node to serve a user's request by retrieving the relevant blobs from the store which can be on any node in the store. A distributed blob store usually does a good job of replicating blobs to each geography so the compute node simply gets the blobs from the same geography. The scheme of processing the data where the data is may not be the best approach given the large number of concurrent users involved. With 100 million users and even using 1,000 data nodes in the store (eBay uses far fewer than that), would simply mean 100 K users per node and the node will be a severe bottleneck for compute for those users.

For the rest of the paper, we'll use the term distributed blob store, or simply blob store, to differentiate from a distributed key-value in-memory hash where a value is relatively small and is always cached in memory. We'll also use the term blob to refer to the value.

This paper describes an approach to ingest data into blobs in such an architecture. We'll use the example of 1,000,000,000,000, or 1 trillion, real time impression per day, 100,000,000, or 100 million users on the site, with 1,000,000,000 listings where each listing is an item to sell, a property to rent, cars and drivers to provide rides, ads to show).

Impression stream is used as basis for discussion in this paper as impression tends to have the largest volume of all the data. The mechanisms discussed can obviously be used for other types streams of smaller volumes.

This paper focuses on the core concepts involved in solving the problem at hand. Given the challenges involved in implementing a reliable distributed stream processing application, topics such as fault-tolerance, exactly-once delivery semantics, etc., are outside the scope of this paper.

2 Ingestion to Blob Store

The ingestion of every impression from the stream will involve using the user (and type impression for current month) as a key to retrieve one of the existing blob containing rows of impressions data already ingested, create a new blob by merging all the existing impressions and the new impression, delete the existing blob, and associate the key with the newly created blob object. Further query from this point on using the composite key of user-type-month will retrieve the new blob containing the new impression data.

This process will repeat for the blob representing the current month (or week, or quarter). At the beginning of the next month, a new blob will be created to start the month with the first impression of the month and the process will repeat only for the new blob created for the new or current month, but the past month (was the current month) is left untouched, unless if some of the data arrives later in the current month but with a time stamp indicating they are for the past month.

We'll use the term write, or write to blob store, update the blob, to refer to the process of creating a new blob and replacing an existing blob with the new.

3 Challenges of Ingestion

The architecture uses a key-value, distributed blob store as the backend storage system and this presents a unique challenge. For a blob that contains a user's impression data for the current month to add a new impression, it will require merging the existing impression data with the new impression to generate a new blob and replace the existing blob with the new blob for the composite key of user-type-month. To ingest one trillion real time impressions per day, that will mean generating 11,574,074 new blobs to replace existing blobs every second, or more than 11 million writes to the blob store per second. To further compound the problem, users are likely to have impressions at the same moment (may be for different listings), but processed by different threads of the system, thus triggering a race condition. The high volume and concurrency are beyond the realm of feasibility for any blob store on the market.

The rest of the paper discusses how to bring the writes to the blob store per second down to reasonable level and do it without concurrent writes.

4 Partition of the Stream

Assuming impressions will arrive in the form a stream (there are many streaming platforms, e.g., Kafka [2]) and consumer of the stream will be the first touch point to the impression data. Consumer of the stream comprises of many threads. Impressions of the same user can potentially be consumed by many threads concurrently, each trying to perform a write to the same monthly blob for the same user to ingest an impression. This will trigger a high frequent race condition on the monthly blob.

To avoid the problem, we need to divert the stream in an organized way and partition the impressions by user id. For simplicity, let's assume there is a unique numerical ID for each user, or user ID, and the impression will be partitioned by mod the user ID by 1,000 (of course, the number 1,000 can be fine-tuned depending on the actual size of the user community). A user ID generated through by hash code of the user identifier, if that is alphanumeric in nature, will serve the purpose.

The partition of the stream can be achieved through Kafka. The consumer at the first touch point simply add user ID to the impression and produce the impression message down the pipeline for a second consumer. Kafka can be configured to dispatch the impression with the same mod based on user ID to the same thread of the second consumer. We'll refer to this second consumer as partition consumer, or partition thread.

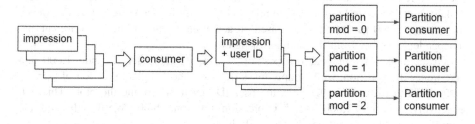

Given it is mod 1,000, there will be 1,000 threads each consuming a sub-stream of impression data based on user ID mod 1,000. Each thread will consume 1 billion impression a day, or 11,574 impressions per second.

Partition by user ID does not reduce the overall number of impressions to be ingested, or the number of writes required to the blob store, but it does eliminate the race condition because partitioning ensures that all impressions of the same user will be ingested by only one partition consumer thread.

5   Ingestion Interval

To reduce the number of writes, an ingestion interval is introduced. The partition consumer thread receiving impression will accumulate impressions in memory and only ingest those impressions to a blob at end of the interval. To make further discussion easier, we assume the impressions in the memory are organized as a hash of user id to a list of impressions of the user, or in Java notation `Map<User, List<Impression>>`.

For the purpose of this paper, we use the interval of 1 min, or 60 s. Since on average, each user can potentially have some impressions every minute and can trigger a write to the blob store. So, it is 100 million writes to the blob store every minute, or 1,666,667 writes per second. Much better than 11 million writes per second, but still too high.

Note that impressions cannot be evenly distributed for each user and for each minute. The effect of uneven distribution on the overall architecture and algorithm will be discussed in a later section.

It is worthwhile to point out that lengthening the interval will reduce the number of writes to the blob store but will have the undesirable side effect increasing the memory footage of each partition, as it will have to store more impressions before flushing them out. We will first continue to tackle the problem of reducing writes to the blob store, then we will discuss this side effect.

6 Daily Blob for Multiple Users

Each partition thread is still processing impressions for 100 millions/1,000 = 100,000 users. To reduce overall number of writes to blob store, we'll introduce a daily blob. A daily blob contains 1,000 users' impression but only for the current day. Of course, the number 1,000 can be adjusted based on actual environment and is a different and independent number from the number of partitions.

Given a user ID of nnnnnnxxxppp, where ppp is user ID mod 1,000 and determines the partition, we'll simply mask out xxx, the 3 digits to the left of ppp, to 000, and use nnnnnn000ppp as the user ID key to determine the blob. This will effectively combine 1,000 users' impression into one blob. We'll only add the impressions of the current day to this daily blob.

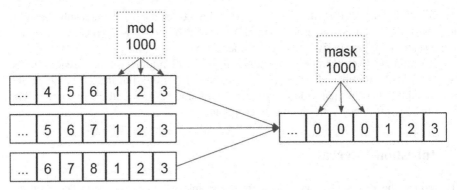

The partition thread will write 1,000 users' impression to one shared daily blob, instead of writing 1,000 users' impression to 1,000 monthly blob (one for each user). This effectively reduces the number of writes to the blob store by a factor of 1,000, from 1,666,667 writes per second to 1,667 writes per second.

Although the daily blob now contains multiple users' impression, adding a filter condition on user id will eliminate other users' impression from the result of a user's query.

The algorithm to compress and de-compress data, and the algorithm to execute a SQL like query as described in [1] can be further optimized to speed up this special filtering by user id.

7 Merging the Daily Blob with Monthly Blob

At the beginning of a new day, a new daily blob is created, and we need to merge the daily blob for the previous day with the current monthly blob. Given there are 100 million users, there should be 100 million monthly blobs, each blob needs one write to add the impression from the previous day, that is 100 million writes. This, of course, cannot be accomplished in an instance when the day switches. Assume we'll spread the writes to an entire day, that is, 100 million writes in 24 h, or 1157 writes per second to merge previous day's daily blob with the current monthly blobs.

Add the writes per second to merge yesterday's daily blob to monthly blobs to the writes per second to update the current daily blob, we arrive at 2,824 writes per second, and that's easily within the scalability of any distributed blob store today.

Note, the two daily blobs for previous day and the current day, and blobs for the current month are all instantly available for queries. it is not difficult to determine if a monthly blob contains the impression from previous day or not, and if not, to include the previous day's daily blob as an extra blob, the same way the current daily blob is included in a query.

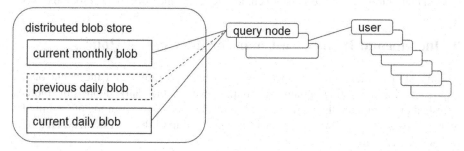

This mechanism ensures that up to date impressions are included in the online users' real-time aggregation. In fact, this mechanism makes the existence of daily blobs entirely transparent to the online users.

Thus, an impression will go through two stages of ingestions, first to the current daily blob and then from the previous daily blob to the current monthly blob. The impressions are available for query and aggregation the moment it gets into one of the blobs.

8 Hourly Impression Count Per Listing

Given 1 trillion impressions per day and 100 million users, each user should have on average 10,000 impressions per day. The number of impressions in a daily blob of 1,000 users will be 10,000 * 1,000 = 10,000,000, or 10 million.

However, given the volume of data for impression, user is usually not interested at each and every impression but rather at an aggregated value, for example, hourly impression per listing.

With the assumption of 1 billion listings 100 million users, each user will have on average 10 listings. Assuming each listing will have at least one impression at each hour, each listing's impression data will only need at most 24 rows per day, one for each hour. A given user will have 10 * 24 = 240 rows for a day. A daily blob of 1,000 users will have 240 * 1,000 = 240,000 rows.

The monthly blob for a single user will contain on average 240 * 31 = 7,440 rows.

The scan and query algorithm described in [1] can process millions of rows in a second.

Now, hourly impression count absorbs more impressions by simply adding to the count, but maintain the same number of rows in a blob. Though we used 1 trillion impressions per day, metrics are very much the same if the daily impression volume is 10 trillion.

As pointed earlier, combining users into a partition has the undesirable side effect of increasing memory footage for the partition, but we can now counter that with a desirable side effect of aggregation to the current hour by the minute in the partition and in memory. The hourly impression aggregation dramatically reduces the number of rows in daily blob and by the same token, it reduces the memory footage for a partition.

9 Incremental Impression Count for the Current Hour

A daily blob contains one row per user per listing per hour with the count of the impression for the listing within the hour. However, the daily impression blob is refreshed every minute, thus a row for a given listing and at a given minute within the hour will only have its impression count incremented until the next hour, when a new row will be created.

user ID	listing	hour	count
978567123	1234567890	2018-03-20 04:00:00	2

+

Impression:
{user ID: 978567123, listing:1234567890, hour:2018-03-20 04:20:19}
⇩

user ID	listing	hour	count
978567123	1234567890	2018-03-20 04:00:00	3

This enables a dashboard that polls the daily blob at a rate, say once per minute, to see the incremental impression count for a given listing minute by minute for the current hour and chart it in a graph. Once time advances to the next hour, the past hour will only have a single hourly count, but once can still chart the daily impression hour by hour. This should be sufficient for most of the analytical purpose.

While most of the parameters in this paper can be changed based on application we believe it is essential to mainly hourly impression count even for historical data.

As communication becomes cheaper and faster covering ever longer distance, a business owner often finds their customers not only in a different time zone by in an entirely different country (and often speaking a different language). For a US owner of a short-term rental in Australia with an ad in Britain hoping to catch potential customers visiting Australia for a Commonwealth Rugby game, it is quite useless to show a daily impression based on any US time zone.

Hourly impression count enables the user to view the daily impression trend based on any of the time zone. by simply assembling the relevant hours of the local day.

10 Implementation

This paper outlines an approach to solve the problem at hand using an example of ingesting one trillion real time impressions per day without delving into the details of implementation. Our overall Big Data system involves three major components, i.e., a real-time ingestion pipeline, blob store and a query engine. This paper focuses on a design approach for a highly scalable ingestion pipeline for extremely large number of impressions and very large number of users.

Our current solution involves a home-grown implementation based on Apache Kafka [2] Java consumer and custom in-memory aggregator. We have also validated the design ideas using stream processing frameworks, Apache Flink [4] and Kafka Streams [5]. However it should be certainly possible to leverage other stream processing frameworks such as open source Apache Storm [3], Apache Spark [6] or commercial solutions like Google Cloud Dataflow [7], Amazon Kinesis [8] and Azure Stream Analytics [9] to implement an ingestion pipeline based on the ideas presented here.

11 Uneven Distribution

The paper so far assumed an even distribution of the impressions among the users and at each time interval down to the second which is certainly not the case in real life application. It is quite reasonable to assume that not all of the users will have impressions every minute for every listing. In fact, it is quite possible that some of the users will not have any impression for any of the listings in a whole day. The more realistic and uneven distribution will involve more complex mathematics and obviously will have impact on how the system work.

To keep it simple, let's just look some examples. If a user happens not to have any impression in a single day, the daily process to merge daily blobs with monthly blob can certainly skip the step for that user, thus reducing the number of writes to the blob store by 1. Given the constant overall number of 1 trillion impressions per day, it simply means some other users will have more impressions in their respective daily impression blobs. However, the process of merging will need 1 write to replace the user's monthly blob with a new monthly blob which contains the increment daily impressions. This is just 1 write to the blob store, regardless of the number of hourly listing impressions involved. So, it can be generally stated that this process requires fewer number of writes to the blob store in the case of uneven distribution among users.

As an extreme example, all that 1 trillion impressions in a day are for just one big user, then there is just only one daily blob to be updated, and only 24 h * 60 min = 1,440 writes in a whole day to the blob store.

The process of writing impressions to daily impression blob at the end of every one minute interval however can potentially see less variation with regarding to the number of writes. Given the sample configuration of 1,000 users sharing one single daily file, it is unlikely that none of the 1,000 users will have any impression. As long as one user has a single impression in that one minute interval, a write to the combined daily blob will be required. However, the algorithm described in this paper has already taken that into consideration.

While the number of listings by a user and their impressions varies in a great degree from a single mom-and-pop user to an enterprise user exploring the site as an additional commerce channel, combing 1,000 users into a single daily file absorbs some of the up and downs. This implies that the load on each partition will somewhat even out on its own. The daily impression file is no longer needed once merged into the monthly blob, thus providing the opportunity to fine tune the partitioning of users for future days.

By now it is obvious that though even distribution is unlikely in reality, it actually presents the worst case scenario in theory in some cases. While even distribution does bring some additional challenges in some other parts of the algorithm, the nature of the design seems to reduce the effect of the uneven distribution.

We have shown that the approach can work for the worst case. and there is opportunity to optimize by further fine tuning the parameters.

12 System at a Glance

The system at eBay which more or less follows the general architecture described in this paper and [1] currently has ~400 TB of data covering a span of 5+ years and growing. The system has ~10 billions of key-value pairs, serving close to 20 millions of queries per day using less than 50 VM for compute, and still managing an average query time under 140 ms. The longest query joins 7 types of data, 17 selects, 15 group by, and is expressed by more than 10,000 chars in text. Depending on the type, data is updated as frequent as 1 min.

13 Summary

An approach is described to ingest trillions of real time impression per day for analytics by a community of 100 million users with a delay of just minutes. The blueprint laid out in this paper stretches to a scale that is beyond that of the most of the ecommerce web site today. It is quite straightforward to come up with a spreadsheet with formula, and plug in the numbers to fine tune the set up for any given real-life production environment.

With this approach, we believe our architecture of decoupling compute and store can scale to 100 million online users end-to-end, from data ingestion to serving online real-time aggregation.

Acknowledgement. We would like to express our gratitude to Sami Ben-Romdhane, VP and eBay Fellow, for his advices and guidance over the years in evolving the architecture. We want to thank Adithya Ramakrishnan for his contributions to this product.

We also want to thank Sujeet Varakhedi, Vijeya Anbalagan, and Hardik Patel for their invaluable support on an internal eBay Key-value distributed blob store that made this architecture a reality.

References

1. Liu, B., Ponnusamy, T., et al.: Distributed data aggregation at scale for large community of users. In: International Conference on Big Data and Education (2018)
2. Kafka, A distributed streaming platform. https://kafka.apache.org/
3. Apache Storm: http://storm.apache.org/
4. Apache Flink: https://flink.apache.org/
5. Kafka Streams: https://docs.confluent.io/current/streams/index.html
6. Spark Streaming: https://spark.apache.org/streaming/
7. Google Cloud Dataflow: https://cloud.google.com/dataflow/
8. Amazon Kinesis: https://aws.amazon.com/kinesis/
9. Azure Stream Analytics: https://azure.microsoft.com/en-us/services/stream-analytics/

Convolutional Neural Network Ensemble Fine-Tuning for Extended Transfer Learning

Oxana Korzh, Mikel Joaristi, and Edoardo Serra[✉]

Computer Science Department, Boise State University, Boise, USA
{oxanakorzh,mikeljoaristi}@u.boisestate.edu,
edoardoserra@boisestate.edu

Abstract. Nowadays, image classification is a core task for many high impact applications such as object recognition, self-driving cars, national security (border monitoring, assault detection), safety (fire detection, distracted driving), geo-monitoring (cloud, rock and crop-disease detection). Convolutional Neural Networks(CNNs) are effective for those applications. However, they need to be trained with a huge number of examples and a consequently huge training time. Unfortunately, when the training set is not big enough and when re-train the model several times is needed, a common approach is to adopt a transfer learning procedure. Transfer learning procedures use networks already pretrained in other context and extract features from them or retrain them with a small dataset related to the specific application (fine-tuning). We propose to fine-tuning an ensemble of models combined together from multiple pretrained CNNs (AlexNet, VGG19 and GoogleNet). We test our approach on three different benchmark datasets: Yahoo! Shopping Shoe Image Content, UC Merced Land Use Dataset, and Caltech-UCSD Birds-200-2011 Dataset. Each one represents a different application. Our suggested approach always improves accuracy over the state of the art solutions and accuracy obtained by the returning of a single CNN. In the best case, we moved from accuracy of 70.5% to 93.14%.

Keywords: Image classification · CNN · Deep learning
Transfer learning

1 Introduction

One of the most promising technologies in machine learning is the concept of transfer learning [31]. Currently, deep learning models require large scale data for training. With transfer learning revolution we can use a relatively small data set for training a deep learning model for a particular application while simultaneously keeping the same performance and reducing the execution time of the training procedure. This method is based on the assumption that current deep learning methods can train the model on a very large and general data set that

F. Y. L. Chin et al. (Eds.): BIGDATA 2018, LNCS 10968, pp. 110–123, 2018.
https://doi.org/10.1007/978-3-319-94301-5_9

includes patterns from different application areas. For a particular application, you do not need to retrain this large model from scratch. You can modify the existing model to be specialized for a particular application while still presenting general knowledge that came from the pretrained model.

There are two main options for using pretrained models for transfer learning [11]. The first one is fine-tuning the model: short-term additional training is applied to the original model to add a particular training set to the model's knowledge base. The second one is to use of pretrained Convolutional Neural Networks (CNN) as a feature extractor to transform images into feature vectors for classification.

For transfer learning for convolutional neural networks [14] it is very popular to use general pretrained networks such as AlexNet [14], GoogleNet [26] and VGG [25] for solving the image classification task for a particular application. In this paper, we propose a method of increasing image classification accuracy by using transfer learning of pretrained CNNs combined into an ensemble. We implement transfer learning using the fine-tuning method [11].

The main advantage of our method is fine-tuning of CNN ensemble when general features from different pretrained networks are shared and applied for a particular application. Our method includes the following steps: (1) each pre-trained CNN is fine-tuned independently for a particular application; (2) weights and biases from fine-tuned networks are used for initialization of the ensemble model; (3) CNN ensemble model is fine-tuned for a particular application; (4) fine-tuned ensemble model can be used as a classifier by itself or as a feature extractor for an external classifier.

The contributions of this paper are:

1. A method of applying fine-tuning procedure to ensemble approach;
2. huge image classification accuracy improvement in three different benchmarks: Yahoo! Shopping Shoes Image Content (93.14% vs best known 70.5% [12]), UC Merced Land Use Dataset (99.76% vs best known 96.90% [11]) and The Caltech-UCSD Birds-200-2011 Dataset (81.91% vs best known 75% [4]);
3. investigation of four different CNN ensembles and summary of their main features with recommendations for improving classification accuracy in particular applications;
4. investigation of fine-tuned CNN ensemble in the task of feature extraction for image classification with external classifiers (SVM, ExtraTrees, Logistic regression).

The remainder of this paper is organized as follows. Section 2 describes related work in the field of neural network ensemble processing. In Sect. 3 we describe data sets used in experiments. In Sect. 4, we propose our approach of CNN ensemble fine-tuning. Section 5 includes details, experiments, result's analysis and discussion. Finally, we have conclusions for this paper with some remarks.

2   Related Work

The ensemble of classifiers is a well known technique to increase classification accuracy [27]. Different approaches for combining classifiers into ensembles already exist. In [19] the approach trains ensembles that directly construct diverse hypotheses using additional artificially-constructed training examples. In the paper [18] general questions of joint loss functions are discussed. In terms of the image classification task, ensemble approach is most commonly used for solving the multi classification task. For example in [16] multiple outputs are extracted as a learning problem over an ensemble of deep networks using a stochastic gradient descent based approach to minimize the loss with respect to an oracle. In [21] authors investigate the problem of pedestrian detection with an ensemble method using histograms of oriented gradients and local receptive fields, which are provided by a convolutional neural network and classified by multi layer perceptrons and support vector machines. The final choice is done by using majority vote and fuzzy integral.Pretrained convolutional neural network fine-tuning technique is successfully used in different applications. Recent research shows that pretraining on general data followed by application-specific fine-tuning yields significant performance improvement in the image classification task. In [10] authors analyze the performance of different fine-tuned CNNs for classification of paintings into art epochs. Paper [23] describes fine-tuning strategy to transfer recognition capabilities from general domains to the specific challenge of plant identification. Authors in [24] used fine-tuning process for CNN models pretrained on natural image dataset to solve medical image processing tasks. Many approaches are also base on synthetic data generation for improve the robustness of the classifier. [8,9] are two works specialized on synthetic data generation In our previous paper [13] we described a stacking approach for improving deep CNN transfer learning for processing low quality remote sensing images. CNN ensemble is used to produce a combination of features, extracted from different CNNs and to combine them in a feature vector for further classification with an external classifier. Paper [15] proposes an ensemble of fine-tuned convolutional neural networks for medical image classification. It describes a method for classifying the modality of medical images using an ensemble of different CNN architectures. The various CNNs in the ensemble allow extracting image features at different semantic levels, thereby enabling the characterization of the varying distinct and subtle differences among modalities. The ensemble of fine-tuned CNNs allows adapting the generic features learned from natural images to be more specific for different medical imaging modalities. In [4], when given a test image, authors use groups of detected keypoints to compute multiple warped image regions that are aligned with prototypical models. Each region is fed through a deep convolutional network, and features are extracted from multiple layers. Then features are concatenated and used as a feature vector for classification. One more paper that we want to mention is [6] where authors use Trunk-Branch Ensemble Convolutional Neural Networks (TBE-CNN) for video-based face recognition. TBE-CNN is composed of one trunk network that learns representations for holistic face images and two branch networks that learn representations for image patches cropped around facial components. The output feature maps of the trunk network and branch networks are fused by concatenation and then last fully connected layer is applied for classification.

The main contribution of out method proposed in this paper in comparison with existing approaches is a fine-tuning procedure for pretrained model ensemble based on the joint loss function. Each ensemble member is a pretrained CNN(AlexNet, VGG19 or GoogleNet) that is prior independently fine-tuned for the specific application domain. Fine-tuned ensemble can be used as image classifier or as feature extractor for further image processing.

3 Data Sets Used in Experiments

We selected three different known benchmark data sets for testing our solution. Datasets belong to different application areas and contain images of different quality and resolution. The first data set is **Yahoo! Shopping Shoes Image Content** [2]. This data set provides a new benchmark for the problem of fine grained object recognition using shoes as an example and contains a diverse collection of types of shoe photos. This dataset contains 107 classes, each corresponding to a type and brand of shoe. Images are in RGB format stored in JPEG format, each of three channels contains 8bit information. Image resolution is 640×480 pixels. The total number of images is 5250. Examples from this data set are shown on the Fig. 1. Paper [12] describes an approach for classification this data set using two-flow model based on usage of pretrained deep neural network for feature extraction. Also, authors extract features directly from the data set using dimensionality reduction. Features from both sources are combined and used in nonlinear classifier to get the final result. In the experiment, 90% of the data is used as train and 10% as test. Achieved classification accuracy is 70.5%. In the paper [3] an approach is proposed for constructing mid-level visual features for image classification. The image is transformed using the outputs of a collection of binary classifiers. These binary classifiers are trained to differentiate pairs of object classes in an object hierarchy. Using this approach authors received 64.7% classification accuracy on random 90/10 split of the Yahoo data set.

The second benchmark used in this paper is well known landscape dataset **UC Merced Land Use Dataset(UCM)** [30]. The images were extracted from the USGS National Map Urban Area Imagery [1] collection for various urban areas around the country. Dataset contains 21 classes and 100 images per each class (2100 images in total). The resolution of this imagery is 1 foot per

Fig. 1. Yahoo! Shopping Shoes Image Content Dataset.

Fig. 2. UC Merced Land Use Dataset.

Fig. 3. The Caltech-UCSD Birds-200-2011 Dataset.

pixel. Each image is 256×256 pixels. Images are in TIF format and contain 8bit three channel (RGB) information. Examples from UCM dataset are shown on the Fig. 2. UCM data set is a widely used benchmark for testing landscape imagery processing methods. Paper [11] describes combination of feature extraction methods using standard image processing methods (such as BOW [17], IFK [22], LLC [29]) and convolutional neural networks. Then feature combination is used for final classification. Best achieved classification accuracy is 96.90%. In the paper [20] fine-tuning process is applied to CNN to achieve better classification accuracy. Fine-tuned CNN is used as a feature extractor for further classification using linear SVM [5]. Achieved accuracy in 5-fold cross-validation process is 96.47%. The most accurate model is based on using fine-tuned GoogleNet [26] as feature extractor in combination with linear SVM for final classification.

The third data set is called **The Caltech-UCSD Birds-200-2011 Dataset** [28]. The dataset contains 11,788 images of 200 bird species. Images are in JPEG format(8bits per channel, RGB). This benchmark data set is used for testing different image processing algorithm: bird species categorization, detection, and part localization. Examples from Birds dataset are shown in Fig. 3. In [7] authors propose a nonparametric approach for part detection which is based on transferring part annotations from related training images to an unseen test image. Feature extraction step is focused on those parts of images where discriminative features are likely to be located. This approach achieves 57.8% classification accuracy. Paper [4] proposed classification methods based on estimating of the object's pose. The features are computed by applying deep convolutional nets to image patches that are located and normalized by the pose. Authors used deep convolutional feature implementations and fine-tuning feature learning for fine-grained classification. Achieved classification accuracy rate is 75%.

All of the three data sets are challenging. Data has large inner class variability and the image resolution is small for traditional feature extraction and fine-grain classification methods. All data sets are widely used for testing convolutional networks approaches for image classification task.

4   Methodology

In this paper, we proposed a fine-tuning procedure for pretrained model ensemble based on the joint loss function. This approach combines the power of different pretrained networks and yields to image classification accuracy increasing. The proposed method includes four main steps. First, each pretrained CNN is fine-tuned independently for a particular application. We include three different networks to test our approach: AlexNet, GoogleNet and VGG19. The second step is ensemble model initialization using weights and biases from single fine-tuned networks. In the third step, CNN ensemble model is fine-tuned for a particular application using joint loss function. The fourth step is final image classification: proposed fine-tuned ensemble model can be used as a classifier by itself or as a feature extractor for an external classifier. For fine-tuning process in each pretrained CNN we replace the last fully connected layer with a new fully connected layer with the number of perceptrons equal to the number of classes in

the dataset. The new layer is randomly initialized. After that standard training procedure with law, learning rate is started. After fine-tuning each CNN
produces a feature vector with the number of elements equal to the number of
classes in output data set. This vector can be processed with some function (for
example softmax) to obtain probabilities for test image to be in the appropriate
class or this vector can be used as an input of external classifier to obtain test
pattern class. In this paper, we investigate four different ensemble models. The
first model is AVnet and it is shown in Fig. 4. This model is a combination of
AlexNet and VGG19 net. To initialize this ensemble we use weights and biases
form single pretrained networks up to fc7 layer. Then a new fully connected layer
is added after concatenation. This layer is randomly initialized before starting
fine-tuning process. Next model is called AGnet (Fig. 5) and is an ensemble of
AlexNet and GoogleNet. In case if GoogleNet is participating in the ensemble we
use its last fully connected layer for concatenation with fc7 layer from AlexNet or
VGG19. Figure 6 shows VGnet that is a combination of VGG19 and GoogleNet.
And finally, we combine all three networks into AVGnet (Fig. 7). The joint loss
function is a cross entropy loss function, which is defined as follows:

$$L = -\sum_{j=1}^{C} y_j \log p_j$$

where C is the number of target classes, y_j is the $j - th$ value of the ground
truth probability (0 or 1 in our case), p_j is the $j - th$ output value of softmax
applied after joint fully connected layer of ensemble network. After ensemble
model fine-tuning process is finalized the model can be used as a classifier by
itself or features from different layers can extracted and used as an input of
external classifier.

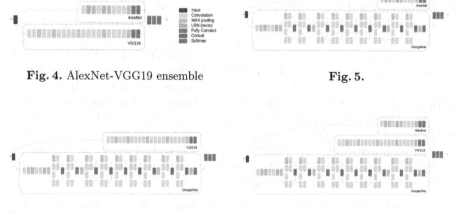

Fig. 4. AlexNet-VGG19 ensemble

Fig. 5.

Fig. 6. VGG19-GoogleNet ensemble
(VGnet).

Fig. 7. AlexNet-VGG19-GoogleNet
ensemble (AVGnet).

Table 1. Single network fine-tuning test classification accuracy

	Yahoo	UCM	Birds
AlexNet	76.56	96.90	65.14
VGG19	87.99	97.85	74.75
Google	83.05	97.85	77.68

Table 2. Ensemble network fine-tuning test classification accuracy

	Yahoo	UCM	Birbs
AVnet	90.45	98.81	79.71
AGnet	88.43	99.76	79.96
VGnet	89.66	99.76	79.88
AVGnet	90.35	99.76	80.56

Table 3. 10 cross fold validation test for CNN fine-tuning on Yahoo data set

	Alexnet	VGG	AVnet
Mean	73.92	84.82	87.83
ST Dev	2.69	1.90	1.92

5   Experiments

In our experiments we implement CNN ensemble using caffe [14]. We use three data sets mentioned in the part 3 for testing our methods: Yahoo! Shopping Shoes Image Content (Yahoo), UC Merced Land Use Dataset (UCM) and The Caltech-UCSD Birds-200-2011 Dataset (Birds). For each data set at first we fine-tune single networks (AlexNet, VGG19 and GoogleNet). Then weights from fine-tuned networks are used to initialize ensemble models for fine-tuning. We process fine-tuning of four ensembles for each data set. Then we compare the classification accuracy of fine-tuned networks and usage of fine-tuned networks as feature extractors in combination with external classifiers. In addition for Yahoo data set we provide the results of ten cross fold validation process for AlexNet, VGG19 and AVnet fine-tuning to estimate the stability of the model.

5.1   Yahoo! Shopping Shoes Image Content

In the Yahoo dataset the total number of images is 5250. We a make random 90/10 split to make our results comparable with experiments in [3,12].

Single Network Fine-Tuning. For single network fine-tuning we transfer images into lmdb database format for faster access. Also, we use fixed batch size in the fine-tuning process for all networks: 32 for train mode and 16 for the test. For AlexNet and GoogleNet we use 10000 iterations and for VGG19 - 20000 iterations. Learning rate starts from 0.0001 and decreases 10 times every 10000 steps. For the last, fully connected layer learning rate is ten times higher than for the other layers in the network. The best classification accuracy result with a single network for this data set was achieved for VGG19 network (87.99%). Even AlexNet shows 76.56% accuracy that is more than the best known accuracy achieved with non neural network fine-tuning methods. Fine-tuned GoogleNet classification accuracy is 83.05%. Classification accuracy for fine-tuned networks for all datasets is summarized in Table 1.

Ensemble Network Training. We are fine-tuning four ensembles: AVnet, AGnet, VGnet, AVGnet. Weight and bias initialization is computed from single fine-tuned networks using the layer by layer copy. Last fully connected layer is initialized randomly. We use batch size 40 in training mode for the first three

Table 4. Classification accuracy when fine-tuned CNN is used for feature extraction for further input to external classifier

Yahoo data set			
Feature extractor	SVM	ExtraTrees	LogReg
AlexNet fc6	74.09%	77.90%	77.33%
AlexNet fc7	75.23%	77.90%	76.19%
AlexNet fc8	76.95%	78.09%	77.52%
VGG19 cf6	88.76%	88.19%	89.14%
VGG19 cf7	88.95%	88.95%	89.71%
VGG19 cf8	89.14%	88.76%	89.90%
GoogleNet	84.95%	85.33%	85.33%
AlexNet + GoogleNet	85.14%	85.71%	86.28%
AlexNet + VGG19	87.42%	86.85%	87.80%
GoogleNet + VGG19	87.23%	88.57%	87.42%
AlexNet + GoogleNet + VGG19	86.28%	87.23%	87.42%
AVnet	90.66%	91.23%	90.85%
AGnet	89.90%	90.66%	91.42%
VGnet	90.47%	91.80%	92.00%
AVGnet	**92.00%**	**93.14%**	**92.57%**

UCM data set			
Feature extractor	SVM	ExtraTrees	LogReg
AlexNet fc6	96.90%	97.14%	97.62%
AlexNet fc7	96.90%	97.38%	97.85%
AlexNet fc8	97.14%	97.62%	98.09%
VGG19 cf6	97.85%	98.33%	98.33%
VGG19 cf7	98.33%	98.81%	98.57%
VGG19 cf8	97.85%	98.33%	98.33%
GoogleNet	98.33%	98.81%	98.81%
AlexNet + GoogleNet	99.52%	99.52%	99.52%
AlexNet + VGG19	99.28%	99.28%	99.28%
GoogleNet + VGG19	99.52%	99.52%	99.52%
AlexNet + GoogleNet + VGG19	99.05%	99.05%	99.05%
AVnet	99.52%	99.52%	99.28%
AGnet	**100.00%**	**100.00%**	**100.00%**
VGnet	99.76%	99.76%	99.76%
AVGnet	99.76%	99.76%	99.76%

Birds data set			
Feature extractor	SVM	ExtraTrees	LogReg
AlexNet fc6	67.14%	67.82%	67.99%
AlexNet fc7	68.67%	68.42%	68.25%
AlexNet fc8	67.99%	67.82%	67.91%
VGG19 cf6	76.40%	76.57%	76.91%
VGG19 cf7	78.18%	77.92%	78.09%
VGG19 cf8	77.75%	77.24%	77.07%
GoogleNet	78.26%	78.52%	78.35%
AlexNet + GoogleNet	78.69%	78.86%	78.94%
AlexNet + VGG19	78.13%	78.35%	78.69%
GoogleNet + VGG19	79.88%	80.05%	79.79%
AlexNet + GoogleNet + VGG19	79.28%	79.45%	79.45%
AVnet	80.64%	80.64%	80.73%
AGnet	80.98%	81.06%	80.98%
VGnet	81.23%	81.15%	80.89%
AVGnet	**81.57%**	**81.91%**	**81.74%**

networks and 32 for AVGnet. In test mode batch size is 16 for all networks. We use the same learning rate decreasing strategy as for single network fine-tuning: learning rate starts with 0.0001 and decreases every 10000 steps, for the last fully connected layer the value is 10 times bigger. Each network is trained with 30000 iterations. Best classification accuracy (90.45%) was achieved with AVnet - ensemble based on the combination of AlexNet and VGG19. Also, AVGnet ensemble has good classification accuracy (90.35%). Final classification accuracy for ensemble fine-tuning for all three data sets is shown in the Table 2.

Fine-Tuned CNN as a Feature Extractor. In this set of experiments, we use fine-tuned CNNs as feature extractors for further classification with an external classifier. For classification we use three classifiers: linear SVM (SVM), Extra-Trees classifier (ExtraTrees) and Logistic Regression (LogReg). Classifier implementation is based on sklearn library. For AlexNet and VGG19 we use feature vectors from three last fully connected layers: fc6, fc7 and fc8. In GoogleNet experiment we use features from last fully connected layer with ReLu and softmax transformations. In experiment AlexNet+GoogleNet we combine features from fc7 layer of AlexNet and last fully connected layer from GoogleNet in one feature vector. AlexNet+VGG19 combines fc7 from AlexNet and fc7 from VGG19. Similarly we do for GoogleNet + VGG19 and AlexNet + GoogleNet + VGG19 experiments. For AVnet, AGnet, VGnet and AVGnet we use a vector from concatenation layer for classification. Classification accuracy results are summarized in Table 4. In most of the experiments the best accuracy is achieved using ExtraTrees classifier. The best classification accuracy is 93.14% for AVGnet ensemble features in combination with ExtreTrees classifier. Also, it is interesting to compare classification results for the combination of feature vectors from different networks and features from ensemble model. Ensemble model gives in average 5% improvement in comparison with the combination of feature vectors (Table 3).

Ten Cross Fold Validation. We have selected three models for 10 folds cross validation process. It is a time consuming procedure because ten different models should be fine-tuned for each evaluation. We used stratified random 10 folder split for this experiment. AlexNet and VGG19 are fine-tuned independently for each of 10 splits. Then fine-tuned networks are used for initialization of appropriate AVnet. Ten cross-fold validation classification accuracy result for proposed ensemble model AVnet is 87.83% and the standard deviation is 1.92%. It means that model is stable and we are obtaining close results for any split. Figures 8, 9, 10, 11, 12 and 13 shows fine-tuning process test accuracy and test loss for AlexNet, VGG19 and AVnet fine-tuning.

5.2   UC Merced Land Use Dataset

Landscape imagery dataset UC Merced Land Use Dataset is the smallest one in our paper. It contains 2100 images for test and train split. We use 80/20 randomly stratified split where 80% of images go to training part and 20% are testing part (420 images for the test in total).

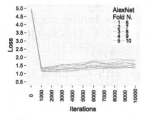
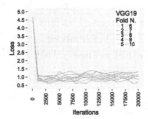
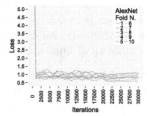

Fig. 8. Alexnet ten cross-fold validation fine-tuning test accuracy for Yahoo data set.

Fig. 9. VGG19 ten cross-fold validation fine-tuning test accuracy for Yahoo data set.

Fig. 10. AV net ten cross-fold validation fine-tuning test accuracy for Yahoo data set.

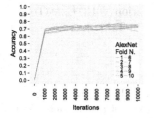
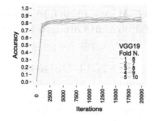
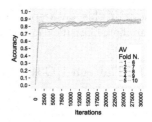

Fig. 11. Alexnet ten cross-fold validation fine-tuning test loss for Yahoo data set.

Fig. 12. VGG19 ten cross-fold validation fine-tuning test loss for Yahoo data set.

Fig. 13. AV net ten cross-fold validation fine-tuning test loss for Yahoo data set.

Single Network Fine-Tuning. In single CNN fine-tuning experiment we use the same methodology as declared in part 5.1 for Yahoo data set. Fine-tuned AlexNet shows 96.90% accuracy that is equal to best known result for this data set. Fine-tuned VGG19 and GoogleNet improve classification accuracy to 97.85%. Classification accuracies for fine-tuned networks are shown in Table 1.

Ensemble Network Training. Ensemble fine-tuning protocol for UCM data set is the same as for Yahoo data set. We fine-tune four ensembles using 40 batch size for train mode and 16 for test mode. As test and train part are not of a big size we use just 10000 iterations for fine-tuning without changing learning rate. After fine-tuning, most of the model shows 99.76% classification accuracy that means for this particular data set just one wrong classified image per 420 images in the test set.

Fine-Tuned CNN as a Feature Extractor. For this data set usage of external classifier after feature extraction is not reasonable because most of the model gives classification accuracy more than 99% without additional processing. The improvement is in half percent range. We achieve 100% classification accuracy for this data split using AGnet model in combination with ExtraTrees classifier. Classification accuracy results are summarized in Table 4.

5.3   Caltech-UCSD Birds-200-2011 Dataset

The Caltech-UCSD Birds-200-2011 Dataset is the most challenging of three datasets represented in this paper. Traditionally some kind of object detection method is applied for this kind of images to improve classification rate. But for us in this paper, the purpose was to show the benefits of ensemble model fine-tuning in comparison with single network fine-tuning. So we use image re-sampling instead of object detection. Images are downsampled to 256 pixels in the smallest dimension.

Single Network Fine-Tuning. Dataset is randomly split into 90/10 ratio of train and test part. Test set contains 1178 images. The fine-tuning protocol is the same as in part 5.1 for Yahoo data set. For single network fine-tuning the best classification accuracy rate (77.68%) is achieved using GoogleNet. This result is 7% more than known classification accuracy rate for this data set.

Ensemble Network Training. Best classification result with ensemble fine-tuning is achieved after fine-tuning AVGnet and is equal to 80.50%. During the fine-tuning process, we use learning rate decreasing strategy similar to the one used for Yahoo data set. Maximum iteration number is 30000 for all four models.

Fine-Tuned CNN as a Feature Extractor. Classification accuracy results are summarized in Table 4. We have in average 3% improvement for ensemble training in comparison with classification of combinations of feature vectors. Best image classification accuracy rate (81.91%) is shown by AVGnet model in combination with ExtraTrees classifier. We summarize best classification accuracy results in Fig. 14. Base is the best known classification accuracy result for each data set (70% for Yahoo,

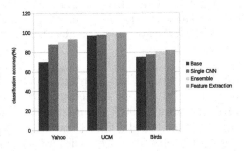

Fig. 14. Best classification accuracy results comparision

96.90% for UCM, 75% for Birds). Single CNN is the best accuracy achieved with single network fine-tuning (87.99%, 97.85%, 77.68% accordingly). Ensemble shows the best result of ensemble CNN fine-tuning (90.45%, 99.76%, 80.56% accordingly). The best results in classification are achieved using fine-tuned ensemble for feature extraction and classification using ExtraTrees classifier (93.14%, 100%, 81.91% accordingly).

6 Conclusions

In this paper, we proposed a transfer learning approach based on the returning of CNN ensemble models combining multiple already pre-trained convolutional neural networks. We tested this approach on three benchmark datasets: Yahoo! Shopping Shoe Image Content, UC Merced Land Use Dataset and The Caltech-UCSD Birds-200-2011 Dataset. We observed that in all of the experiments our approach is able to classify better than the method present in literature and better than the standard transfer learning approach that fine-tunes only a single network at a time. In addition, we show that in terms of accuracy our approach works even better if it is used as a feature extractor. We obtained a maximum accuracy among all the ensemble models of 93.14% for Yahoo! Shopping Shoe Image Content, 100% for UC Merced Land Use Dataset and 81.91% for The Caltech-UCSD Birds-200-2011 Dataset. Our approach always improves the accuracy of all the state of art with 23% of classification accuracy improvement in the best case.

References

1. Usgs national map urban area imagery: 2017. https://nationalmap.gov/ortho.html. Accessed 14 Nov 2017
2. Yahoo webscope datasets: 2016. http://webscope.sandbox.yahoo.com/. Accessed 14 Nov 2017
3. Albaradei, S., Wang, Y., Cao, L., Li, L.-J.: Learning mid-level features from object hierarchy for image classification. In: Proceedings of 2014 IEEE Winter Conference on Applications of Computer Vision (WACV), pp. 235–240, November 2014
4. Branson, S., Van Horn, G., Belongie, S.J., Perona, P.: Bird species categorization using pose normalized deep convolutional nets. CoRR, abs/1406.2952 (2014)
5. Cortes, C., Vapnik, V.: Support-vector networks. Mach. Learn. **20**(3), 273–297 (1995)
6. Ding, C., Tao, D.: Trunk-branch ensemble convolutional neural networks for video-based face recognition. CoRR, abs/1607.05427 (2016)
7. Göring, C., Rodner, E., Freytag, A., Denzler, J.: Nonparametric part transfer for fine-grained recognition. In: Proceedings of the IEEE Conference on Computer Vision and Pattern Recognition (CVPR), pp. 2489–2496, June 2014
8. Guzzo, A., Moccia, L., Saccà, D., Serra, E.: Solving inverse frequent itemset mining with infrequency constraints via large-scale linear programs. ACM Trans. Knowl. Discov. Data (TKDD) **7**(4), 18 (2013)
9. Guzzo, A. Saccà, D., Serra, E.: An effective approach to inverse frequent set mining. In: Ninth IEEE International Conference on Data Mining, ICDM 2009, pp. 806–811. IEEE (2009)
10. Hentschel, C., Wiradarma, T.P., Sack, H.: Fine tuning cnns with scarce training data - adapting imagenet to art epoch classification. In: 2016 IEEE International Conference on Image Processing (ICIP), pp. 2381–8549, September 2016
11. Fan, H., Xia, G.-S., Jingwen, H., Zhang, L.: Transferring deep convolutional neural networks for the scene classification of high-resolution remote sensing imagery. Remote Sens. **7**(11), 14680–14707 (2015)

12. Ilyukovich-Strakovskaya, A., Dral, A., Dral, E.: Using pre-trained models for fine-grained image classification in fashion field. In: Proceedings of the First International Workshop on Fashion and KDD, KDD 2016, August 2016
13. Korzh, O., Cook, G., Andersen, T., Serra, E.: Stacking approach for cnn transfer learning ensemble for remote sensing imager. In: Proceedings of Intelligent Systems Conference 2017, September 2017
14. Krizhevsky, A., Sutskever, I., Hinton, G.E.: Imagenet classification with deep convolutional neural networks. In: Proceedings of the NIPS 2012: Neural Information Processing Systems (2012)
15. Kumar, A., Kim, J., Lyndon, D., Fulham, M., Feng, D.: An ensemble of fine-tuned convolutional neural networks for medical image classification. IEEE J. Biomed. Health Inform. **21**(1), 31–40 (2017)
16. Lee, S., Purushwalkam, S., Cogswell, M., Ranjan, V., Crandall, D.J., Batra, D.: Stochastic multiple choice learning for training diverse deep ensembles. CoRR, abs/1606.07839 (2016)
17. Li, F.-F., Perona, P.: A bayesian hierarchical model for learning natural scene categories. In: Proceedings of the 2005 IEEE Computer Society Conference on Computer Vision and Pattern Recognition (CVPR 2005), pp. 524–531, June 2005
18. Liu, Y., Yao, X.: Ensemble learning via negative correlation. Neural Netw. **12**(10), 1399–1404 (1999)
19. Melville, P., Mooney, R.J.: Creating diversity in ensembles using artificial data. Inf. Fus. **6**(1), 99–111 (2005)
20. Nogueira, K., Penatti, O.A.B., dos Santos, J.A.: Towards better exploiting convolutional neural networks for remote sensing scene classification. In CoRR, volume abs/1602.01517 (2016)
21. Oliveira, L., Nunes, U., Peixoto, P.: On exploration of classifier ensemble synergism in pedestrian detection. IEEE Trans. Intell. Transp. Syst. **11**(1), 16–27 (2010)
22. Perronnin, F., Sánchez, J., Mensink, T.: Improving the fisher kernel for large-scale image classification. In: Daniilidis, K., Maragos, P., Paragios, N. (eds.) ECCV 2010, Part IV. LNCS, vol. 6314, pp. 143–156. Springer, Heidelberg (2010). https://doi.org/10.1007/978-3-642-15561-1_11
23. Reyes, A.K., Caicedo, J.C., Camargo, J.E.: Fine-tuning deep convolutional networks for plant recognition. In: Working Notes of CLEF 2015, September 2015
24. Shin, H., Roth, H., Chen, M., Lu, L., Xu, Z., Nogues, I., Yao, J., Mollura, D.: Deep convolutional neural networks for computer-aided detection: Cnn architectures, dataset characteristics and transfer learning. IEEE Trans. Med. Imaging **35**, 1285–1298 (2016)
25. Simonyan, K., Zisserman, A.: Very deep convolutional networks for large-scale image recognition. CoRR, abs/1409.1556 (2014)
26. Szegedy, C., Liu, W., Jia, Y., Sermanet, P., Reed, S., Anguelov, D., Erhan, D., Vanhoucke, V., Rabinovich, A.: Going deeper with convolutions. volume abs/1409.4842 (2014)
27. Tumer, K., Ghosh, J.: Error correlation and error reduction in ensemble classifiers. Connect. Sci. **8**(3–4), 385–404 (1996)
28. Wah, C., Branson, S., Welinder, P., Perona, P., Belongie, S.: The caltech-ucsd birds-200-2011 dataset. California Institute of Technology, (CNS TR 2011 001) (2011)
29. Wang, J., Yang, J., Yu, K., Lv, F., Huang, T., Gong, Y.: Locality-constrained linear coding for image classification. In: Proceedings of the 2010 IEEE Conference on Computer Vision and Pattern Recognition (CVPR), pp. 3360–3367, June 2010

30. Yang, Y., Newsam, S.: Bag-of-visual-words and spatial extensions for land-use classification. In: Proceedings of the 18th SIGSPATIAL International Conference on Advances in Geographic Information Systems, pp. 270–279, March 2010
31. Yosinski, J., Clune, J., Bengio, Y., Lipson, H.: How transferable are features in deep neural networks. In: Proceedings of Neural Information Processing Systems (NIPS 2014), November 2014

GLDA-FP: Gaussian LDA Model for Forward Prediction

Yunpeng Xiao[1(✉)], Liangyun Liu[1], Ming Xu[2], Haohan Wang[3], and Yanbing Liu[1]

[1] Chongqing Engineering Laboratory of Internet and Information Security,
Chongqing University of Posts and Telecommunications,
Chongqing 400065, China
xiaoyp@cqupt.edu.cn
[2] Research Institute of Information Technology, Tsinghua University,
Beijing 100084, China
[3] School of Computer Science, Language Technologies Institute,
Carnegie Mellon University, Pittsburgh, PA, USA

Abstract. In social networks, information propagation is affected by diversity factors. In this work, we study the formation of forward behavior, map into multidimensional driving mechanisms and apply the behavioral and structural features to forward prediction. Firstly, by considering the effect of behavioral interest, user activity and network influence, we propose three driving mechanisms: interest-driven, habit-driven and structure-driven. Secondly, by taking advantage of the Latent Dirichlet allocation (LDA) model in dealing with problems of polysemy and synonymy, the traditional text modeling method is improved by Gaussian distribution and applied to user interest, activity and influence modeling. In this way, the user topic distribution for each dimension can be obtained regardless of whether the word is discrete or continuous. Moreover, the model can be extended using the pre-discretizing method which can help LDA detect the topic evolution automatically. By introducing time information, we can dynamically monitor user activity and mine the hidden behavioral habit. Finally, a novel model, Gaussian LDA, for forward prediction is proposed. The experimental results indicate that the model not only mine user latent interest, but also improve forward prediction performance effectively.

Keywords: Multidimensional driving mechanisms · Forward prediction
LDA

1 Introduction

From the perspective of information propagation, forwarding is viewed as an atomic behavior. Published messages are visible to the followers, and a follower can quickly share information that he or she is interested in. Considering the direct driving force on information diffusion, forward prediction has been applied in many fields [1, 2]. The relevant research can help us explore the direction of information dissemination [3], and has a positive significance to public opinion control [4]. Although improved

© Springer International Publishing AG, part of Springer Nature 2018
F. Y. L. Chin et al. (Eds.): BIGDATA 2018, LNCS 10968, pp. 124–139, 2018.
https://doi.org/10.1007/978-3-319-94301-5_10

methods have achieved positive results in current research on forward prediction, some challenges still remain.

On the one hand, user behavior in social networks is caused by complex factors [5]. Users are the core media of online social network information dissemination and directly affect the breadth and depth of information dissemination. Although the current research concerns the impact of user attributes on information forwarding, it only considers the basic attributes, such as the number of fans or friends [6], and ignores the intrinsic mechanisms [7] affecting forward behavior such as interests and habits [8]. Moreover, user interests tend to be multidimensional. This may make users participate in different social actions and be influenced by different users. However, there are few studies that consider the multidimensional interests of users. We need to explore more in the future research.

On the other hand, a lot of methods for forward prediction in social networks consider only static features and attributes [9, 10], and few works take time factor into consideration [11]. Not only nodes and edges are changing with time, the information forwarding behavior also change with time. Incorporating time factor into forward prediction methods would be promising.

In order to analyze the complexity of user forward behavior, it is mapped into multiple mechanisms. Both internal driving mechanisms such as interests, habits and external driving mechanism of network structure are considered. By introducing time information, we propose a model to dynamically monitor user forward behavior. To verify our proposed model, we choose real data of the Sina Weibo for evaluation. Experimental results indicate that the model not only mine user latent topic in multiple dimensions, but also improve forward prediction performance.

Our contribution can be summarized as follows:

- In order to analyze the complexity of user forward behavior, it is mapped into multiple mechanisms: interest-driven, habit-driven and structure-driven. By analyzing and quantifying these driving mechanisms, we can effectively predict user forward behavior.
- Owing to the continuity of some user attributes, the traditional LDA text modeling method is improved by Gaussian distribution and applied to user interest, activity and influence modeling. In this way, the user topic distribution for each dimension can be obtained regardless of whether the word is discrete or continuous.
- Given the insufficiency of current model consider only static features and the advantage of pre-discretizing method on helping LDA detect the topic evolution automatically. Our model can be extended using the pre-discretizing method. By introducing time information, we can dynamically monitor user activity and mine the hidden behavioral habit.

The remainder of this paper is organized as follows. Section 2 introduces related work. Section 3 formulates the problem and gives the necessary definitions. Section 4 explains the proposed model and describes the learning algorithm. Section 5 presents and analyzes the experimental results. Finally, Sect. 6 concludes the paper.

2 Related Work

In online social networks, information dissemination is mainly depended on forward behavior. Forward prediction is achieved by learning the user interests and behavior patterns. According to different assumptions, we structure the discussion of related work onto two broad previously mentioned categories: user behavior and interest modeling, dynamic modeling.

Forward Prediction with User Behavior and Interest Models. There are some prior works that focused on predicted forwarding by similar interests [12, 13]. These approaches treat forwarding as the way people interact with the messages. It is critical in understanding user behavior patterns and modeling user interest. Qiu et al. [14] proposed an LDA-based behavior-topic model which jointly models user topic interests and behavioral patterns. Bin et al. [15] proposed two novel Bayesian models which allow the prediction of future behavior and user interest in a wide range of practical applications. Comarela et al. [16] studied factors that influence a users' response, and found that the previous behavior of user, the freshness of information, the length of message could affect the users' response. However, most of the existing work is difficult to take into account the complex drivers of user behavior and neglected the intrinsic mechanisms such as habits.

Modeling and Predicting Forward Behavior Dynamically. Many studies propose dynamic modeling of user behavior. Ahmed et al. [17] proposed a time-varying model. They assumed that user actions are fully exchangeable, and that users' interest are not fixed over time. The paper divided user actions into several epochs based on the time stamp of the action and modeled user action inside each epoch using LDA. Liu et al. [18] proposed a fully dynamic topic community model to capture the time-evolving latent structures in such social streams. Moreover, some researches [19–21] assumed data in similar space are exchangeable and effectively capture the dynamics of topics in message. Zhao et al. [22] proposed a dynamic user clustering topic model. The model adaptively tracks changes of each user's time-varying topic distribution based both on the short texts the user posts during a given time period and on the previously estimated distribution. Most of the methods implement dynamic modeling by quantifying user action or interest.

This paper models user interest and behavior by using GLDA which integrate the Gaussian distribution into LDA. The internal driving mechanisms and external driving mechanisms are jointed into our model. Meanwhile, time factor is introduced by pre-discretizing method, which can help GLDA detect the topic evolution dynamically. Since then, we can obtain user interest and mine the hidden behavioral habit, as well as predict forward behavior dynamically.

3 Problem Definition

3.1 Related Definitions

We use $G = (V, E)$ to denote the structure of a social network, where V is the set of all users and E is an $N \times N$ matrix, with each element $e_{m,n} = 0$ or 1 indicating whether user v_m has a link to user v_n. The cardinality $|V| = N$ is used to denote the total number of whole network users. For predicting the forward behavior, some basic concepts and related definitions are introduced.

Definition 1. Interest following vector $\vec{e}_m^{(a)} = \left[e_{m,1}^{(a)}, e_{m,2}^{(a)}, \ldots, e_{m,N_m}^{(a)} \right]$.

We hold the view that users are more likely to follow a user they are interested in. Therefore, the user following behavior is used to define the interest following vector. $e_{m,n}^{(a)} (n = 1, 2, \ldots, N_m)$ is a user followed by user v_m, referred to here as a followed user. N_m is the number of followed users.

Definition 2. Interest interacting vector $\vec{e}_m^{(i)} = \left[e_{m,1}^{(i)}, e_{m,2}^{(i)}, \ldots, e_{m,N_m'}^{(i)} \right]$.

The following relationship only indicates the possibility of interaction between users and reflects the static user interests. By analyzing the historical interaction, we can mine active user interests. $e_{m,n}^{(i)} (n = 1, 2, \ldots, N_m')$ is a user interacted with user v_m, referred to here as an interacted user. N_m' is the number of interacted users.

Definition 3. Interest-driven vector $I(v_m) = \left[e_{m,1}^{(a)}, \ldots, e_{m,N_m}^{(a)}, e_{m,1}^{(i)}, \ldots, e_{m,N_m'}^{(i)} \right]$.

The interest-driven vector is referred to as a user interest document, which can also be expressed as the superposition of interest followed users and interacted users. Each followed user or interacted user can be referred to here as a behavioral user.

Definition 4. Habit-driven vector $A(v_m, t) = [x_{m,t,1}, x_{m,t,2}]$.

The activity is divided into post activity $x_{m,t,1}$ and forward activity $x_{m,t,2}$. Considering the characteristics of user daily routine, we divide a day into four six-hour slices and map the activity related attributes into multiple time slices, which are defined as:

$$\begin{cases} x_{m,t,1} = n_{m,t}^{pos} / n^{pos} \\ x_{m,t,2} = n_{m,t}^{ret} / n^{ret} \end{cases} \tag{1}$$

Where $n_{m,t}^{pos}$ and $n_{m,t}^{ret}$ represent the average post or forward number of user v_m at time t, n^{pos} and n^{ret} are the average post or forward number per day.

Definition 5. Structural-driven vector $S(v_m) = [d_{m,1}, d_{m,2}, d_{m,3}]$.

Network provides substrate for information propagation, thus forward behavior strongly depends on network structure. Based on the network influence related attributes, we can define structural-driven vector, where $d_{m,1}, d_{m,2}, d_{m,3}$ are the in-degree, out-degree, and node degree centrality respectively.

3.2   Problem Formulation

To formally formulate the problem of our research, let $G = (V, E)$ be the whole network, $B = \{(b, v, t)|v \in V\}$ represents the behavior information of all users. Firstly, the cause of forward behavior can be mapped into multidimensional vectors: I, A, S. If a user published a message at time slice t, then we can use our method to predict fans forward behavior Y. Specifically, the problem is formulated as follows:

$$\left.\begin{array}{c} G, B \to I, A, S \\ t \end{array}\right\} \Rightarrow f : (I, A, S, t) \to Y \qquad (2)$$

4   Proposed Model

To solve the above problems, we propose a novel prediction model, GLDA, based on user behavior and relationships. The details of the model framework are introduced in three modules: driving mechanisms quantification, user interest, activity and influence modeling and forward prediction, as shown in Fig. 1. In the first module, related attributes are considered for driving mechanisms quantification, and multiple driven vectors are proposed to represent them. In the second module, the user topic distribution for each dimension can be obtained based on improved LDA. In the third module, using Gibbs sampling method to get the probability distribution of forward behavior, then the model can be proposed to do forward prediction.

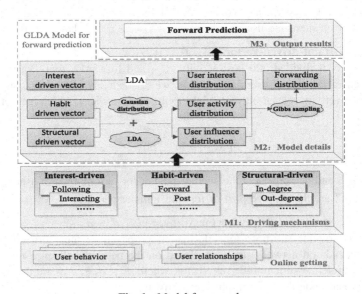

Fig. 1. Model framework.

4.1 Model Details

Given the three driven vectors defined in Sect. 3.1, the problem to be solved becomes how to incorporate those vectors into multiple prediction features and behavior modeling. This module presents the process of modeling, includes: interest-driven simulation analysis, habit-driven simulation analysis, structural-driven simulation analysis. Corresponding to different driving mechanisms, the relevant distributions of prediction features can be obtained.

Interest-Driven Simulation Analysis. User interest is reflected primarily in user behavior. We focus on the analysis of following behavior and interacting behavior. Taking advantage of the LDA topic model in dealing with polysemy and synonym problems, the traditional text modeling method is used to model user interest. Each user can be understood as the component of followed users and interacted users, which can also be expressed as its interest-driven vector. Given parameter Z as the number of interest topics, the simulated interest-driven vector generative process is:

1. For each interest topic z, draw $\vec{\xi}_z \sim Dir(\lambda)$;
2. Given the *mth* user v_m, in whole network G, draw $\vec{\varphi}_m \sim Dir(\alpha)$;
3. For the *nth* behavioral user in the *mth* user $e_{m,n}$:
   a. Draw an interest topic $z = z_{m,n} \sim Mult(\vec{\varphi}_m)$;
   b. Draw a behavioral user $e_{m,n} \sim Mult(\vec{\xi}_{z_{m,n}})$;
   Here, $Dir(.), Mult(.)$ denotes Dirichlet distribution and Multinomial distribution. The graphic model is shown in Fig. 2 and the symbols are described in Table 1.

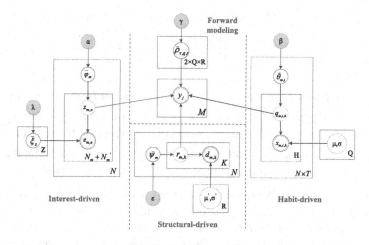

Fig. 2. Graphic model.

Actually, the aim of user interest modeling is to compute Multinomial distributions $\Phi = \left[\vec{\varphi}_1, \vec{\varphi}_2, \ldots, \vec{\varphi}_N\right]$ and $\Sigma = \left[\vec{\xi}_1, \vec{\xi}_2, \ldots, \vec{\xi}_Z\right]$. Owing to the coupling of Φ and Σ, we

Table 1. Description of symbols in graphic model.

Symbols	Descriptions	Symbols	Descriptions
$\alpha, \beta, \lambda, \gamma, \varepsilon$	Dirichlet priors	N	Number of the users in G
N_m, N'_m	Number of followed or interacted users about v_m	Z, Q, R	Number of interest or activity or influence topics
T	Number of time slices	$\vec{\varphi}_m$	Interest distribution of v_m
$\vec{\xi}_z$	Behavioral distribution of z	$e_{m,n}$	The nth behavioral user of v_m
$\vec{\psi}_m$	Influence distribution of v_m	$z_{m,n}$	Interest topic assigned to $e_{m,n}$
$\vec{\theta}_{m,t}$	Activity distribution of v_m at time slice t	$d_{m,k}$	The kth influence related attribute of v_m
$q_{m,t,h}$	Activity topic assigned to $x_{m,t,h}$	$r_{m,k}$	Influence topic assigned to $d_{m,k}$
$x_{m,t,h}$	The hth activity related attribute of v_m at time slice t	H, K	Number of activity or influence related attributes
$\mu, \sigma, \mu', \sigma'$	Parameters of Gaussian distributions	$\vec{\rho}_{\tau,q,r}$	Forward behavior distribution of multiple prediction features
y_j	The jth forward behavior to be predicted	M	Number of forward behaviors to be predicted

cannot compute them directly and Gibbs sampling [23] is applied to indirectly get Φ and Σ. The principle of Gibbs sampling in terms of extracting topic z_i of behavior user e_i is as follows:

$$p\left(z_i = z | \vec{z}_{-i}, E\right) \propto p\left(z_i = z, e_i = e | \vec{z}_{-i}, E\right) = \widehat{\varphi}_{m,z} \times \widehat{\xi}_{z,e}$$

$$= \frac{n_{m,\neg i}^{(z)} + \alpha}{\sum_{z=1}^{Z} n_{m,\neg i}^{(z)} + \alpha} \times \frac{n_{z,\neg i}^{(e)} + \beta}{\sum_{e=1}^{N} n_{z,\neg i}^{(e)} + \beta} \tag{3}$$

Where $\vec{z}_{-i}$ represents the topic of behavioral users except for the current behavioral user; $\vec{e}_{-i}$ represents behavioral users except for the current behavioral user; $n_{z,\neg i}^{(e)}$ is the number of behavioral user e assigned to interest topic z except for the current behavioral user; and $n_{m,\neg i}^{(z)}$ is the number of interest topic z assigned to user v_m except for the current behavioral user. When the sampling converges, Φ and Σ can be obtained.

Habit-Driven Simulation Analysis. The user behavioral habit can be analyzed based on historical behavior. Here, we focus on post behavior and forward behavior. Considering the dynamics of user behavioral habit, the past behavioral data are pre-discretized to get habit-driven vector. In other words, a user at every time slice is regarded as a document and the activity related attributes are regarded as words. By introducing activity as topics, we can mine potential user behavioral habits.

Unlike the value of $e_{m,n}$, the activity related attributes are both continuous. Owing to the useless for dealing with continuous attributes modeling, Gaussian distribution is used to replace Multinomial distribution in standard LDA. In the improved model, the activity related attributes obey the following Gaussian distribution:

$$f(x_{m,t,h}; \mu_{q,h}, \sigma_{q,h}) = \frac{1}{\sqrt{2\pi}\sigma_{q,h}} \exp\left[-(x_{m,t,h} - \mu_{q,h})^2 / 2\sigma_{q,h}^2\right] \tag{4}$$

Where $x_{m,t,h}$ is the hth attribute of user v_m at time slice t, and $\mu_{q,h}, \sigma_{q,h}$ are parameters of Gaussian distribution that $x_{m,t,h}$ obeys. The simulated habit-driven vector generative process can be described as follows:

1. Given the mth user v_m at any time slice t, draw $\vec{\theta}_{m,t} \sim Dir(\beta)$;
2. For the hth activity related attribute of the mth user $x_{m,t,h}$:

    a. Draw an activity topic $q = q_{m,t,h} \sim Mult(\vec{\theta}_{m,t})$;
    b. Draw an attribute value $x_{m,t,h} \sim N(\mu_{q_{m,t,h},h}, \sigma_{q_{m,t,h},h})$;

Where $N(.)$ denotes Gaussian distribution. The purpose of user activity modeling is to learn the distribution set $\Pi = [(\mu_{1,h}, \sigma_{1,h}), (\mu_{2,h}, \sigma_{2,h}), \ldots, (\mu_{Q,h}, \sigma_{Q,h})]$ $(h \in [1, H])$ and $\Theta = [\vec{\theta}_{1,t}, \vec{\theta}_{2,t}, \ldots, \vec{\theta}_{N,t}]$ $(t \in T)$. Owing to the existence of hidden variables, the EM algorithm [24] is used to estimate model parameters. E step computes the responsiveness of topic to attribute according to the current model parameters:

$$\chi_{m,t,h,q} = P(q|v_m, t, x_{m,t,h}) = P(x_{m,t,h}|q) \times P(q|v_m, t) = \frac{f(x_{m,t,h}; \mu_{q,h}, \sigma_{q,h})\theta_{m,t,q}}{\sum_{q'=1}^{Q} f(x_{m,t,h}; \mu_{q',h}, \sigma_{q',h})\theta_{m,t,q'}} \tag{5}$$

M step updates the model parameters for the new round of iteration:

$$\mu_{q,h} = \frac{\sum_{m=1}^{N} \sum_{t=1}^{T} \chi_{m,t,h,q} * x_{m,t,h}}{\sum_{m=1}^{N} \sum_{t=1}^{T} \chi_{m,t,h,q}}, \quad \sigma_{q,h} = \sqrt{\frac{\sum_{m=1}^{N} \sum_{t=1}^{T} \chi_{m,t,h,q}(x_{m,t,h} - \mu_{q,h})^2}{\sum_{m=1}^{N} \sum_{t=1}^{T} \chi_{m,t,h,q}}} \tag{6}$$

$$\theta_{m,t,q} = \frac{1}{H} \sum_{h=1}^{H} \chi_{m,t,h,q} \tag{7}$$

Where $\chi_{m,t,h,q}$ is the responsiveness of activity topic q to the attribute $x_{m,t,h}$. $\theta_{m,t,q}$ denotes the probability of user v_m assigned to activity topic q at time slice t. Repeat the above two steps until convergence, Θ and Π can be obtained.

Structural-Driven Simulation Analysis. The network structure contains many user attributes, such as in-degree, out-degree, and other attributes, which can be expressed as structural-driven vector. Based on it, we can classify users into clusters. Each cluster can be regarded as an influence role that users play. Each influence role has a set of parameters of distribution that the influence related attributes conform to. Similar with the previous Section, we also use Gaussian distribution. If user v_m play influence role r, its kth attribute $d_{m,k}$ conforms to:

$$f\left(d_{m,k}; \mu'_{r,k}, \sigma'_{r,k}\right) = \frac{1}{\sqrt{2\pi}\sigma'_{r,k}} \exp\left[-\left(d_{m,k} - \mu'_{r,k}\right)^2 / 2\sigma'^2_{r,k}\right] \tag{8}$$

Where $\mu'_{r,k}$ and $\sigma'_{r,k}$ are parameters of Gaussian distribution that $d_{m,k}$ obeys. The simulated structural-driven vector generative process is as follows:

1. Given the *mth* user v_m, in the whole network, draw $\vec{\psi}_m \sim Dir(\varepsilon)$;
2. For the *kth* influence related attribute of the *mth* user $d_{m,k}$:

   a. Draw an influence topic $r = r_{m,k} \sim Mult(\vec{\psi}_m)$;
   b. Draw an attribute value $d_{m,k} \sim N(\mu'_{r_{m,k},k}, \sigma'_{r_{m,k},k})$;

Our goal is to learn distributions $\Pi' = [(\mu'_{1,k}, \sigma'_{1,k}), (\mu'_{2,k}, \sigma'_{2,k}), \ldots, (\mu'_{R,k}, \sigma'_{R,k})]$ $(k \in [1, K])$ and $\Psi = [\vec{\psi}_1, \vec{\psi}_2, \ldots, \vec{\psi}_N]$. And EM algorithm is also used to estimate model parameters. E step computes the responsiveness as:

$$\chi'_{m,k,r} = P(r|v_m, d_{m,k}) = P(d_{m,k}|r) \times P(r|v_m) = \frac{f(d_{m,k}; \mu'_{r,k}, \sigma'_{r,k})\psi_{m,r}}{\sum_{r'=1}^{R} f(d_{m,k}; \mu'_{r',k}, \sigma'_{r',k})\psi_{m,r'}} \tag{9}$$

M step updates the model parameters for the new round of iteration:

$$\mu'_{r,k} = \frac{\sum_{m=1}^{N} \chi'_{m,k,r} d_{m,k}}{\sum_{m=1}^{N} \chi'_{m,k,r}}, \ \sigma'_{r,k} = \sqrt{\frac{\sum_{m=1}^{N} \chi'_{m,k,r}(d_{m,k} - \mu'_{r,k})^2}{\sum_{m=1}^{N} \chi'_{m,k,r}}} \tag{10}$$

$$\psi_{m,r} = \frac{1}{K} \sum_{k=1}^{K} \chi'_{m,k,r} \tag{11}$$

Where $\chi'_{m,k,r}$ is the responsiveness of influence role r to the attribute $d_{m,k}$. $\psi_{m,r}$ denotes the probability of user v_m assigned to influence role r. Repeat the two steps until convergence, Ψ and Π' can be obtained.

4.2   Comprehensive Forward Behavior Modeling and Prediction

Based on the previous modeling process, the relevant probability distributions of prediction features are obtained. By combining these distributions, the forward behavior distribution can be computed to predict the forward action that a user may take. Assume that $Y = \{y_1, y_2, \ldots, y_M\}$ is the behavior set to be predicted and its generative process can be described as follows:

1. Draw $\vec{\rho} \sim Dir(\gamma)$;
2. For the *jth* behavior y_j:

   a. Draw an interest topic $z_m \sim Mult(\vec{\varphi}_m)$ for post user v_m;
   b. Draw an interest topic $z_n \sim Mult(\vec{\varphi}_n)$ for fans v_n;
   c. Draw an activity topic $q = q_{n,t} \sim Mult(\vec{\theta}_{n,t})$ for fans v_n;

d. Draw a influence role $r = r_m \sim Mult(\vec{\psi}_m)$ for post user v_m;

e. Draw the behavior $y_j \sim Mult(\vec{\rho}_{\tau,q,r})$;

Where τ is an indicator function of interest, if $z_m = z_n$, $\tau = 1$, otherwise $\tau = 0$. Behavior y_j only contains two cases ($y_j = 1$ indicates establish forward action, $y_j = 0$ indicates not establish forward action), so we can use a Bernoulli distribution $\vec{\rho}_{\tau,q,r}$ to represent the probability distribution of multiple features over forward actions and the parameter $\Omega = \left[\vec{\rho}_{0,q,r}, \vec{\rho}_{1,q,r}\right]$ ($q \in [1, Q], r \in [1, R]$). By using Gibbs sampling, the principle of extracting features τ, q, r of behavior y_j is as follows:

$$p\left(\tau_j = \tau, q_j = q, r_j = r | \vec{\tau}_{-j}, \vec{q}_{-j}, \vec{r}_{-j}, Y\right) \propto p\left(\tau_j = \tau, q_j = q, r_j = r, y_j = y | \vec{\tau}_{-j}, \vec{q}_{-j}, \vec{r}_{-j}, Y_{-j}\right)$$
$$= p(\tau|\Phi)p(q|\Theta)p(r|\Psi)\widehat{\rho}_{\tau,q,r,y}$$
$$= \left(\vec{\varphi}_m \vec{\varphi}_n^T\right)\theta_{n,t}^{(q)}\psi_m^{(r)} \times \frac{n_{\tau,q,r,-j}^{(y)} + \gamma}{\sum_{y=0}^{1} n_{\tau,q,r,-j}^{(y)} + \gamma}$$

(12)

Where $\vec{\tau}_{-j}, \vec{q}_{-j}, \vec{r}_{-j}$ represents the prediction features of behavioral except for the current behavior; Y_{-j} represents behavior to be predicted except for the current behavior; $n_{\tau,q,r,-j}^{(y)}$ is the number of behavior y assigned to prediction features τ, q, r except for the current behavior; When the sampling converges, Ω can be obtained.

Given a message, we can predict the forward behavior based on the trained model. Firstly, get the features τ, q, r for model input by probability sampling method, and then the user's forward probability $\rho_{\tau,q,r,1}$ and non-forward probability $\rho_{\tau,q,r,0}$ are calculated according to the parameter Ω. If $\rho_{\tau,q,r,1} > \rho_{\tau,q,r,0}$, we predict the fans will forward the message, $y = 1$; Otherwise it will not, $y = 0$, formally expressed as:

$$y = \text{argmax}_{\Omega} p(|y|\tau, q, r)$$

(13)

5   Experiments and Analysis

5.1   Experimental Data and Evaluation Metrics

The experimental data used in this paper is collected from Sina micro-blog, a popular social networking platforms in China. In the process of data collection, we randomly selected a user (user ID: 2312704093) as the starting point. Some users and their micro-blogs are captured based on breadth-first-search, forming a sub-network containing 49,556 users and 61,880 user relationships for the 2011/08/21-2012/02/22. The statistics of the dataset is shown in Table 2.

In this paper, Accuracy, Precision, Recall, F1-Measure, and ROC curve were used to verify the prediction results. We assumed that the forward behavior of fans is a positive example "1", and non-forward behavior is a negative example "0". Meanwhile, the dataset

Table 2. Statistics of the dataset

Item	Users	Relationships	Post	Forward	Review
Count	49,556	61,880	3,057,635	506,765,237	185,079,821

needs to be partitioned into training set and test set. Here, we set the proportion of training set and test set to be 8:2. The better prediction results have greater Accuracy, Precision, Recall, F1-Measure, and their ROC curves are close to the upper left corner.

5.2   Prediction Performance Analysis

In this section, the performances of our model are evaluated from three viewpoints. Firstly, we show the results of user latent interest distribution and analyze the overall interest distribution of network. Then, the impact of interest number and the proportion of training set on forward prediction can be verified. Finally, we evaluate the performance by comparing our model with other baseline methods. According to the above three viewpoints, the superiority of our model can be verified.

Firstly, the result of user latent interest distribution is analyzed. We select several representative users to show their latent interests in Fig. 3. Meanwhile, latent interest distributions in network can be shown in Fig. 4. Both in the figures, the x-axis represents the interest ID and the y-axis represents the probability value. The highest interest focus values are provided in parentheses in the legend.

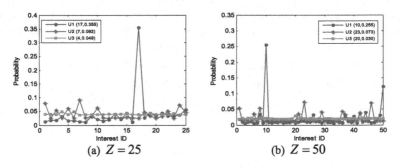

(a) $Z = 25$ (b) $Z = 50$

Fig. 3. User latent interest distributions.

As shown in Fig. 3, the distribution of each user interest is different. When latent interest number $Z = 25$, user $U1$ interest is obvious and prefer Interest $ID = 17$. The range of user $U2$ interest is relatively wide and the proportion interest of user $U3$ is average. And from Fig. 4, we can observe that the network interest distribution is uniform, although user preferences in the entire network have some differences. Next we will verify the latent interest has a driving effect on forward prediction.

Secondly, considering the excellent classification effect of LR and SVM, they are applied to the forward prediction problem and compared with our model. And the effects of the proportion of training set on forward prediction are shown in Fig. 5.

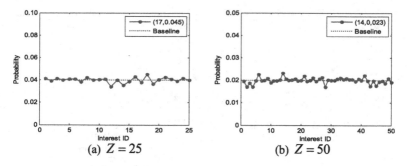

Fig. 4. Latent interest distributions in network.

In addition, by reducing the dimension of driving mechanism in our model, three sub-models are obtained: Sub-IA, Sub-IS, and Sub-I. Comparing our model with the sub-models, the effects of the interest topic number on each model can be shown as Fig. 6.

As shown in Fig. 5, the performance of GLDA is better than LR and SVM. And the performance of our model is least affected by the training set proportion. Overall, as the proportion of training set increases, the prediction effect of each method improves. From Fig. 6, we also can see our model has better prediction performance than its sub-models. It indicates that the extraction of multidimensional driven vector can improve the effect of forward prediction. With the increase of interest topic number, the Precision increases gradually, while the Recall decreases rapidly. From the change of F1-Measure, we can see that when Z is 10–20, the model performs well. In addition, the

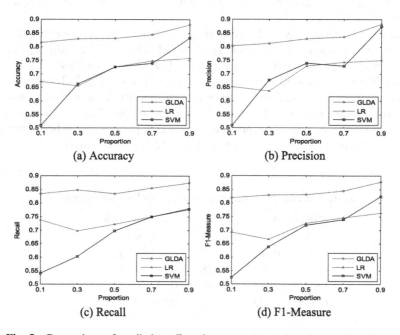

Fig. 5. Comparison of prediction effects between proposed model and classifiers.

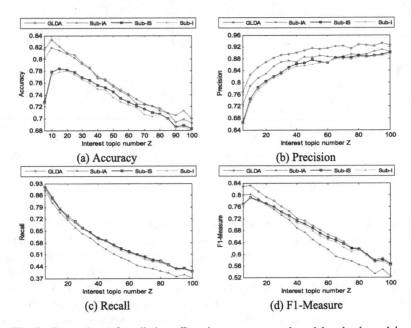

Fig. 6. Comparison of prediction effects between proposed model and sub-models.

number of activity topics and influence topics are proposed in the literature [25, 26]. It proposes to classify the activity level as inactive, generally active and very active and points out that users can be divided into three types: ordinary user, opinion leader and structural hole spanner.

Finally, the performance of our model is evaluated by comparing with some baseline methods, such as probabilistic graph model LDA [27] and CRM [6], classical forward prediction methods CF [28] and VSM [29]. The performances of them are shown in Table 3. And ROC curves comparison are shown in Fig. 7. The results show that our model plays optimal performance in Accuracy, Recall and F1-Measure compared with baseline methods. And it can be seen that the ROC curve of our proposed model is closest to the upper left corner, and the overall performance is best. Therefore, our model can improve the forward prediction performance effectively.

Table 3. Comparison between our model and baseline methods.

Methods	Accuracy	Precision	Recall	F1-Measure
LDA	0.774	0.734	0.805	0.768
CRM	0.784	0.785	0.822	0.803
CF	0.805	**0.848**	0.746	0.794
VSM	0.665	0.703	0.742	0.722
GLDA	**0.833**	0.827	**0.838**	**0.833**

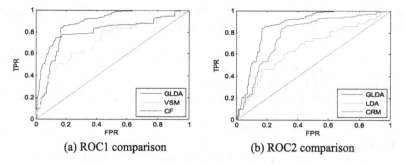

(a) ROC1 comparison   (b) ROC2 comparison

Fig. 7. Comparison of different methods in ROC.

6 Conclusion

In this study, a novel forward prediction model GLDA is proposed, and it can effectively predict forward behavior by analyzing user behavior and relationships. Firstly, we mapped the cause of forward behavior into three driving mechanisms: interest-driven, activity-driven and structure-driven. Secondly, the traditional LDA was improved by Gaussian distribution and applied to user interest, activity and influence modeling. Finally, the model was extended with the pre-discretizing method and we can dynamically monitor user activity and mine the hidden behavioral habit.

The experimental results showed that our model can improve forward prediction performance in comparison to other baseline methods. By studying forward prediction in social networks, we can acquire a better understanding of information propagation mechanism. And the model can provide support for public opinion management and control. In the future work, it would be intriguing to integrate nonparametric methods into our model to base parameter value choices on the data itself. It is also interesting to explore a method to alleviate the time complexity of training algorithms.

Acknowledgements. This paper is partially supported by the National 973 Key Basic Research Program of China (Grant No.2013CB329606), the National Natural Science Foundation of China (Grant No.61772098), Chongqing Science and Technology Commission Project (Grant No.cstc 2017jcyjAX0099) and Chongqing key research and development project (Grant No.cstc2017 zdcy-zdyf0299, cstc2017zdcy-zdyf0436).

References

1. Yu, H., Bai, X.F., Huang, C.Z., et al.: Prediction of users retweet times in social network. Int. J. Multimed. Ubiquit. Eng. **10**(5), 315–322 (2015)
2. Huang, D., Zhou, J., Mu, D., et al.: Retweet behavior prediction in Twitter. In: 7th International Symposium on Computational Intelligence and Design, pp. 30–33. IEEE, Hangzhou (2015)
3. Xu, Q., Su, Z., Zhang, K., et al.: Epidemic information dissemination in mobile social networks with opportunistic links. IEEE Trans. Emerg. Top. Comput. **3**(3), 399–409 (2017)

4. Yoo, E., Rand, W., Eftekhar, M., et al.: Evaluating information diffusion speed and its determinants in social media networks during humanitarian crises. J. Oper. Manag. **45**, 123–133 (2016)

5. Xiao, Y., Li, N., Xu, M., et al.: A user behavior influence model of social hotspot under implicit link. Inf. Sci. **396**, 114–126 (2017)

6. Han, Y., Tang, J.: Probabilistic community and role model for social networks. In: Proceedings of the 21th ACM SIGKDD International Conference on Knowledge Discovery and Data Mining, Sydney, NSW, Australia, pp. 407–416. ACM (2015)

7. Lee, S.: What makes twitterers retweet on Twitter? Exploring the roles of intrinsic/extrinsic motivation and social capital. J. Korea Acad.-Ind. **15**(6), 3499–3511 (2014)

8. Jin, Y., Zhai, L.H.: An Investigation and Analysis of the Impact of User's Forwarding Behavior on the Quality of Information in Social Media Environment. Libr. Theor. Pract. September 2016

9. Wang, C., Li, Q., Wang, L., et al.: Incorporating message embedding into co-factor matrix factorization for retweeting prediction. In: International Joint Conference on Neural Networks, Anchorage, AK, USA, pp. 1265–1272. IEEE (2017)

10. Zhang, Y., Lyu, T., Zhang, Y.: Hierarchical community-level information diffusion modeling in social networks. In: Proceedings of the 40th International ACM SIGIR Conference on Research and Development in Information Retrieval, Shinjuku, Tokyo, Japan, pp. 753–762. ACM (2017)

11. Zarrinkalam, F., Kahani, M., Bagheri, E.: Mining user interests over active topics on social networks. Inf. Process. Manag. **54**(2), 339–357 (2018)

12. Chen, W., Wang, Y., Yang, S.: Efficient influence maximization in social networks. In: Proceedings of the 15th ACM SIGKDD International Conference on Knowledge Discovery and Data Mining, Paris, France, pp. 199–208. ACM (2009)

13. Kempe, D., Kleinberg, J., Tardos, E.: Maximizing the spread of influence through a social network. In: Proceedings of the Ninth ACM SIGKDD International Conference on Knowledge Discovery and Data Mining, Washington, D.C., pp. 137–146. ACM (2003)

14. Qiu, M., Zhu, F., Jiang, J.: It is not just what we say, but how we say them: LDA-based behavior-topic model. In: Proceedings of the 2013 SIAM International Conference on Data Mining, pp. 794–802. Society for Industrial and Applied Mathematics (2013)

15. Bi, B., Cho, J.: Modeling a retweet network via an adaptive Bayesian approach. In: Proceedings of the 25th International Conference on World Wide Web, Canada, pp. 459–469. International World Wide Web Conferences Steering Committee (2016)

16. Comarela, G., Crovella, M., Almeida, V., et al.: Understanding factors that affect response rates in Twitter. In: Proceedings of the 23rd ACM Conference on Hypertext and Social Media, Milwaukee, Wisconsin, USA, pp. 123–132. ACM (2012)

17. Ahmed, A., Low, Y., Aly M, et al.: Scalable distributed inference of dynamic user interests for behavioral targeting. In: Proceedings of the 17th ACM SIGKDD International Conference on Knowledge Discovery and Data Mining, San Diego, California, USA, pp. 114–122. ACM (2011)

18. Liu, Z., Zheng, Q., Wang, F., et al.: A dynamic nonparametric model for characterizing the topical communities in social streams. In: Proceedings of the 2014 SIAM International Conference on Data Mining, pp. 379–387. Society for Industrial and Applied Mathematics (2014)

19. Ahmed, A., Xing, E.P.: Timeline: a dynamic hierarchical Dirichlet process model for recovering birth/death and evolution of topics in text stream. arXiv preprint arXiv:1203.3463 (2012)

20. Ahmed, A., Xing, E.P.: Dynamic non-parametric mixture models and the recurrent chinese restaurant process: with applications to evolutionary clustering. In: Proceedings of the 2008 SIAM International Conference on Data Mining, pp. 219–230. Society for Industrial and Applied Mathematics (2008)
21. Blei, D.M., Frazier, P.I.: Distance Dependent Chinese Restaurant Processes. J. Mach. Learn. Res. **12**(1), 2461–2488 (2009)
22. Zhao, Y., Liang, S., Ren, Z., et al.: Explainable user clustering in short text streams. In: Proceedings of the 39th International ACM SIGIR Conference on Research and Development in Information Retrieval, Pisa, Italy, pp. 155–164. ACM (2016)
23. Zhao, F., Zhu, Y., Jin, H., et al.: A personalized hashtag recommendation approach using LDA-based topic model in microblog environment. Future Gener. Comput. Syst. **65**, 196–206 (2016)
24. Zanetti, M., Bovolo, F., Bruzzone, L.: Rayleigh-rice mixture parameter estimation via EM algorithm for change detection in multispectral images. IEEE Trans. Image Process. **24**(12), 5004–5016 (2015)
25. Zhu, Y., Zhong, E., Pan, S.J., et al.: Predicting user activity level in social networks. In: Proceedings of the 22nd ACM International Conference on Information and Knowledge Management, San Francisco, California, USA, pp. 159–168. ACM (2013)
26. Yang, Y., Tang, J., Leung, C.W., et al.: RAIN: social role-aware information diffusion. In: AAAI, pp. 367–373 (2015)
27. Li, L., He, J., Wang, M., et al.: Trust agent-based behavior induction in social networks. IEEE Intell. Syst. **31**(1), 24–30 (2016)
28. Jiang, B., Liang, J., Sha, Y., et al.: Retweeting behavior prediction based on one-class collaborative filtering in social networks. In: Proceedings of the 39th International ACM SIGIR Conference on Research and Development in Information Retrieval, Pisa, Italy, pp. 977–980. ACM (2016)
29. Waitelonis, J., Exeler, C., Sack, H.: Enabled generalized vector space model to improve document retrieval. In: NLP-DBPEDIA@ ISWC, pp. 33–44 (2015)

Tracking Happiness of Different US Cities from Tweets

Bryan Pauken[1], Mudit Pradyumn[2(✉)], and Nasseh Tabrizi[2]

[1] The University of Alabama, Tuscaloosa, AL 35401, USA
[2] Department of Computer Science, East Carolina University, Greenville
NC 27858, USA
pmudit90@gmail.com

Abstract. Research into the possibilities of Twitter data has grown greatly over the past few years. Studies have shown its potential in identifying and managing disasters, predicting flu trends, predicting the success of movies at the box office, and analyzing people's emotions. In this study, tweets from Twitter were collected and analyzed from nine different cities across America. East Carolina University's Hadoop cluster was used to run our application and the Stanford CoreNLP was then used to give the sentiment of each statement in the tweets. Although our research reviled small distinction between nine individual cities in the percentage of positive, negative, and neutral statements, but however, there were significant differences in overall statements, where up 47.88% of all the statements were neutral, positive statements only 14.95%, while 37.16% of the statements were negative.

Keywords: Twitter · Sentiment analysis · Happiness levels
Stanford CoreNLP

1 Introduction

Sentiment analysis has become a topic of growing interest in recent years, especially through the use of the Internet and social media. It is commonly studied in data mining, web mining, text mining, and is one of the most prevalent research areas in the field of natural language processing [1]. The data acquired through sentiment analysis is extremely useful. For example, sentiment analysis can help businesses assess the public opinion of a specific product and predict the success of movies at the box office [1–4]. The study aims to use sentiment analysis from tweets posted in specific cities on Twitter and compare the results to the list of happiest cities in the United States of America.

Twitter is an online micro-blogging website that allows people to post statuses called tweets. These tweets are short messages of no more than 140 characters. Each tweet contains information like who the tweet belongs to, the text of the tweet, the number of favorites, the location, the hashtags, and more. Much research has been done using Twitter [5] due to the easy to use application program interface (API) and the amount of potential data Twitter has to offer. A quick search on GoogleScholar using the keyword "Twitter" yields over 6 million results. With over 300 million monthly

© Springer International Publishing AG, part of Springer Nature 2018
F. Y. L. Chin et al. (Eds.): BIGDATA 2018, LNCS 10968, pp. 140–148, 2018.
https://doi.org/10.1007/978-3-319-94301-5_11

users and according to [6, 7] Twitter produces on average 6000 tweets per second, 350,000 tweets a minute, and 500 million tweets a day. Four main motivations for tweeting are reaching information quickly, gaining higher visibility, feeling connected, and being aware of what is happening [6]. Furthermore, Twitter users most commonly tweet about themselves [8], which makes Twitter a valuable tool for sentiment analysis.

Currently, there are two common ways to gauge happiness level. One way to assess a city's happiness is to have researchers conduct interviews like Gallup-Healthways [9] held 350,000 interviews from 2015 to 2016 in order to help rank the happiness 189 communities. These interviews are time intensive and can often give an unrealistic image of a person's happiness [10] as the interviews may contain bias. The problem with the interview or survey method of assessing happiness is that it only captures a small snapshot of that person's life. A person could be having a terrible day or a great day at the time of the interview, which would skew the data and not give an accurate overall depiction of how that person feels on a day to day basis. To be more effective, interviews would have to take place over a longer period of time. However, that would cause the process to take even longer and to be costlier.

The second method [11] of assessing a city's happiness is to look at key statistics like depression rate, average income, unemployment rate, and life expectancy. Once again, this method of evaluation may not give an accurate depiction of a city's happiness. A city could have good statistics in these areas, and its people may still be unhappy. Collecting data from Twitter [12] could give a more accurate depiction of a location's emotional wellbeing. Evidence suggests that social network users share emotional states such as mood, happiness, depression, and suicidality online.

In this study, Twitter data was collected from nine different and diverse group of cities (San Francisco, San Diego, Washington D.C., New York, Chicago, Dallas, Detroit, Cleveland, and Philadelphia) over a 3-day time frame in the United States of America. Three cities were chosen from the top ten happiest cities list provided by Wallethub [13] San Diego, San Francisco, and Washington D.C. Three cities were also chosen from the bottom of the list: Detroit, Philadelphia, and Cleveland. Then three cities were chosen from the middle of the list in order to have data to compare to New York, Chicago, and Dallas. The Stanford CoreNLP [14] was then used to analyze the sentiment of the tweets. The results were compared to the Wallethub's list of happiest cities to see the correlation.

2 Related Work

Twitter has become a heavily researched field recently. One big topic of interest is using Twitter in disaster management situations. Sakaki et al. used Twitter to detect target events by looking at keywords in a tweet, the number of words, and the context of the tweet. Their proposed method could accurately detect an earthquake with a seismic scale of 3 or more 96% of the time [12]. Also, the researchers were able to send emails to registered users quicker than the Japan Meteorological Agency could report the earthquakes. The designed system took about a minute, sometimes just 20 s, to deliver warning emails to users after an earthquake was detected. In comparison, JMA takes 6 min on average after an earthquake occurs to report it [15].

The authors [16] studied the usage of Twitter in Padang Indonesia in order to help in disaster management situations. Five different ways of obtaining data were proposed and evaluated. Bound-boxing, setting the coordinates of where the tweets to be collected from, was determined to be the best method. The study determined due to the high and widespread usage of Twitter, high amounts of geo-information, and accurate language markers, that Twitter would be effective in helping warn about and manage disaster situations in Padang, Indonesia.

Companies have been interested in opinion mining and sentiment analysis for quite some time as they can see the public's reaction to a new product. It is valuable for marketers to understand how the public is responding on social media about xyz. Most of the research done on Twitter and sentiment analysis is focused on a specific event or product. Daniel et al. [17] used sentiment analysis of tweets created by the financial community on Twitter to detect event polarity. Tweets were collected from thirty companies, filtered, and then sentiment analysis was performed to detect the importance of events in the company. Four different text analysis tools were used to perform the sentiment analysis. Financial events were able to be detected successfully through the model designed in the study. Baek et al. [18] investigated how various opinions posted on different websites impacts box office revenue. Twitter was found to be influential in the early stages of the movies being released, while Yahoo!Movies were more influential later after the movie was released.

Yu and Wang [19] collected tweets from U.S. soccer fans from five different games during the 2014 FIFA World Cup. A keyword search was used to collect tweets with hashtags like #FIFA, #Football, #Worldcup, and #Soccer. The Natural Language Toolkit was used to analyze the sentiments of the tweets after preprocessing was done. The results of the experiments matched the expected results. U.S. fans experienced an increase in fear and anger after the opponent scored a goal. Also, U.S. fans showed a lower amount of negative emotions during non-U.S. games. One of the limitation of this study was the way in which the sentiment analysis was done. Each word in the tweet was examined, but the overall semantic meaning of the words was not taken into consideration.

Durahim and Coskun [10] developed a sentiment analysis model to calculate the Gross National Happiness of Turkey. In 2003, Turkey implemented a Life Satisfaction Survey to gauge the happiness of the country. The goal of the study was to create a more accurate way to calculate the happiness of Turkey. SentiStrength V2.2 was used to provide the sentiments of tweets. The sentimental analysis provided similar results to the GNH by Province survey results of Turkish Statistical Institute.

The study by Nguyen et al. [12] collected geotagged Twitter data from Salt Lake County, San Francisco County, and New York County in order to examine neighborhood happiness, diet, and physical activity. The Stanford Natural Language Processing Group was used to tokenize the tweets. In order to perform the sentiment analysis of the tweets, the study utilized the Language Assessment by Mechanical Turk word list and computed the average happiness for each word in a tweet throwing out words like "is", "the", and "a". The automated scores from the algorithm were compared to happiness scores from humans, and the kappa statistics for agreement between the two techniques was 73%. Neutral and positive tweets were found to be more frequent than negative sentiment tweets. New York County had the highest prevalence of happy tweets, San Francisco had the second highest, and Salt Lake County had the lowest. For San Francisco County, the highest happiness scores were recorded near the bay. In New York County, the highest happiness scores were found near Central Park, the Meatpacking District, and other areas.

3   Methodology

The cluster that was used to collect, process, and analyze data, is made up of sixteen Macintosh desktop computers, each with 16 GB of memory, 2 quad core CPU's, a 1 Tb hard disk, and they are all connected with a 1 GB per second network. Hadoop allows for distributed processing of data-intensive analytics through a Java-based software framework [20, 21].

Tweets were collected by using Twitter's streaming API, which allows the live streaming of 1% of all tweets worldwide. However, if filters are applied to what tweets are being collected, then the percentage of tweets being collected can be increased greatly [16]. The Hosebird Client [22, 23], a Java HTTP client for consuming Twitter's Streaming API, was used to collect tweets from nine cities across the United States of America. Only tweets that were geotagged were looked at in this study. Coordinates for New York, Chicago, Dallas, Washington D.C., San Francisco, San Diego, Philadelphia, Detroit, and Cleveland were input into the Hosebird Client in order to create a bounding box around each city, allowing producer program to distribute the tweets between nodes on the Kafka [22] Cluster. Kafka is a distributed streaming platform that has a low latency platform good for handling real-time streaming data.

The data from the producer program stored in the Kafka was then passed to a consumer program, that was written in Spark [21], was then used to process the tweets in real time. Spark is a framework for parallel processing that keeps data in memory instead of on the Hadoop Distributed File System. The consumer program cleaned the text of the tweets, where links, hashtags, and mentions were all removed from the Tweets. Certain short hands like "idk" "idc" and "smh" were replaced with "I don't know" "I don't care" and "Shake my head" respectively in order for a better more accurate analysis of the tweets. The coordinates were pulled and cleaned from the tweet to only give each longitude and latitude coordinate. The consumer program also extracted the language of the tweets, the user string id, the location name, and the tweet id.

Sentiment grades of positive, negative, and neutral were given to each sentence in a tweet at the time of processing. The Stanford CoreNLP was used to provide the sentiment analysis of each tweet. The Stanford CoreNLP sentiment model differs from most sentiment prediction systems as it takes into account the order of the words in a sentence. Other systems generally give a positive or negative score to each word in a sentence and the sum up the scores to give the overall sentiment score of the sentence. Ignoring the order of the words can lead to loss of information and inaccurate classification. For example, the words funny and witty are positive, but the StanfordCoreNLP can correctly label the following sentence as negative based on the words around "funny" and "witty" whereas other sentiment programs label this sentence as positive: "This movie was actually neither that funny nor super witty."

The deep learning model of the Stanford CoreNLP is based on a Recursive Neural Network [24], the model increases the accuracy of positive and negative classification of a sentence from 80% to 85.4% [25].

The analyzed tweets were passed back to the Kafka in real time that in turn passed the processed data to Flume [26]. Flume is a simple and flexible distributed service that can be used for collecting, moving, and aggregating Big Data [26]. Flume saves the messages being written to the Kafka topic into the Hadoop Distributed File System. Once the data collection was completed, another Spark program was used to group the data in batch mode. The total number of tweets, statements, positive statements, negative statements, and neutral statements were counted. This data was then imported into Tableau to better analyze and visualize the data.

As shown in Fig. 1, Tweets are collected using Twitter's Streaming API. The producer program collects tweets from the 9 specific cities and passes the raw twitter information to the nodes on the Kafka. Raw Twitter data is passed to the consumer program. The data is extracted, cleaned, classified using the Standford CoreNLP then sent back to the Kafka cluster. Flume saves the data written to the Kafka nodes to the Hadoop Distributed File System. Spark program groups the data and then the data is visualized using Tableau.

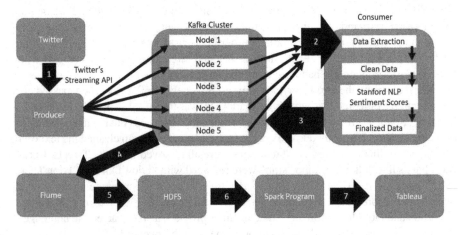

Fig. 1. Tweets collection using Twitter's streaming API.

4   Results

Data was collected from: San Francisco, San Diego, Washington D.C., New York, Chicago, Dallas, Detroit, Cleveland, and Philadelphia over a three day time frame. A total of 229,733 tweets were collected, and they contained a total of 309,743 sentences or statements. New York City produced significantly more tweets and statements than the other cities as expected since it is the largest city by population. Detroit and Cleveland produced the least number of tweets. Neutral statements made up the majority of data with 148,322 neutral statements. There were 115,112 negative statements and 46,309 positive statements. The data for each city can be viewed in Table 1.

Table 1. Number of tweets and statements per city

City	Total tweets	Total statements	Positive	Negative	Neutral
New York	81496	110534	15851	41520	53163
Chicago	31538	43023	6680	15629	20714
Washington DC	26283	35642	5522	13372	16748
Dallas	25787	33677	4933	12357	16387
San Francisco	18064	25006	3813	9220	11973
Philadelphia	17812	23392	3489	8693	11210
San Diego	12707	17733	2849	6461	8423
Detroit	8208	10721	1659	4128	4934
Cleveland	7841	10015	1513	3732	4770
Overall	229733	309743	46309	115112	148322

The percentage shown in Table 2 of positive, negative, and neutral tweets were close in every city. The percentage in each sentiment category was within 2% of the other cities. Overall, 48% of the statements from tweets were neutral, 37% were negative, and 15% were positive. Although the cities percentages are close, San Diego, the second happiest city used in this city behind San Francisco, had the highest percentage of happy tweets by 0.5%. Detroit, the saddest city used in this study, had the highest percentage of negative tweets by 0.94%.

Table 2. Percent of positive, negative and neutral statements per city

City	Percent positive	Percent negative	Percent neutral
San Diego	16.07	36.43	47.50
Chicago	15.53	36.33	48.15
Washington DC	15.49	37.52	46.99
Detroit	15.47	38.50	46.02
San Francisco	15.25	36.87	47.88
Cleveland	15.11	37.26	47.63
Philadelphia	14.92	37.16	47.92
Dallas	14.65	36.69	48.66
New York	14.34	37.56	48.10
Overall	14.95	37.16	14.89

5 Discussion

The results of this study were unexpected. There was only a small variance (2%), between all the cities in positive, negative, and neutral statements. Nguyen et al. [12] were able to successfully find trends in happiness at a neighborhood level and found that most of the tweets that were collected from February to August of 2015 were of positive or neutral sentiment.

In contrast, the majority of the statements in this study were neutral making up 47.88% of all the statements. Negative statements were over twice as prevalent as positive statements. Only 14.95% of the statements were positive while 37.16% of the statements were negative. In the following paragraphs, different theories as to why the results were similar are discussed.

One reason the results may have been similar is due to the way the Stanford CoreNLP classifies the statements. Only saying if a statement is positive, negative, or neutral without giving the statement a number grade may have been the cause of small variance in the results. One statement may have a greater positive meaning than another positive statement, but the Stanford CoreNLP would rank both statements as positive. To be more precise and accurate, the statements should be given a number value. For example, Nguyen et al. used a bag of words algorithm with the Language Assessment by Mechanical Turk to give words a score from 1 (sad) to 9 (happy) [12]. Without a number grading system, different levels of happiness and sadness cannot be measured.

The way in which people tweet and use Twitter could be another reason as to why the results were too close. While using Twitter, users may tweet messages and communicate in similar styles. Unlike in real life communication, a person's location may not affect how he or she talks on Twitter.

Also, the small sample size could have been a factor. Other studies that have attempted to analyze the sentiment of a city or country using Twitter collect data over months at a time [10, 12]. Using a longer time period would allow for more positive or negative events to happen in each city and could cause the percentages to change and spread out more.

6   Conclusion

Twitter is a popular research topic and has seen much interest in recent years. Although disaster management, event detection, and sentiment analysis are common research fields in the area of Twitter, companies have been doing sentiment analysis online for years in order to gauge the public's response to new products or advertising. This study was intended to use Twitter and sentiment analysis to evaluate the levels of happiness in nine different cities across the United States of America.

Using Twitter's streaming API and East Carolina University's cluster, tweets were collected and cleaned in real time. Stanford CoreNLP was then used to grade the statements of the tweets as either positive, negative, or neutral. Neutral statements made up the largest percent of statements collected at 47.89%. Negative statements made up 37.16%, and positive statements only accounted for 14.95% of all statements. Across the nine cities, the percentage of positive, negative, and neutral statements varied by less than 2% in each category. The small sample size, the way the Stanford CoreNLP classifiers tweets or the way people across America interact on Twitter are all possible reasons as to why the data shows small variance.

Much work can be done in the future to further the research done in this study. When running the data collection again, data can be collected over a longer time period (1 to 2 months), also, more cities should be added for more comparison. If the results do not change after collecting data over a longer time frame, other countries can be looked at next, knowing that England's sentiment on Twitter should differ from the United States of America's sentiment. If the results are still similar to the results found in this study, a different method of classifying a tweet's sentiment should be considered. The result of the research can also be improved by analyzing happiness related to certain aspect of life. This can be done by collecting data according to right keywords and suiting the interest of the person.

Acknowledgement. This research is supported in part by grant #1560037 from the National Science Foundation.

References

1. Liu, B.: Sentiment analysis and opinion mining. Synth. Lect. Hum. Lang. Technol. **5**(3), 1–167 (2002)
2. Pang, B., Lee, L.: Opinion mining and sentiment analysis. Found. Trends Inf. Retr. **2**(1–2), 1–135 (2012)
3. Tan, H., Tan, S., Cheng, X.: A survey on sentiment detection of reviews. Expert Syst. Appl. **36**(7), 10760–10773 (2009)
4. Tsytsarau, M., Palpanas, T.: Survey on mining subjective data on the web. Data Min. Knowl. Discov. **24**(3), 478–514 (2012)
5. Yadranjiaghdam, B., Yasrobi, S., Tabrizi, N.: Developing a real-time data analytics framework for Twitter streaming data. In: Proceedings of BigData Congress, pp. 329–336 (2017)
6. Laylavi, F., Rajabifard, A., Kalantari, M.: Event relatedness assessment of Twitter messages for emergency response. Inf. Process. Manag. **53**, 266–280 (2017)
7. Twitter Usage Statistics - Internet Live Stats, Internetlivestats.com (2017). http://www.internetlivestats.com/twitterstatistics. Accessed 2017
8. Naaman, M., Boase, J., Lai, C.: Is it really about me? Message content in social awareness streams. In: Proceedings of the 2010 ACM Conference on Computer Supported Cooperative Work, pp. 189–192 (2010)
9. Johnson, D.: These Are the Happiest and Healthiest Cities in America, Time.com (2017). http://time.com/4691862/bestcities-us-happiest-healthiest/. Accessed 2017
10. Durahim, A.Q., Coskun, M.: #iamhappybecause: gross national happiness through Twitter analysis and big data. Technol. Forecast. Soc. Change **99**, 92–105 (2015)
11. Bernardo, R.: 2017s Happiest Places to Live. WalletHub (2017). https://wallethub.com/edu/happiest-places-tolive/32619/. Accessed 2017
12. Nguyen, Q.C., Kath, S., Meng, H., Li, D., Smith, VanDerslice, J.A., Wen, M., Li, F.: Leveraging geotagged Twitter data to examine neighborhood happiness, diet, and physical activity. Appl. Geogr. **73**, 77–88 (2016)
13. Wallethub. https://wallethub.com/
14. Stanford CoreNLP. https://stanfordnlp.github.io/CoreNLP/
15. Sakaki, T., Okazaki, M., Matsuo, Y.: Earthquake shakes Twitter users: real-time event detection by social sensors. In: Proceedings of the 19th International Conference on World Wide Web, pp. 851–860 (2010)
16. Carley, K.M., Malik, M., Landwehr, P.M., Pfeffer, J., Kowalchuck, M.: Crowd sourcing disaster management: the complex nature of Twitter usage in Padang Indonesia. Saf. Sci. **90**, 48–61 (2016)

17. Daniel, M., Neves, R.F., Horta, N.: Company event popularity for financial markets using Twitter and sentiment analysis. Expert Syst. Appl. **71**, 111–124 (2017)

18. Baek, H., Oh, S., Yang, H., Ahn, J.: Electronic word-of-mouth, box office revenue and social media. Electr. Commer. Res. Appl. **22**, 13–23 (2017)

19. Yu, Y., Wang, X.: World cup 2014 in the Twitter world: a big data analysis of sentiments in U.S. sports fans' tweets. Comput. Hum. Behav. **48**, 392–400 (2015)

20. Chen, H., Chiang, R.H., Storey, V.C.: Business intelligence and analytics: from big data to big impact. MIS Q. **36**(4), 1165–1188 (2012)

21. Yasrobi, S., Alston, J., Yadranjiaghdam, B., Tabrizi, N.: Performance analysis of sparks machine learning library. Trans. Mach. Learn. Data Min. **10**(2), 67–77 (2017)

22. Kafka, A.: A distributed streaming platform. Kafka. https://kafka.apache.org/. Accessed 2017

23. Yadranjiaghdam, B., Pool, N., Tabrizi, N.: A survey on real-time big data analytics: applications and tools. In: 2016 International Conference on Computational Science and Computational Intelligence, pp. 404–409 (2016)

24. Sentiment Analysis, Deeply Moving: Deep Learning for Sentiment Analysis, Stanford (2017). https://nlp.stanford.edu/sentiment/. Accessed 2017

25. Socher, R., Perelygin, A., Wu, J.Y., Chuang, J., Manning, C.D., Ng, A.Y., Potts, C.: Recursive deep models for semantic compositionality over a sentiment treebank. In: Conference on Empirical Methods in Natural Language Processing, pp. 1631–1642 (2013)

26. Welcome to Apache Flume. Apache Flume (2017). https://flume.apache.org/

An Innovative Lambda-Architecture-Based Data Warehouse Maintenance Framework for Effective and Efficient Near-Real-Time OLAP over Big Data

Alfredo Cuzzocrea[1(✉)], Rim Moussa[2], and Gianni Vercelli[3]

[1] ICAR-CNR, University of Trieste, Trieste, Italy
alfredo.cuzzocrea@dia.units.it
[2] LaTICE Laboratory, University of Tunis, Tunis, Tunisia
rim.moussa@enicarthage.rnu.tn
[3] University of Genoa, Genoa, Italy
gianni.vercelli@unige.it

Abstract. In order to speed-up query processing in the context of *Data Warehouse Systems*, *auxiliary summaries*, such as *materialized views* and *calculated attributes*, are built on top of the data warehouse relations. As changes are made to the data warehouse through *maintenance transactions*, summary data become stale, unless the refresh of summary data is characterized by an expensive cost. The challenge gets even worst when near *real-time environments* are considered, even with respect to emerging *Big Data features*. In this paper, inspired by the well-known *Lambda architecture*, we introduce *a novel approach for effectively and efficiently supporting data warehouse maintenance processes in the context of near real-time OLAP scenarios*, making use of so-called *big summary data*, and we assess it via an empirical study that stresses the complexity of such OLAP scenarios via using the popular *TPC-H benchmark*.

1 Introduction

Usually, *Data Warehouse Systems* are deployed as part of a decision support system separated from the system of records (a.k.a. production databases). *On-Line Analytical Processing* (OLAP) queries, which execute at the data warehouse level, are long-running and complex query meant to extract *actionable knowledge*, but they are typically resource-consuming (e.g., [1–4]). Hence, in order to speed-up query processing, *auxiliary summaries*, such as *materialized views* and *calculated attributes*, are built on top of the data warehouse relations (e.g., [5]). As changes are made to the data warehouse through *maintenance transactions*, summary data become stale, unless the refresh of summary data is characterized by an expensive cost. Due to performance concerns, fresh data propagate down to the data warehouse system episodically (yearly

F. Y. L. Chin et al. (Eds.): BIGDATA 2018, LNCS 10968, pp. 149–165, 2018.
https://doi.org/10.1007/978-3-319-94301-5_12

or monthly refreshes), periodically (night refreshes) or, at-the-best, after some lag. Performance concerns are mainly resulting from complex data integration workflow executions and ACID (Atomicity, Coherency, Isolation, Durability – [6]) properties enforcement during the transaction maintenance task. Traditional data warehouse systems, with episodic or periodical data refreshing, foster *Retrospective Analytics* (e.g., [7]). The latter provides a look at what has already happened, and allows us to analyze past activities of an organization. On the other hand, in order to provide insights and actionable decisions at the right time, it is important to analyze *real-time data* (e.g., [8]). *Real-Time analytics* use cases occur in multiple instances, e.g.: *(i)* intelligent road-traffic management, *(ii)* remote health-care monitoring, *(iii)* complex event processing systems, and *(iv)* inventory management. Henceforth, novel approaches and paradigms are required in order to deal with OLAP query processing in dynamic environments, as the case of (data) changes at the data source level. In this paper, inspired by the well-known *Lambda architecture* [9], we introduce *a novel approach for effectively and efficiently supporting data warehouse maintenance processes in the context of near real-time OLAP scenarios*, with emerging *Big Data features* (e.g., [10]), and we assess it via an empirical study that stresses the complexity of such OLAP scenarios via using the popular *TPC-H benchmark* [11]. The paper extends a previous introduction short paper [12].

Refresh functions of data warehouse systems represent a common solution to face-of the described problem. They are commonly referred as *batch-incremental update processing* or *maintenance transactions* (e.g., [13]). Indeed, in standard data warehouse systems, *refresh procedures* load data into the data warehouse in a bulk mode. Prior to data loading, data are transformed. Data transformations are modeled in terms of complex *data integration workflows* and are part of the whole *data integration process*. Therefore, *big data processing for data warehouse maintenance* is becoming more and more relevant, as combined with *summary data management*, as also confirmed by recent research initiatives (e.g., [14]). Our research work lies in this specific scientific area, and predicts a new instance of *big data warehouse data*, the so-called *big summary data*, i.e. summary data structures that aid big data warehouse maintenance processes in emerging big data (e.g., *Cloud-based* – [15]) environments.

On a larger extent, data warehouse refresh can be modeled in terms of an eight-step process, as follows:

(Step 1) *Coping fresh data to the staging area.* The *staging area* is an intermediate storage area used for data processing during the data integration process; it sits between the data source(s) and the data target(s) [7,16].

(Step 2) *Running transformations on fresh data.* Transformations include: cleaning, de-duplication, data format conversion, derivation of new calculated values from existing data, filtering, joining, splitting, and so forth.

(Step 3) *Preparing transformed fresh data.* Prepare the insertion of transformed fresh data by usually disabling reference constraints and entity constraints, thus making indexes able to accelerate data warehouse insertion performance.

(Step 4) *Inserting fresh data into the data warehouse.* In some cases, it is necessary to merge fresh and stale data, indicate the time of last data update or maintain multiple data versions in order to handle suitable *Change Data Capture* (CDC).

(Step 5) *Validating inserted data.* Validate inserted data and processing different alerts (e.g., constraint violations). Alerts may need human solutions.

(Step 6) *(Re-)Setting-up constraints.* Re-enable reference constraints, entity constraints and other kinds of constraint over inserted data.

(Step 7) *Refreshing indexes.* Refresh indexes over inserted data.

(Step 8) *Refreshing data summaries.* Refresh auxiliary structures, such as materialized views, over inserted data.

Our research proposal aims at improving the big data warehouse maintenance problem via interacting with the process above. In order to better describe the proposed solution, we first need to focus on state-of-the-art data warehouse benchmarks (e.g., [17]), and how they are exploited to assess the performance of data warehouse systems.

The most prominent benchmarks for evaluating decision support systems are the various benchmarks issued by the *Transaction Processing Council* (TPC). The TPC-H benchmark [11] and its successor *TPC-DS* [18] assess the performance of the system under test according to two different ways, namely: (*i*) *Power Test*, and *Throughput Test*. The *Power Test* measures the query execution power of the system when connected with a single user. It runs the analysis in a *serial manner*, i.e. queries and update functions run one at a time and the elapsed time is measured separate for query run and refresh run. The *Throughput Test* measures the ability of the system to process concurrent queries and update functions in a multi-user environment.

Looking at refreshing operations in greater details, TPC-H benchmark exposes two refresh functions, namely *RF1*, for loading new sales (TPC-H introduces a classical product-order-supplier multidimensional model), and *RF2* for purging obsolete sales. In order to publish a TPC-H compliant performance result, the system needs to support ACID properties, by also specifying one of the four *isolation levels* (namely serializable, repeatable reads, read committed, read uncommitted) as well as the snapshot isolation level [19]. Initially, TPC-DS benchmark follows the TPC-H benchmark. In order to satisfy big data analytics requirements, *TPC-DS 2.0* reverted to a simpler model, in which analytic queries and data maintenance procedures are strictly distinct. Big data solutions are inherently not ACID-compliant, while most systems are indeed BASE-compliant (BASE – Basically Available, Soft state, Eventual consistency) compliant [20].

By inspecting the nature of tests implemented by popular data warehouse benchmarks, we can observe that *the overlapping of ACID-compliant OLAP queries and BASE-compliant data maintenance functions (e.g., refresh) overall turns to be extremely costful.* Hence, ad-hoc solutions must be devised in order to circumvent the target problem.

Specifically, we propose (and experimentally assess) a new big data warehouse maintenance methodology that pursues the idea of *first* performing a shortcut from (Step 1) to (Step 8), in order to improve both query performance as well as the query accuracy with respect to fresh data, and *then* performing the remaining steps from (Step 2) to (Step 7). The proposed approach is based on so-called *delta computations*, inspired by the well-known Lambda architecture [9], which allow us to (1) serve queries from both stale (big) data summaries in the data warehouse system and fresh (big) data in the staging area. In addition to this, the proposed approach applies *factorization* of streams processing for (2) fast computation of so-called *delta views*, in order to capture dynamic properties of big data systems. Finally, the proposed approach also (3) applies *postponement of the data warehouse maintenance transaction* to an opportune time (still based on a cost-aware analysis).

2   Related Work

Several proposals have been presented to address the management of real-time data in Data Warehouses. Proposals address fast data processing in OLAP systems using different approaches. Hereafter, we overview related work. Authors in [21–23] propose inserting real time data into OLAP multidimensional cube structures, instead of into the Data Warehouse itself. They argue that insertions in the data cubes would occur faster, due to the fact that they are not executed over highly indexed tables that contain a large amount of historical data.

In [24], authors foster data fragmentation of the data warehouse over a shared-nothing architecture, to accelerate the data integration process. The *maintenance transaction* becomes distributed and more complex to manage with well admitted commit distributed protocols (2-PC and 3-PC).

Dehne et al. propose *CR-OLAP* [25–27], a Real-time OLAP system based on a distributed index structure for OLAP, refered to as a *distributed PDCR tree*. *R-Store* [28,29] -A Scalable Distributed System for Supporting Real-time Analytics, which periodically materializes real-time data into a data cube. *R-Store* uses HBase for data storage and MapReduce for query processing, and implements MVCC (Multi-version concurrent control) to support real-time OLAP. In *CR-OLAP*, [28], summary data maintenance is not investigated.

In [30–33], Ferreiran, Cuzzocra et al., demonstrate propose near-real time data warehouses and propose a Rewrite/Merge Approach for Real-Time Data Warehousing. The proposed architecture in [32,33] implements a real-time data warehouse without data duplication. It is composed of three main components: the Dynamic Data Warehouse (D-DW), the Static Data Warehouse (S-DW) and the Merger. The Integration between the D-DW and the S-DW is performed offline.

Real-time new systems were deployed at Google and LinkedIn. The latters have different workloads. In [34], authors propose *Mesa* a highly scalable analytic data warehousing system that stores critical measurement data related to Google's Internet advertising business. *Mesa* satisfies near real-time data ingestion and query-ability requirements. It supports continuous updates which should be available for querying consistently across different views within minutes. *Pinot* [35] is a real-time distributed OLAP datastore designed to scale horizontally and open sourced by LinkedIn. The database is column oriented and implements bitmaps and inverted indexes. It is suited for analytical use cases on immutable append-only data with exclusively selection, aggregation, filtering, group by, order by, distinct queries on fact data. *Pinot* does not suit TPC DSS benchmarks like workloads, which show complex join operations. *Druid* [36] is an open source data store designed for real-time exploratory analytics on large data sets. The system combines a column-oriented storage layout, a distributed, shared-nothing architecture, and an advanced indexing structure allowing fast data ingestion and analytics of events.

Materialized views (a.k.a. MVs, summary tables, aggregate tables), store pre-computed results, and are widely adopted to facilitate fast queries on large data sets. As update batches arrive at a high rate, it is infeasible to continuously update MVs and a common solution is to group and defer maintenance transactions. Meanwhile, the MVs become stale, which leads to inaccurate query results. Multiple papers investigated Materialized views' refresh mechanisms and optimizations [37–41].

The *Lambda Architecture* [42–45] targets Big Data processing at scale and involves both batch and stream systems. Indeed, batch and stream workloads run in parallel on the same incoming data. The Lambda Architecture is made up of three layers, namely, (*i*) the *Batch Layer* which ingests and stores large quantities of immutable data and calculates batch views, (*ii*) the *Speed Layer* which processes stream data into views and deploys the views on the *Serving Layer*, and (*iii*) the *Service Layer* which queries the batch and real-time views and merges them into serving up views. Challenging problems are handling large quantities of data at the batch layer, unknown and time-varying data streams at the speed layer, as well as fast merge of views at the service layer.

Our paper is inspired by Lambda architecture principles for architecting near-real time OLAP scenarios in big data systems.

3   Big Summary Data: Definition and Management

Data warehousing is based on (1) collecting, cleansing, and integrating data from a variety of operational systems; (2) calculating summary data to address performance leaks, and (3) performing data analysis for decision support.

In order to address performance leaks related to complex OLAP queries' processing over big data, data warehouses build data summaries. Next, we overview *derived attributes* and *materialized views* cycle life from design to refresh. Data Summaries' management has three costs, namely (*i*) building cost, (*ii*) storage cost, (*iii*) refresh cost and have an age which indicates how old are these data snapshots.

Examples in this section, are based on TPC-H benchmark [11]. Figure 1 illustrates the Database Schema of TPC-H Benchmark.

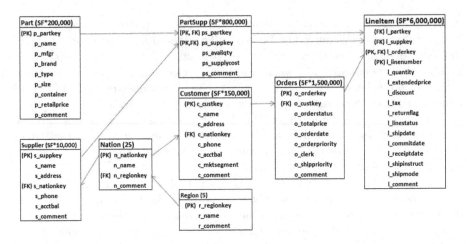

Fig. 1. Database schema of TPC-H benchmark.

3.1   Design of Data Summaries

Given, (*i*) a relational data warehouse schema, (*ii*) a workload composed of OLAP queries, (*iii*) refresh streams, we propose simple recommendations for proposing both of (*i*) derived attributes and (*ii*) materialized views which allow to achieve high performance OLAP over relational database management systems. Hereafter, we briefly recall definitions of *materialized views* and *derived attributes*, and motivate their usage for each type of business query of TPC-H workload.

Materialized Views. A *materialized view* summarizes large number of detail rows into information that has a coarser granularity. As the data is precomputed, an aggregate table allows faster cube processing. Research work propose cost models assessing materialized view's recommendations [46–49]. In this paper, we propose a materialized view for each query, then a grouping along full inclusion of dimensions. That's if the MV -MV_i of a query Q_i is included in the MV -MV_j recommended for query Q_j, then -MV_j is proposed for both Q_i and Q_j. The size of an MV is the number of rows in the MV. The latter is simply derives as the product of all attributes cardinalities, except attributes which are in functional dependency with an other attribute (as customer name is in functional dependency with customer key), and attributes to which refer other attributes (as customer key is referred in Orders table). We investigate Materialized Tables for two types of OLAP queries. First, for OLAP queries which MVs sizes are scale factor independent, and OLAP queries having very sparse cubes. Next, we overview by example, these two types of OLAP queries.

Business query Q12 of TPC-H benchmark is illustrated in Fig. 2(a). Q12, the *Shipping Modes and Order Priority Query* counts, by *ship mode*, for lineitems actually received by customers in a given year, the number of lineitems belonging to orders for which the *L_receiptdate* exceeds the *L_commitdate* for two different specified ship modes. Only lineitems that were actually shipped before the *L_commitdate* are considered. The aggregate table for Q12 is shown in Fig. 2(b) while the MV12 recommended for Q12 is illustrated in Fig. 2(c). Whether is the scale factor of TPC-H benchmark, MV12 has fixed number of rows equal to 49 computed as follows, ♯ *Line Receipt Years* = 7 × ♯ *line ship modes* = 7.

```
SELECT l_shipmode,
SUM(case when o_orderpriority in ('1-URGENT','2-HIGH')
     then 1 else 0 end) as high_line_count,
SUM(case when o_orderpriority not in ('1-URGENT','2-HIGH')
     then 1 else 0 end) as low_line_count
FROM orders, lineitem
WHERE o_orderkey = l_orderkey
AND l_shipmode in ('[SHIPMODE1]', '[SHIPMODE2]')
AND l_commitdate < l_receiptdate
AND l_shipdate < l_commitdate
AND l_receiptdate >= date '[DATE]'
AND l_receiptdate < date '[DATE]' + '1' year
GROUP BY l_shipmode
ORDER BY l_shipmode;
```
(a)

```
CREATE TABLE mv_q12 AS (
SELECT year(l_receiptdate) as years, l_shipmode,
sum(case when o_orderpriority in ('1-URGENT','2-HIGH')
     then 1 else 0 end) as high_line_count,
sum(case when o_orderpriority not in ('1-URGENT','2-HIGH')
     then 1 else 0 end) as low_line_count
FROM orders, lineitem
WHERE o_orderkey = l_orderkey
AND l_commitdate < l_receiptdate
AND l_shipdate < l_commitdate
GROUP BY year(l_receiptdate), l_shipmode) ;
```
(b)

```
SELECT l_shipmode,sum(high_line_count),sum(low_line_count)
FROM mv_q12
WHERE l_shipmode in ('[SHIPMODE1]', '[SHIPMODE2]')
AND year = '[YEAR]'
GROUP BY l_shipmode ;
```
(c)

```
SELECT c_name, c_custkey, o_orderkey, o_orderdate,
     o_totalprice, sum(l_quantity)
FROM customer, orders, lineitem
WHERE o_orderkey IN ( SELECT l_orderkey
     FROM lineitem
     GROUP BY l_orderkey
     HAVING sum(l_quantity) > [QUANTITY] )
AND c_custkey = o_custkey
AND o_orderkey = l_orderkey
GROUP BY c_name,c_custkey,o_orderkey,....,o_totalprice
ORDER BY o_totalprice desc, o_orderdate;
```
(d)

```
CREATE TABLE mv_q18 AS (
SELECT c_name, c_custkey, o_orderkey, o_orderdate,
     o_totalprice, sum(l_quantity) as sum_qty
FROM customer, orders, lineitem
WHERE c_custkey = o_custkey
AND o_orderkey = l_orderkey
GROUP BY c_name, c_custkey, o_orderkey, ..., o_totalprice
HAVING sum(l_quantity) > 300);
```
(e)

Fig. 2. SQL statement of TPC-H business query *Q*12 (a), aggregate table *Agg_Q*12 for *Q*12 (b), *MV_Q*12 for *Q*12 (c), *Q*18 (d), *MV_Q*18 for *Q*18 (e).

Business query Q18 of TPC-H benchmark is illustrated in Fig. 2(d). Q18 -the *Large Volume Customer Query* finds all customers who have ever placed large quantity orders. The query lists the customer name, customer key, the order key, date and total price and the quantity for the order. Q18 return few rows as 3.8 ppm (parts per million) of orders are big orders. Hence, it's recommended to calculate an MV for Q18 (see Fig. 2(e)).

Derived Attributes. *Derived Attributes* are calculated from other attributes. We recommend derived attributes for OLAP cubes which dimensionality is scale factor dependent, as Q10 (see Fig. 3).

Indeed, for this type of business queries, *derived attributes* are much less space consuming than *aggregate tables*. Q10 identifies customers who might be having problems with the parts that are shipped to them, and have returned

```
SELECT n_name, c_custkey, ..., c_comment,
       sum(l_extendedprice*(1-l_discount))
FROM customer, orders, lineitem, nation
WHERE o_orderdate >= ['DATE']
AND o_orderdate < ['DATE'] + interval '3' month
AND l_returnflag = 'R'
AND o_orderkey = l_orderkey
AND c_custkey = o_custkey
AND c_nationkey = n_nationkey
GROUP BY n_name, c_custkey, ..., c_comment
ORDER BY revenue desc;
```

(a)

```
SELECT n_name, c_custkey, ..., c_comment
       sum(o_sum_lost_revenue) lost_revenue
FROM customer, orders, nation
WHERE o_orderdate >= ['DATE']
AND o_orderdate < ['DATE'] + interval '3' month
AND c_custkey = o_custkey
AND c_nationkey = n_nationkey
GROUP BY n_name, c_custkey, ..., c_comment ;
ORDER BY revenue desc;
```

(b)

```
SELECT n_name, c_custkey, c_name, ..., c_comment
       sum(lost_revenue) as lost_revenue
FROM (
SELECT n_name, c_custkey, c_name, ..., c_comment,
       sum(o_sum_lost_revenue) as lost_revenue
FROM customer, orders, nation
WHERE c_custkey = o_custkey
AND o_orderdate >= date ['DATE']
AND o_orderdate < date ['DATE'] + interval '3' month
AND c_nationkey = n_nationkey
GROUP BY n_name, c_custkey, c_name, ..., c_comment
UNION ALL
SELECT n_name, c_custkey, c_name, ..., c_comment,
       sum(o_sum_lost_revenue) as lost_revenue
FROM customer, orders_temp, nation
WHERE c_custkey = o_custkey
AND o_orderdate >= date ['DATE']
AND o_orderdate < date ['DATE'] + interval '3' month
AND c_nationkey = n_nationkey
GROUP BY n_name, c_custkey, c_name, ..., c_comment)
GROUP BY n_name, c_custkey, c_name, ..., c_comment;
```

(c)

```
SELECT n_name, c_custkey, c_name, ..., c_comment,
       sum(lost_revenue) as lost_revenue
FROM (
SELECT n_name, c_custkey, c_name, ..., c_comment,
       sum(o_sum_lost_revenue) as lost_revenue
FROM customer, orders, nation
WHERE c_custkey = o_custkey
AND o_orderdate >= date ['DATE']
AND o_orderdate < date ['DATE'] + interval '3' month
AND c_nationkey = n_nationkey
GROUP BY n_name, c_custkey, c_name, ..., c_comment
UNION ALL
SELECT n_name, c_custkey, c_name, ..., c_comment,
       -sum(l_extendedprice*(1-l_discount)) as lost_revenue
FROM customer, v_join_rf2_streams, nation, time
WHERE c_custkey = o_custkey
AND l_orderkey = o_orderkey
AND o_orderdate >= date ['DATE']
AND o_orderdate < date ['DATE'] + interval '3' month
AND l_returnflag = 'R'
AND c_nationkey = n_nationkey
GROUP BY n_name, c_custkey, c_name, ..., c_comment)
GROUP BY n_name, c_custkey, c_name, ..., c_comment;
```

(d)

Fig. 3. SQL statement of TPC-H business query $Q10$ (a), with $o\_sum\_lost\_revenue$ immutable derived attributes (b), serving layer processing RF1 refresh function (c), serving layer processing RF2 refresh function (d).

them, for so, it calculates the *lost revenue* for each customer for a given quarter of a year. In order to improve the response time of Q10, we propose the following alternatives, (1) Either add 28 derived attributes $c\_sum\_lost\_rev$ /year/quarter to CUSTOMER relation, or (2) add one attribute $o\_sum\_lost\_rev$ to ORDERS relation. Notice that, the second alternative is better than the first with respect to both storage overhead and cost of refresh of stale derived attributes. Indeed, following inserts or deletes of orders (respectively TPC-H refresh functions RF1 and RF2), the 28 derived attributes are stale, while refreshes do not render stale the attribute $o\_sum\_lost\_rev$. Attribute $o\_sum\_lost\_rev$ will enable a gain in performance results from saving the cost of the join of LINEITEM and ORDERS tables.

Derived attributes alter the data warehouse schema, which is not costful for column-oriented storage systems, and allow a gain in performance through reducing both temporal and spatial complexities. The main point is to choose attributes which are not stale after data refresh or refresh cost is not costful.

TPC-H Benchmark Analysis. Following our directives for recommending on when to recommend Materialized Views and Derived Attributes, We concluded that TPC-H business queries fall into three categories (see Table 1).

Table 1. TPC-H workload taxonomy.

Dimensionality	Sparsity	TPC-H Business Queries	Recommendation
SF dependent	very sparse	Q15, Q18	Materialized Views
SF dependent	dense enough	Q2, Q9, Q10, Q11, Q20, Q21	Derived Attributes
SF independent	very sparse	—	Materialized Views
SF independent	dense enough	Q1, Q3, Q4, Q5, Q6, Q7, Q8, Q12, Q13, Q14, Q16, Q17, Q19, Q22	Materialized Views

Table 2 illustrates a total storage cost of all materialized views equal to 0.88 GB for all scale factors. The cost of derived attributes scales with TPC-H scale factor and is 0.5 GB for SF =10 (see Table 3).

Table 2. Materialized views data for TPC-H benchmark for any scale factor.

MV-Qi	♯Rows	Volume (MB)
mv-q1	129	0.008
mv-q3	2210908	52.712
mv-q4	135	2.241
mv-q5	175	2.563
mv-q6	3850	0.088
mv-q7	4375	0.067
mv-q8	131250	2.128
mv-q12	49 1813	0.002
mv-q13	47 3196	0.003
mv-q14	84 1344	0.001
mv-q15	28 728	0.001
mv-q16	187495	2.861
mv-q17	1000	0.011
mv-q18	624	0.023
mv-q19	39859141	836.278
mv-q22	500018	7.630
	total	*901.817*

Table 3. Tables' volumes respectively before and after adding new derived attributes for SF = 10.

Table	Volume (MB)	New Volume (MB)
Supplier	4.578	4.959
PartSupp	366.211	427.246
Orders	1001.358	1115.799
Lineitem	9381.974	9839.631

3.2   Refresh of Data Summaries

In this section, we first overview different data integration strategies, namely *Lazy integration* or *Eager integration*, which handle differently the processing of fresh data. Then, we detail data summaries management from inception to maintenance.

Data Integration [50]. Data Integration is the process of integrating data from multiple sources. Integration is either *Lazy* or *Eager* [51]. *Lazy Integration* keeps data at the sources and requires a mid-tier for query processing. The mid-tier determines the sources which answer the query and devises the execution tree. Once queries' answer sets obtained, the mid-tier performs required post-processing and returns a result set to the user-application. Summarizing, *Lazy integration* leaves data at sources, integrates data on-demand i.e. at query time, and queries' answer sets is accurate. The system is out-of-service when the sources are unavailable. The data sources are queried by the decision support system as well as the transactional system. This might degrade queries' performances.

Eager integration, is based on data warehousing. Thus, information of each source of interest is extracted in-advance and processed as appropriate, then merged with information from other sources and stored. The data warehouse is a database that is designed for query and analysis and is operational even when sources are unavailable. High query performance is achieved by building data summaries and local processing at sources is unaffected by the decision support system workload. Summarizing, with *Eager integration* a business query answer set might be stale. In order to overcome staleness, the data warehouse is refreshed episodically, periodically, or at best after some time. The maintenance transaction is also costful.

The combination of eager and lazy integration approaches is challenging and will enable OLAP over fresh data and load balancing between the transactional and the decision support system.

Refresh Operations. Given, (*i*) a relational data warehouse schema, (*ii*) a workload -a set of queries, (*iii*) refresh streams, (*iv*) calculated attributes and (*v*) materialized views; we have to determine when and how data summaries are refreshed. Two refresh strategies are proposed for the refresh of data summaries, and are detailed hereafter,

- *Eager refresh*: derived attributes and materialized views are refreshed with in the *maintenance transaction*. Hence, the data warehouse is coherent at the expense of costful maintenance.
- *Lazy refresh*: the refresh of calculated attributes and materialized views is delayed and is not part of the *maintenance transaction*. Thus, the data warehouse is incoherent for better performances.

Data Summaries refresh processing is performed is either *incremental*, requires *full reprocessing*, or both,

- *Incremental processing*: an *incremental refresh* executes first a sophisticated merge of the old snapshot and a new snapshot built over fresh data and if needed relations in the warehouse, and second integrates fresh data in the data warehouse.
- *Full reprocessing*: a *full reprocessing* integrates fresh data in the data warehouse, then recomputes data summaries.
- *Hybrid processing*: some parts require full reprocessing, while others can be incrementally refreshed.

In Eqs. (1)–(4), we analyze relational algebra operations involving different types of relations, namely relations which undergo inserts or deletes and immutable relations.

For R1 and R2 concerned by new inserts,

$$(R1 \cup \Delta R1) \bowtie (R2 \cup \Delta R2)$$
$$= (R1 \bowtie R2) \cup (R1 \bowtie \Delta R2) \cup (\Delta R1 \bowtie R2) \cup (\Delta R1 \bowtie \Delta R2)$$

IF fresh data does not refer to stale data as in TPC-H RF1

THEN $R1 \bowtie \Delta R2 =$ AND $\Delta R1 \bowtie R2 =$

$$= (R1 \bowtie R2) \cup (\Delta R1 \bowtie \Delta R2)$$

(1)

For R1 concerned by new inserts and R2 immutable,

$$(R1 \cup \Delta R1) \bowtie R2$$
$$= (R1 \bowtie R2) \cup (\Delta R1 \bowtie R2)$$

(2)

For R1 and R2 concerned by deletes,

$$(R1 - \Delta R1) \bowtie (R2 - \Delta R2)$$
$$= (R1 \bowtie R2) - (R1 \bowtie \Delta R2) - (\Delta R1 \bowtie R2) \cup (\Delta R1 \bowtie \Delta R2)$$

IF $(\Delta R1 = R1 \bowtie \Delta R2)$ THEN

$$\Delta R1 \bowtie R2 = R1 \bowtie \Delta R2 \bowtie R2 = R1 \bowtie \Delta R2$$
$$AND \ \Delta R1 \bowtie \Delta R2 = R1 \bowtie \Delta R2 \bowtie \Delta R2 = R1 \bowtie \Delta R2$$

$$= (R1 \bowtie R2) - (\Delta R1 \bowtie R2)$$

(3)

For R1 concerned by deletes and R2 immutable,

$$(R1 - \Delta R1) \bowtie R2$$
$$= (R1 \bowtie R2) - (\Delta R1 \bowtie R2)$$

(4)

The analysis of TPC-H benchmark workload reveals that all TPC-H benchmark business questions are concerned by refresh functions, except the following business queries Q2, Q11, Q13 and Q16 (i.e. 4 over 22 queries).

4   Near Real-Time OLAP Scenarios

In this section, we detail near real-time OLAP scenarios querying fresh data for an OLAP query improved using derived attribute, and a second for an OLAP query tuned using a materialized view.

4.1   Stream Workflow Management

Each stream has (1) a unique identifier *streamID*, typically a *timestamp*, (2) a *stream type* -for TPC-H benchmark, *stream type* is either *RF*1 or *RF*2, and (3) is composed of a sequence of operations (inserts, updates and deletes). All data entering the system is dispatched to both the *batch layer* and the *speed layer* for processing. The *batch layer* has two functions: (*i*) to manage the data warehouse, and (*ii*) to compute the *batch materialized views*. The *speed layer* has also two functions: (*i*) to manage the incremental updates i.e. streams, and (*ii*) to compute the *speed materialized views*. The *serving layer* merges *batch materialized views* and the *speed materialized views*, and indexes the resulting views i.e. *serving views*; so that they can be queried in low-latency and ad-hoc way.

In Table 4, we enumerate the different processing related to TPC-H business queries, and regroup them in order to depict which queries perform the same relational operations.

Table 4. TPC-H business queries' processing requirements implied by the batch updates.

Processing	Business queries: Q_i
σ LineItem	*l_shipdate* [?][?]: Q1,Q3,Q6,Q7,Q14,Q15,Q20
	l_discount [?][?]: Q6
	l_quantity < [?]: Q6,Q19
	l_commitdate < *l_receiptdate*: Q4,Q12
	l_returnflag =' R': Q10
	l_shipmode [= |in][?]: Q12, Q19
	l_shipdate < *l_commitdate*: Q12
	l_shipdate [?][?]: Q1
	l_receiptdate [?][?]: Q12
	sum(l_quantity) > [?]: Q18
	l_shipinstruct = [?]: Q19
	l_quantity [?]: Q19
	l_receiptdate < *l_commitdate*: Q21
σ Orders	*o_orderdate* [?][?]: Q3,Q4,Q5,Q8,Q10
	o_orderpriority [?][?]: Q12
	o_commentlike[?]: Q13
	o_orderstatus =' F': Q21
LineItem ⋈ R	R = *Orders*: Q3,Q4,Q5,Q7,Q8,Q9,Q10,Q12,Q21
	R = *Part*: Q8,Q9,Q14,Q17,Q19
	R = *PartSupp*: Q9,Q20
	R = *Supplier*: Q5,Q7,Q8,Q9,Q15,Q20,Q21
Orders ⋈ R	R = *Customer*: Q3,Q5,Q7,Q8,Q10,Q13,Q18,Q22

In Fig. 4(a), we show general processing as dictated by TPC-H queries. Then, in Fig. 4(b), we propose the join of *Orders' stream* and *LineItems' stream*, since the join operation is required by multiple queries.

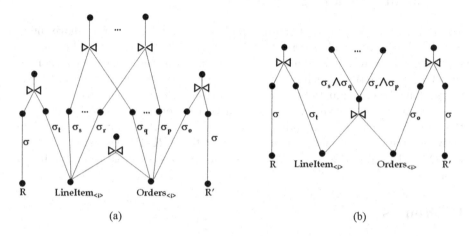

(a)   (b)

Fig. 4. Non-optimized stream processing (a) - Factorized stream processing (b).

4.2   Materialized View Management

The discounted revenue query Q19 of TPC-H benchmark [11], finds the gross discounted revenue for all orders for three different types of parts that were shipped by air or delivered in person. Parts are selected based on the combination of specific brands, a list of containers, and a range of sizes.

4.3   Derived Attribute Management

Business query Q10 of TPC-H benchmark, illustrated in Fig. 3, identifies customers who might be having problems with the parts that are shipped to them. The query calculates for each customer the lost revenue of returned parts, i.e. lineitems fulfilling ($l\_returnflag = 'R'$). In order, to accelerate the processing of Q10, we propose the derived attribute $o\_sum\_lost\_revenue$ to be calculated for each order. This derived attribute allows to save the join of the two tables *lineitem* and *orders*. The computation of $o\_sum\_lost\_revenue$ is described in Fig. 3(b). Next, we discuss Q10 processing in case there are refreshes, namely process new orders (RF1) and process deletes of orders (RF2).

– Query Q10 is affected by new inserts of orders (RF1). In order to enable real-time Q10, the derived attribute $o\_sum\_lost\_revenue$ is calculated for each new order in each new stream, as illustrated in Fig. 3(c). This processing performed at the speed layer requires different relational operations such as restrictions, join, projection and scalar functions. The service layer is responsible for the merge of Q10 resultsets at the batch layer and at the speed layer.

– Query Q10 is also affected by delete of orders (RF2). The corresponding processing is illustrated in Fig. 3(d).

5   Conclusions and Future Work

Inspired by the well-known Lambda architecture, in this paper we have introduced and experimentally assessed a novel approach for effectively and efficiently supporting data warehouse maintenance processes in the context of near real-time OLAP scenarios, which makes use of innovative big summary data. Experiments have been conducted against the popular TPC-H benchmark.

In future work, we will first conduct further experiments for bigger scales of the TPC-H benchmark. Secondly, we will generate *sketch synopsis* rather than derived attributes for several query classes of the TPC-H benchmark, namely: Q2, Q9, Q10, Q11, Q20 and Q21. Those, in fact, are suitable to sketch-based computations.

References

1. Cuzzocrea, A., Song, I., Davis, K.C.: Analytics over large-scale multidimensional data: the big data revolution! In: Proceedings of DOLAP 2011, pp. 101–104. ACM (2011)
2. Cuzzocrea, A.: Aggregation and multidimensional analysis of big data for large-scale scientific applications: models, issues, analytics, and beyond. In: Proceedings of the 27th International Conference on Scientific and Statistical Database Management, SSDBM 2015, La Jolla, 29 June–1 July 2015, pp. 23:1–23:6 (2015)
3. Cuzzocrea, A., Bellatreche, L., Song, I.: Data warehousing and OLAP over big data: current challenges and future research directions. In: Proceedings of the Sixteenth International Workshop on Data Warehousing and OLAP, DOLAP 2013, San Francisco, 28 October 2013, pp. 67–70 (2013)
4. Cuzzocrea, A.: Analytics over big data: Exploring the convergence of data warehousing, OLAP and data-intensive cloud infrastructures. In: 37th Annual IEEE Computer Software and Applications Conference, COMPSAC 2013, Kyoto, 22–26 July 2013, pp. 481–483 (2013)
5. Gupta, H., Mumick, I.S.: Selection of views to materialize in a data warehouse. IEEE Trans. Knowl. Data Eng. **17**(1), 24–43 (2005)
6. Härder, T., Reuter, A.: Principles of transaction-oriented database recovery. ACM Comput. Surv. **15**(4), 287–317 (1983)
7. Kimball, R., Ross, M.: The Data Warehouse Toolkit: The Definitive Guide to Dimensional Modeling. John Wiley, New York (2013)
8. Cuzzocrea, A.: CAMS: OLAPing multidimensional data streams efficiently. In: Pedersen, T.B., Mohania, M.K., Tjoa, A.M. (eds.) DaWaK 2009. LNCS, vol. 5691, pp. 48–62. Springer, Heidelberg (2009). https://doi.org/10.1007/978-3-642-03730-6_5
9. Marz, N.: Big Data: Principles and Best Practices of Scalable Realtime Data Systems. O'Reilly Media, [S.l.] (2013)
10. Cuzzocrea, A., Saccà, D., Ullman, J.D.: Big data: a research agenda. In: 17th International Database Engineering & Applications Symposium, IDEAS 2013, Barcelona, 09–11 October 2013, pp. 198–203 (2013)

11. Transaction Processing Council: TPC-H Benchmark (2013). http://www.tpc.org/tpch
12. Cuzzocrea, A., Moussa, R.: Towards lambda-based near real-time OLAP over big data. In: 42nd IEEE International Conference on Computers, Software and Applications, Tokyo, 23–27 July 2018
13. Gupta, A., Mumick, I.S.: Maintenance of materialized views: problems, techniques, and applications. IEEE Data Eng. Bull. **18**(2), 3–18 (1995)
14. Krishnan, K.: Data Warehousing in the Age of Big Data. Morgan Kaufmann, Waltham (2013)
15. Agrawal, D., Das, S., El Abbadi, A.: Big data and cloud computing: current state and future opportunities. In: Proceedings of the 14th International Conference on Extending Database Technology, EDBT 2011, Uppsala, Sweden, 21–24 March 2011, pp. 530–533 (2011)
16. Inmon, W.H.: Building the Data Warehouse. Wiley, New York (2005)
17. Transaction Processing Council: TPC-DS Benchmark (2013). http://www.tpc.org/tpcds
18. Nambiar, R.O., Poess, M.: The making of TPC-DS. In: Proceedings of the 32nd International Conference on Very Large Data Bases, Seoul, 12–15 September 2006, pp. 1049–1058 (2006)
19. Berenson, H., Bernstein, P., Gray, J., Melton, J., O'Neil, E., O'Neil, P.: A critique of ANSI SQL isolation levels. In: Proceedings of the ACM SIGMOD International Conference on Management of Data, SIGMOD, pp. 1–10 (1995)
20. Pritchett, D.: Base: an acid alternative. Queue **6**(3), 48–55 (2008)
21. Nguyen, T.M., Tjoa, A.M., Schiefer, J.: Towards the stream analysis model in grid-based zero-latency data stream warehouse. In: Professional Knowledge Management - Experiences and Visions, Contributions to the 3rd Conference Professional Knowledge Management - Experiences and Visions, WM, pp. 630–635 (2005)
22. Nguyen, T.M., Brezany, P., Tjoa, A.M., Weippl, E.R.: Toward a grid-based zero-latency data warehousing implementation for continuous data streams processing. IJDWM **1**(4), 22–55 (2005)
23. Doka, K., Tsoumakos, D., Koziris, N.: Efficient updates for a shared nothing analytics platform. In: Proceedings of the Workshop on Massive Data Analytics on the Cloud, MDAC, pp. 7:1–7:6 (2010)
24. Pereira, D., Azevedo, L.G., Tanaka, A.K., Baião, F.A.: Real time data loading and OLAP queries: living together in next generation BI environments. JIDM **3**(2), 110–119 (2012)
25. Dehne, F., Zaboli, H.: Parallel real-time OLAP on multi-core processors. In: Proceedings of the 12th IEEE/ACM International Symposium on Cluster, Cloud and Grid Computing, CCGRID 2012, pp. 588–594. IEEE Computer Society (2012)
26. Dehne, F.K.H.A., Kong, Q., Rau-Chaplin, A., Zaboli, H., Zhou, R.: A distributed tree data structure for real-time OLAP on cloud architectures. In: Proceedings of the IEEE International Conference on Big Data, pp. 499–505 (2013)
27. Dehne, F., Zaboli, H.: Parallel real-time OLAP on multi-core processors. IJDWM **11**(1), 23–44 (2015)
28. Li, F., Özsu, M.T., Chen, G., Ooi, B.C.: R-store: a scalable distributed system for supporting real-time analytics. In: IEEE 30th International Conference on Data Engineering, ICDE, pp. 40–51 (2014)
29. Li, F., Özsu, M.T., Chen, G., Ooi, B.C.: R-Store - Source Code (2015). https://github.com/lifeng5042/RStore

30. Ferreira, N., Martins, P., Furtado, P.: Near real-time with traditional data ware-house architectures: factors and how-to. In: 17th International Database Engineer-ing & Applications Symposium, IDEAS, pp. 68–75 (2013)
31. Ferreira, N., Furtado, P.: Real-time data warehouse: a solution and evaluation. IJBIDM **8**(3), 244–263 (2013)
32. Cuzzocrea, A., Ferreira, N., Furtado, P.: Enhancing traditional data warehousing architectures with real-time capabilities. In: Foundations of Intelligent Systems - 21st International Symposium, ISMIS Proceedings, pp. 456–465 (2014)
33. Cuzzocrea, A., Ferreira, N., Furtado, P.: Real-time data warehousing: a rewrite/merge approach. In: 16th International Conference on Data Warehousing and Knowledge Discovery, DaWaK, pp. 78–88 (2014)
34. Gupta, A., Yang, F., Govig, J., Kirsch, A., Chan, K., Lai, K., Wu, S., Dhoot, S.G., Kumar, A.R., Agiwal, A., Bhansali, S., Hong, M., Cameron, J., Siddiqi, M., Jones, D., Shute, J., Gubarev, A., Venkataraman, S., Agrawal, D.: Mesa: Geo-replicated, near real-time, scalable data warehousing. PVLDB **7**(12), 1259–1270 (2014)
35. LinkedIn: Pinot - A Realtime Distributed OLAP Datastore (2015). https://github.com/linkedin/pinot/
36. Yang, F., Tschetter, E., Léauté, X., Ray, N., Merlino, G., Ganguli, D.: Druid: a real-time analytical data store. In: Proceedings of the ACM SIGMOD International Conference on Management of Data, SIGMOD 2014, pp. 157–168. ACM (2014)
37. Salem, K., Beyer, K., Lindsay, B., Cochrane, R.: How to roll a join: asynchronous incremental view maintenance. SIGMOD Rec. **29**(2), 129–140 (2000)
38. Quass, D., Widom, J.: On-line warehouse view maintenance. In: Proceedings of the ACM SIGMOD International Conference on Management of Data, SIGMOD, pp. 393–404 (1997)
39. Agrawal, D., El Abbadi, A., Singh, A., Yurek, T.: Efficient view maintenance at data warehouses. SIGMOD Rec. **26**(2), 417–427 (1997)
40. Huyn, N.: Multiple-view self-maintenance in data warehousing environments. In: Proceedings of 23rd International Conference on Very Large Data Bases, VLDB 1997, pp. 26–35 (1997)
41. Krishnan, S., Wang, J., Franklin, M.J., Goldberg, K., Kraska, T.: Stale view clean-ing: getting fresh answers from stale materialized views. Proc. VLDB Endow. **8**(12), 1370–1381 (2015)
42. Marz, N., Warren, J.: Principles and Best Practices of Scalable Realtime Data Systems. Manning, New York (2015)
43. Marz, N., Warren, J.: Big Data: Principles and Best Practices of Scalable Realtime Data Systems, 1st edn. Manning Publications Co., Greenwich (2015)
44. Kiran, M., Murphy, P., Monga, I., Dugan, J., Baveja, S.S.: Lambda architecture for cost-effective batch and speed big data processing. In: IEEE International Con-ference on Big Data, pp. 2785–2792 (2015)
45. Piekos, J.: Simplifying the (Complex) Lambda Architecture (2014). https://voltdb.com/blog/simplifying-complex-lambda-architecture
46. Roussopoulos, N.: Materialized views and data warehouses. SIGMOD Rec. **27**(1), 21–26 (1998)
47. Agrawal, S., Chaudhuri, S., Narasayya, V.R.: Automated selection of material-ized views and indexes in SQL databases. In: Proceedings of 26th International Conference on Very Large Data Bases, pp. 496–505 (2000)
48. Aouiche, K., Jouve, P.E., Darmont, J.: Clustering-based materialized view selection in data warehouses. In: Proceedings of the 10th East European Conference on Advances in Databases and Information Systems, ADBIS, pp. 81–95 (2006)

49. Hose, K., Klan, D., Marx, M., Sattler, K.: When is it time to rethink the aggregate configuration of your OLAP server? PVLDB **1**(2), 1492–1495 (2008)
50. Cuzzocrea, A., Moussa, R.: Multidimensional database modeling: literature survey and research agenda in the big data era. In: IEEE ISNCC 2017, pp. 1–6 (2017)
51. Widom, J.: Integrating heterogeneous databases: lazy or eager? ACM Comput. Surv. **28**(4es), 91 (1996)

Application Track: BigData Algorithms

The Application of Machine Learning Algorithm Applied to 3Hs Risk Assessment

Guixia Kang[1,2](✉), Bo Yang[3](✉), Dongli Wei[1,2](✉), and Ling Li[1](✉)

[1] Key Laboratory of Universal Wireless Communications, Ministry of Education, Beijing University of Posts and Telecommunications, Beijing, China
gxkang@bupt.edu.cn, dongli_wei6674@163.com,
LingLiql@163.com
[2] Wuxi BUPT Sensory Technology and Industry Institute CO.LTD,
Wuxi, China
[3] Department of Cardiology, Chinese PLA General Hospital,
Beijing 100853, People's Republic of China
dryangb@aliyun.com

Abstract. Hypertension, Hyperglycemia and Hyperlipidemia (3Hs) are the significant factors of Cardiovascular Disease. Considering indicators related to obesity containing Body Mass Index (BMI), Waist Circumference (WC), Hip Circumference (HC), Waist-to-hip Ratio (WHR), Waist-to-height Ratio (WHtR) and disease history, disease history of family, dietary and etc. obtained conveniently and noninvasively, this article mainly set up two models to study the application of algorithm applied to 3Hs risk assessment. According to different combinations and gender, we build prediction model respectively to test the performance of them. In this article, 10-fold cross-validation was used to verify the model. In model I (HCRI - Logistic Model), the logistic regression algorithm was used to train the RC of Harvard cancer risk index. In model II (Logistic - Cart Model), taking the advantage of Decision Tree dealt with continuous variables, we set the output of CART as the input of logistic. The results show that, in HCRI - Logistic Model, the differences between male and female were not obvious, the accuracies are both only close to 70%, and the prediction of hyperglycemia is better than other 2Hs. In Logistic - Cart Model, the prediction of adult female is superior than men using indicators related to obesity. Especially about hyperglycemia, for model II, the accuracy is as high as 89.85% raised by 19.28% compared with model I, the specificity is 96.62% and the sensitivity is 84.56%. It provides an important reference for the evaluation of 3Hs to reduce the growth of relative chronic diseases.

Keywords: CART · 3Hs · Obesity · Machine learning

G. Kang and B. Yang—Contributed equally.

© Springer International Publishing AG, part of Springer Nature 2018
F. Y. L. Chin et al. (Eds.): BIGDATA 2018, LNCS 10968, pp. 169–181, 2018.
https://doi.org/10.1007/978-3-319-94301-5_13

1 Introduction

Chronic disease has become an important public health issue facing the world now, especially the cardiovascular disease (CVD) which ranks the first cause of death in the world. As the predisposing factors of CVD, Hypertension, Hyperglycemia and Hyperlipidemia (3Hs) has attracted much attention and research. According to statistics, nearly 3 million people die of varying degrees of CVD in China every year, while the number of people suffering from CAD is far more than 300 million. The death toll caused by 3Hs accounts for more than a quarter in China. Literature shows that high blood fatness in adults can lead to long-term risk of coronary heart disease [1].

The link among the 3Hs is well-connected. The increased blood fatness leads to increased blood pressure. The patients with hyperglycemia are always accompanied with the increased blood lipid. Obese hypertensive patients are generally suffering from diabetes.

Obesity relates to 3Hs closely especially hypertension [2]. In fact, the prevalence of obesity is increasing substantially in the world. It is estimated that 35.5% of adult female and 32.2% of adult male in the US are obese [3]. The Brazilian Institute of Geography and Statistics has shown that 50.1% of male and 48% of female in Brazil are overweight, while 12.4% of male and 16.9% of female are suffering from obesity [4].

The report shows that the hyperlipidemia and hyperglycemia is often overlooked, thanks to the inconvenient examination method, and more than half of patients realize their illness till the complications occur showing by kidney or eye. According to the National Health Research [5], over 30% of Indonesian is suffering from hypertension and 76% of them do not realize. To minimize the risk of CVD, an early prevention and treatment are needed. Therefore, it is necessary to adopt a low-cost method to assess the 3Hs risk, especially in underdeveloped and developing countries.

The body parameters such as body mass index (BMI), waist circumference (WC), hip circumference (WHR), waist to height ratio (WHtR) are the most practical and effective indicators for assessment of obesity, which is accessible and noninvasive [3]. Literature shows a positive correlation between WC and WHR and the amount of visceral fat, and combined with them can predict CVD effectively [6]. Young et al. [7] validate the predictive ability of WC, WHR and BMI for hypertension based on 722 Chinese adults. Akdag determined that the risk factors for hypertension are WHR, gender, triglycerides (TC) and family history of hypertension [8]. Fava C validates that the addition of additional variables is more predictive of the hypertensive model [9]. Sandi G uses C4.5 to predict the health risk for the treatment of hypertension [10]. Tayefi M et al. apply to CART of decision tree with multi types of data in prediction of hypertension [11]. Yanrui S. proposes adaptive prediction algorithm which established a blood glucose prediction model based on glucose data, exercise and dietary [12]. However, it is valuable only if the user has a large amount of monitoring data.

Since blood fat detection is not as convenient as blood pressure and glucose, there is no prediction model for blood fat across the world. However, the relationship between blood fat and CVD is even closer than that of high blood pressure and glucose, so the prediction of blood fat is of great significance.

With the powerful ability to extract complex patterns and accurate assessment, machine learning has been widely used in the medical arena such as prediction of obesity, classification of prostate cancer etc. The Microsoft Machine Learning Summit presented new machine-based health science applications in 2013. Classification and regression tree (CART) has a special significance for health research, owing to it can find out a best combination of variables.

This article mainly contrast the Harvard Cancer Risk Index (HCRI) and CART in the 3Hs prediction. We will analyze that which variables or the combinations of them can assess the risk of 3Hs better so that patients have a choice to take an early treatment, which can prevent the occurrence of 3Hs or the complications such as CVD effectively [13].

2 Data

2.1 Data Set

The data are collected from a number of chronic disease surveillance sites in 10 cities of China (Beijing, Shanghai, Wuhan, etc.) from August 2013 to June 2015.

Screening of obese body parameters, including Age, BMI, WC, HC, WHtR, WHR, etc., along with features for 3Hs family history, history of illness, smoking, drinking, dietary, exercise, pressure from data sets. There are 6071 samples in total, age from 19 to 86 years old.

The statistical analysis of features shows in Table 1.

Table 1. Statistics of the 3Hs patient

	Gender (Male 0/Female 1)		Age	BMI	HC	WC	WHR	WHtR
Hypertension	0	AVG	51.37	27.1427	99.197	93.215	0.9390	0.5123
		SD	11.908	3.72884	8.2322	10.1855	0.0548	0.0595
	1	AVG	58.87	25.9811	96.262	85.053	0.8842	0.5621
		SD	9.044	3.80338	9.6045	10.7807	0.0892	0.0623
Hyperlipidemia	0	AVG	38.94	22.0799	75.42	91.28	0.8256	0.4720
		SD	13.449	3.44866	9.861	7.363	0.0788	0.0648
	1	AVG	51.98	24.5448	82.23	94.69	0.8676	0.5191
		SD	12.107	3.70455	10.106	7.779	0.06958	0.0661
Hyperglycemia	0	AVG	51.55	27.0015	93.93	99.09	0.9471	0.5476
		SD	11.452	3.61055	10.773	8.381	0.06001	0.0605
	1	AVG	58.25	26.1691	87.06	95.84	0.9073	0.5551
		SD	9.438	3.64422	11.775	10.434	0.07039	0.0749

It is obvious that the difference between both genders of feature is large, thus we analyze separately. In order to take full use of data resources and make the data balanced, the final data set consist of 1400 males (including 643 hypertensive patients while 757 not) and 800 females (including 421 hypertensive patients while 379 not) for high blood pressure study; 2800 males (including 1421 hyperlipidemia patients while 1379 not) and 2000 females (including 945 hyperlipidemia patients while 1055 not) for high blood fatness study; 1400 males (including 682 hyperglycemia patients while 718 not) and 800 females (including 367 hyperglycemia patients while 433 not) for high blood glucose study.

Each data set is randomly divided into ten groups containing one validation set and nine training set. Then we evaluate the model by averaging the accuracy, specificity, etc.

2.2 Data Analysis

The single factor analysis results of features are as follow:

For hypertension, the three most relevant factors are history of hypertension, family history of hypertension and gender (F-test = 615.232, 51.245 and 50.156 respectively). In addition, age, WC and HC are strong correlation (F-test = 6.215, 5.667 and 2.667 respectively). Fully illustrates the relationship between obesity and high blood pressure is very close. This demonstrates that obesity has a high correlation with hypertension.

For hyperlipidemia, the three most relevant factors are gender, age and WC (F-test = 192.065, 9.578 and 9.409 respectively) in the absence of fat-related medical history data. This demonstrates that obesity has a high correlation with hyperlipidemia. Besides, the F-test value of systolic blood pressure and HC is 6.026 and 5.789 respectively which means there is a relationship between hyperlipidemia and hypertension.

For hyperglycemia, the three most relevant factors are the history of diabetes, gender and family history of diabetes, and gender is more prominent than family history of diabetes. Compared with hypertension and hyperlipidemia, the F-test values of WHR, WHtR, WC and BMI are 16.774, 11.412, 9.667 and 9.435 respectively indicating that the relationship between obesity and hyperglycemia is well-connected. In addition, DBP, SBP and TC also have a significant correlation with hyperglycemia, which shows that the relationship among 3Hs is extremely close.

In order to show the relationship between each feature and the 3Hs more intuitively and clearly, Receiver Operational Characteristic (ROC) curve of male and female were drawn respectively by SPSS, with the body measurement parameters as test variables, and diagnostic as state variable. The area under the curve (AUC) was used to explore the predictive value of each feature for 3Hs.

Compared to the WHtR, WC, WHR of male, the WHtR and WC of female show a better prediction according to the hypertension ROC curve. From hyperlipidemia ROC curve, the feature shows similar predictive value for both male and female. The top three most valuable variables are WC, WHR and WHtR. From the AUC of hyperglycemia ROC curve, we can see that the feature of female show a better prediction than male, and WHtR ranks first (Figs. 1, 2 and 3).

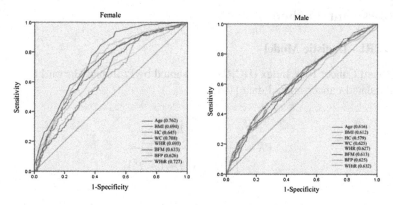

Fig. 1. ROC curve and AUC of the variables in hypertension

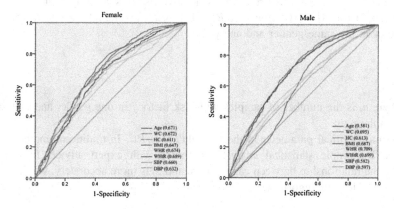

Fig. 2. ROC curve and AUC of the variables in hyperlipidemia

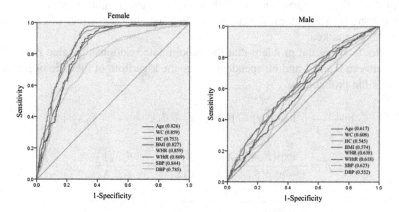

Fig. 3. ROC curve and AUC of the variables in hyperlipidemia

3   Assessment Methods

3.1   HCRI - Logistic Model

The Harvard Cancer Risk Index (HCRI) is presented by Professor Harvard for 10 years of accumulated cancer-related data [14]:

$$RR = \frac{\prod\limits_{i=1}^{N} RI_i}{\prod\limits_{j=1}^{N} (P_j \cdot RC_j + 1 - P_j)} \tag{1}$$

Where N is the number of risk factors, RR is the relative risk of the predicted sick individual compared with the general of same gender age group, and RC is the relative risk of a certain risk factor. RIi refers to the relative risk of sick individual, if one has a certain risk factor, $RI = RC$, otherwise $RI = 1.0$. And P_j is the ratio of those who has a risk factor in the same gender and age [15]:

$$P_j = \frac{n_j}{m} \tag{2}$$

Where n_j is the number of samples with risk factor j in one gender and age group and m is the total.

The most critical parameter in the formula (1) is RC. In traditional methods, it is assigned by simple statistical analysis combined with expert advice. In order to describe the relationship between each feature and 3Hs more accurately, we determine RC by using logistic algorithm.

$$x = \omega_0 + \omega_1 a_1 + \omega_2 a_2 + \cdots + \omega_k a_k \tag{3}$$

Where x is the category, a is the attribute and w is the weight. By training the data, the corresponding weights can be calculated, and the attribute values of each training sample can be marked.

In order to solve the problem that the subordinate relationship value range is not always between 0 and 1 and independence, take the logarithm of the ratio of p and $1-p$, where p is the probability:

$$\ln \frac{p}{1-p} = \log it(p) = \beta_0 + \beta_1 X_1 + \beta_2 X_2 + \cdots + \beta_k X_k + \varepsilon \tag{4}$$

$$\begin{aligned} \hat{p} &= \frac{\exp(\beta_0 + \beta_1 X_1 + \beta_2 X_2 + \cdots + \beta_k X_k)}{1 + \exp(\beta_0 + \beta_1 X_1 + \beta_2 X_2 + \cdots + \beta_k X_k)} \\ &= \frac{1}{1 + \exp[-(\beta_0 + \beta_1 X_1 + \beta_2 X_2 + \cdots + \beta_k X_k)]} \end{aligned} \tag{5}$$

Statistical variables ($P < 0.05$) were obtained, while insignificant attributes were screened out ($P > = 0.05$). Through logistic regression training, the regression coefficient can be get. And $OR = exp\ (\beta)$ is the relative risk (it is a risk factor when OR > 1 while protective factor when OR < 1).

This article uses the Logistic classifier from the tree package in scikit-learn to tune and train the model and for make 10-fold cross-validation by using Kfold package. RC can be get (Fig. 4):

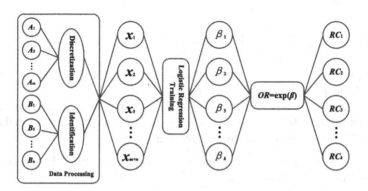

Fig. 4. The Calculation Process of *RC*

3.2 Logistic – Cart Model

Different from the traditional statistics, the CART algorithm is given in the form of a binary tree, which is easy to understand and has a more accurate prediction criterion. And the more complex the data, the more variables, the more significant the superiority of the algorithm. It consists of two parts:

Construction. First, determine the condition attributes and decision attributes in the target data set, and then determine the split criteria. According to the value of the splitting function F_i, the largest corresponding attribute A_i is selected as the splitting attribute.

$$F_i = \prod P_{ij} \sum \left(1 - 2 \times \min\{k = 1 \cup k = 2 | P_{ijk}\}\right) \tag{6}$$

Where P_{ij} is the quantity ratio of the value j whose property is i in data set, P_{ijk} is the quantity ratio of the value j whose property is i and which belongs to decision attribute k in data set.

After determining the best split attribute, we need to select the best split value:

$$\Phi(s/t) = 2 \times P_L P_R \sum |P(C_j| t_L) - P(C_j| t_R)| \tag{7}$$

Where P_L and P_R are the ratio of the number of samples in the left and right sub tree to the total respectively:

$$P_L \cup P_R = \frac{m_{L|R}}{n} \tag{8}$$

Where is the number of samples in the left or right sub tree and n is the total. $P(C_j| t_L)$ and $P(C_j|t_R)$ refer to the probability of belonging to the Cj in the left and right sub tree respectively:

$$P(C_j| t_L) \cup P(C_j| t_R) = \frac{n_j}{N} \tag{9}$$

Where n_j is the number of samples belonging to class j in the left or right sub tree, and N is the total number at the target node.

After selecting the best split attribute and the best split attribute value, the data set of the current node is divided into W_{ijL} and W_{ijR}, and continue to split until all the nodes belong to the same class or the attribute set is empty.

Pruning. Since the generated decision tree is likely over-fitting, it is necessary to prune the decision tree to optimize:

$$R_\alpha(T) = R(T) + \alpha|\tilde{T}| \tag{10}$$

Where $R_\alpha(T)$ represents the cost complexity of T, $R(T)$ is misclassification loss, $|\tilde{T}|$ is the number of nodes, and α is the complexity coefficient. During the increase of α from 0, always choose the sub tree with the smallest value of $R_\alpha(T)$, and make it a node, finally we get the best decision tree. Textual data are pre-processed and tagged. Obesity-related variables $A_1 A_2...A_m$ are input into the decision tree model. The output value is entered into the logistic regression model as a new variable B_1 along with the family history, personal history, and drinking habits $(B_1 B_2...B_n)$ with a threshold of 0.5 (Fig. 5).

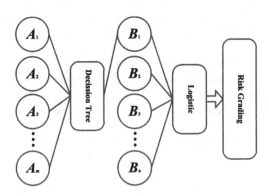

Fig. 5. Decision tree and logistics regression assessment model

3.3   Assessment Indicators

After training, the article evaluates by accuracy, specificity and sensitivity (Fig. 6 and Table 2).

$$Accuracy = (a+b)/(c+d) \tag{11}$$

$$Sensitivity = a/c \tag{12}$$

$$Specificity = b/d \tag{13}$$

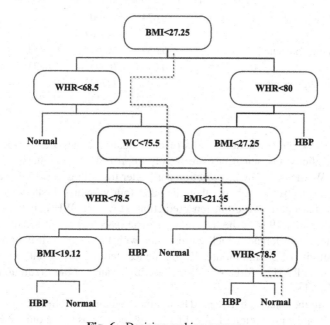

Fig. 6. Decision-making process

Table 2. Risk grading

P of logistic output	Risk level
<0.4	lower than the general
>= 0.4 & <0.6	general
>= 0.6 & <0.8	higher than the general
>= 0.8	higher than the general significantly

Where a and b are the number of patients and the normal which is determined correctly, while c and d are the total number of patients and the normal respectively.

3.4 Result

The result of the model 1 is in Table 3, the accuracy, specificity and sensitivity are taken as the mean of 10-fold cross-validation.

Table 3. Risk grading

	Male 0/Female 1	Accuracy	Specificity	Sensitivity
Hypertension	1	0.6741	0.7003	0.6465
	0	0.6322	0.6587	0.6115
Hyperlipidemia	1	0.6591	0.6685	0.6453
	0	0.6865	0.6942	0.6623
Hyperglycemia	1	0.7057	0.7233	0.6829
	0	0.6983	0.7115	0.6799

From Table 4, the prediction results of individual variables are not ideal, and the accuracy, specificity and sensitivity are greatly improved after synthesis.

The only WHtR is the ideal prediction parameter for both genders whose accuracy, specificity and sensitivity are better than the other four variables. It shows that the only WHtR has high value in hypertension prediction. However, WHtR has a decrease of 3.43%, 2.52% and 1.19% in the accuracy of male, specificity and sensitivity respectively compared with female which corresponds with ROC curve. It is noticeable that the max sensitivity of the individual WC on the test set is 60.61%, and the accuracy is next only to WHtR. Therefore, WC also has a high value in the prediction of hypertension for female.

Considering the relative between 3Hs and the prediction of blood pressure, we add DBP and SBP in the prediction of blood fat and blood glucose. The only WHR and the only WHtR have better accuracy, specificity and sensitivity than other 4 variables in both male and female models. Secondly, the accuracy, specificity and sensitivity of male model 8 and female model 8 have been greatly improved which indicates that the interaction of variables can provide more information for the hyperlipidemia of prediction.

We can see that the prediction of female is significantly better than male which is match to the results derived from ROC curve. In the prediction model of individual variables, WHtR is obviously superior to other variables which fully proves the high value of WHtR in the prediction of female hyperglycemia. The prediction variable next to WHtR is BMI, WC, WHR whose accuracy, specificity and sensitivity are close to the optimal model 8 of male, which means that the obesity index is closely related to

Table 4. Decision

	N	Variable	Female			Male		
			Accuracy	Specificity	Sensitivity	Accuracy	Specificity	Sensitivity
Hypertension	1	BMI	0.6074	0.7113	0.4545	0.5601	0.6134	0.4956
	2	WC	0.6442	0.6701	0.6061	0.5760	0.7153	0.4953
	3	HC	0.6012	0.6598	0.5152	0.5352	0.5547	0.5044
	4	WHR	0.5828	0.6289	0.5125	0.6120	0.6204	0.6018
	5	WHtR	0.6503	0.7113	0.5606	0.6280	0.6934	0.5487
	6	BMI+WC +HC+WHR +WHtR	0.7239	0.7526	0.6818	0.6861	0.7023	0.6283
Hyperlipidemia	1	BMI	0.5854	0.5402	0.6287	0.5902	0.5322	0.6704
	2	WC	0.6385	0.6878	0.5909	0.6181	0.6110	0.6248
	3	HC	0.5842	0.5029	0.6648	0.5525	0.4518	0.6432
	4	WHR	0.6504	0.6712	0.6311	0.6175	0.5727	0.6564
	5	WHtR	0.6308	0.7003	0.5726	0.6488	0.6985	0.6218
	6	DBP+SBP	0.5601	0.5678	0.5534	0.6107	0.5484	0.6679
	7	Age+BMI +WC+HC +WHR +WHtR	0.6769	0.6883	0.6652	0.6821	0.7123	0.6525
	8	Age+BMI +WC+HC +WHR +WHtR +DBP+SBP	0.7103	0.7321	0.6798	0.7065	0.7254	0.6478
Hyperglycemia	1	BMI	0.7801	0.7590	0.7986	0.6271	0.6149	0.6378
	2	WC	0.7913	0.8663	0.7539	0.5507	0.4693	0.6292
	3	HC	0.6788	0.6315	0.7183	0.5271	0.4786	0.5802
	4	WHR	0.7913	0.7919	0.7911	0.6193	0.6021	0.6344
	5	WHtR	0.8150	0.8275	0.8038	0.5964	0.5725	0.6205
	6	DBP+SBP	0.7788	0.7593	0.7967	0.7101	0.5566	0.8572
	7	Age+BMI +WC+HC +WHR +WHtR	0.8750	0.9650	0.7985	0.7793	0.8970	0.6703
	8	Age+BMI +WC+HC +WHR +WHtR +DBP+SBP	0.8875	0.9572	0.8283	0.7836	0.8909	0.6834

the hyperglycemia of female. BMI, WHR and blood pressure are relatively ideal prediction parameters for male, and the accuracy and the sensitivity of the prediction affected by the only blood pressure are the highest which are 71.01% and 85.72% respectively, which means that blood pressure and blood glucose have confidential relation. It can be seen from the table that the accuracy and the specificity of model 8, are higher 7.35% and 27.6% respectively than the maximum of 71.01% (model 7) and 61.49% (model 2) of the single variable model.

According to the predictions of the 3Hs models, WHtR is the most crucial factor, and the more the parameters, the more favorable the prediction results. After adding factors such as 3Hs family history, history of illness, smoking, drinking, dietary, exercise, pressure. The prediction results are shown in Table 5.

Table 5. Decision tree prediction results of 3Hs

	Male 0/Female 1	Accuracy	Specificity	Sensitivity
Hypertension	1	0.7326	0.7555	0.7022
	0	0.7132	0.7395	0.6842
Hyperlipidemia	1	0.7163	0.6852	0.7384
	0	0.7142	0.6882	0.7301
Hyperglycemia	1	0.8985	0.9662	0.8456
	0	0.7925	0.9012	0.7522

4   Discussion

Literature assumes that 500 million adults worldwide are troubled with obesity including fifteen million who are overweight [2]. Obesity is a public health problem, so that its and relative diseases treatment expenditure can be considerable. The prevalence of this chronic non-communicable diseases have been ever-increasing both in developed developing countries. The prevalence of female in the United States was 35.5%, compared with 32.2% for male. And the prevalence of female in the United States was 16.9%, compared with 12.4% for male [16].

Considering certain limitations, it is the most practical and low-cost way to assess obesity by anthropometric variables. The report of WHO in 2008 noted that BMI, WC and WHR are well-connected related to cardiovascular risk, hypertension, hyperlipidemia and other health problems.

The research of this article is that the algorithm of machine learning can predict 3Hs instantly with corresponding accuracy by measuring certain body parameters. Although we can now easily measure blood pressure and various body indicators in hospitals, there are poor supplies in some remote places, such as some parts of Africa and poor villages in China. Therefore, the study of this article will become very practical, because we only need ruler-measured parameters of the body to predict 3Hs. In the present study, we use the classification regression tree model to study the 3Hs prediction model for body measurement parameters. We can see that body measurement parameters such as obesity are more valuable to female than male in prediction. In all body measurements, the waist circumference was the most valuable predictor of 3Hs which is suitable for both male and female.

5   Conclusion and Future Work

This article mainly uses machine learning algorithm of CART to realize the prediction of 3Hs comparing with the HCRI - Logistic model. The study provides a simple classification rule to determine obesity risk factors related to 3Hs, which is potential to help to make management solution of 3Hs.

Acknowledgment. The research is supported by the National Natural Science Foundation of China (no.61471064 and no.81570272), and National Science and Technology Major Project of China (No. 2017ZX03001022-005).

References

1. Navar-Boggan, A., Peterson, E., D'agostino, R.: Hyperlipidemia in early adulthood increases long-term risk of coronary heart disease. Circulation **131**(5), 451–458 (2015). https://doi.org/10.1161/CIRCULATIONAHA.114.012477
2. Flegal, K., Carroll, M., Ogden, C., et al.: Prevalence and trends in obesity among US adults, 1999–2008. JAMA **303**(3), 235–241 (2010)
3. Suchanek, P., Kralova Lesna, I., Mengerova, O., Mrazkova, J., Lanska, V., Stavek, P.: Which index best correlates with body fat mass: BAI, BMI, waist or WHR? Neuro Endocrinol Lett. **2012**(33), 78–82 (2012)
4. Vazquez, G., Duval, S., Jacobs, D., Silventoinen, K.: Comparison of body mass index, waist circumference, and waist/hip ratio in predicting incident diabetes: a meta-analysis. Epidemiol. Rev. **2007**(29), 115–128 (2007)
5. Bergman, R., Stefanovski, D., Buchanan, T., Sumner, A., Reynolds, J., Sebring, N., et al.: A better index of body adiposity. Obesity (Silver Spring) **2011**(19), 1083–1089 (2011)
6. Perichart-Perera, O., Balas-Nakash, M., Schiffman-Selechnik, E., Barbato-Dosal, A., Vadillo-Ortega, F.: Obesity increases metabolic syndrome risk factors in school-aged children from an urban school in Mexico City. J. Am. Diet. Assoc. **107**(1), 81–91 (2007)
7. Yong, L., Guanghui, T., Weiwei, T., Liping, L., Xiaosong, Q.: Can bodymass index, waist circumference, waist-hip ratio and waist-height ratio predict the presence of multiple metabolic risk factors in Chinese subjects?. BMC Public Health **11** (2011). Article 35
8. Akdag, B., Fenkci, S., Degirmencioglu, S., Rota, S., Sermez, Y., Camdeviren, H.: Determination of risk factors for hypertension through the classification tree method. Adv. Ther. **23**(2006), 885–892 (2006)
9. Fava, C., Sjögren, M., Montagnana, M., et al.: Prediction of blood pressure changes over time and incidence of hypertension by a genetic risk score in Swedes. Hypertension **61**(2), 319–326 (2013)
10. Sandi, G., Supangkat, S., Slamet, C.: Health risk prediction for treatment of hypertension. In: 2016 International Conference on Cyber and IT Service Management, pp. 1–6. IEEE (2016)
11. Tayefi, M., Esmaeili, H., Karimian, M., et al.: The application of a decision tree to establish the parameters associated with hypertension. Comput. Methods Programs Biomed. **139**, 83 (2017)
12. Yanrui, S.: Research on Glucose Prediction Model and Hypoglycemia Alarm Technology. Zhengzhou University, Zhengzhou (2014)
13. Appel, L., Brands, M., Daniels, S., et al.: Dietary approaches to prevent and treat hypertension a scientific statement from the American Heart Association. Hypertension **47**(2), 296–308 (2006)
14. Kim, D., Rockhill, B., Colditz, G.: Validation of the Harvard Cancer Risk Index: a prediction tool for individual cancer risk. J. Clin. Epidemiol. **57**(4), 332–340 (2004)
15. Harvard Obesity Prevention Source, Harvard School of Public Health, Cambridge, Mass, USA. http://www.hsph.harvard.edu/obesity-prevention-source/.2017/10/11
16. POF I. POF 2008-2009: desnutrição cai e peso das crianças brasileiras ultrapassa padrão internacional [Internet]. 2010.[citado em 2012 ago 03] (2012)

Development of Big Data Multi-VM Platform for Rapid Prototyping of Distributed Deep Learning

Chien-Heng Wu[1]([✉]), Chiao-Ning Chuang[2], Wen-Yi Chang[1], and Whey-Fone Tsai[1]

[1] National Center for High-Performance Computing, Hsinchu 30076, Taiwan
`garywu@narlabs.org.tw`,
`{c00wyc00,wftsai}@nchc.narl.org.tw`
[2] National Chiao Tung University, Hsinchu 30010, Taiwan
`chiaoning.chuang@gmail.com`

Abstract. The present study utilizes VirtualBox virtual environment technology to develop the personal big data multi-VM platform with four-node Spark and Hadoop cluster that can effectively replicate and provide an environment for developers to easily design and implement the Spark and Hadoop Map/Reduce programming. Before running their Big Data and deep learning applications in physical multi-node Spark and Hadoop Cluster, developers can conduct Map/Reduce programing simply on the proposed multi-VM platform, which is exactly the same as the physical one. To demonstrate its capability and applicability, this study utilizes the deep learning application as an example for function illustration. In this study, the big data multi-VM platform provides the rapid prototyping of distributed deep learning by using a cutting-edge framework TensorFlowOnSpark (TFoS) for AI developers. To look into deep insight, this study performs the deep-learning benchmark in different types of cluster systems including the multi-node big data VM platform, physical standalone system and the physical small-cluster system. The results indicate that InputMode.SPARK can get 3.3 times faster than InputMode.TENSORFLOW on the big data VM platform and even achieve 6.1 times faster on the physical server.

Keywords: Big data multi-VM platform · Deep learning application
Spark · In-memory computing · Hadoop Map/Reduce

1 Introduction

With explosively growing data volume in many application fields, the demand for large-scale computation and storage capacities has been greatly boosted. Under such circumstances, the well-known Hadoop system [1, 2] has been developed and adopted by many enterprises. Generally, building a Spark and Hadoop Cluster at least requires two physical machines for NameNode and DataNode, which enables functionalities of data partition, data replication and data distribution. Hence, for Map/Reduce programming and testing purposes, a Hadoop system with the basis of master-slave architecture is highly recommended. In addition to Hadoop, Spark [3] is the newly

© Springer International Publishing AG, part of Springer Nature 2018
F. Y. L. Chin et al. (Eds.): BIGDATA 2018, LNCS 10968, pp. 182–193, 2018.
https://doi.org/10.1007/978-3-319-94301-5_14

emerging Big Data platform, which boasts the ability of in-memory computing and has been shown to be potentially 10–100 times faster than Hadoop. However, building a physical Hadoop or Spark system may be difficult for entry-level beginners without strong IT background. Furthermore, Hadoop system is built in Linux operation system, so it can also be an obstacle for application users who are only familiar with Windows operating system.

For the considerations described above, this study utilizes the virtual machine VirtualBox [4] to deploy multi-node Big Data VM platform, which can run Spark and Hadoop cluster in Multi-Node (1 name node + 3 data nodes) Virtual Machines (VM) under Windows operating system. Although running the (1+3) Multi-Node VMs would share the same resources (CPU, memory and hard disk) in one physical machine, the developers or users are still able to perform and simulate big data computing by using the (1+3) Multi-Node virtual environment of Spark and Hadoop cluster. In addition, we also provide the solution to eliminate the problems of unstable HBase system and the superfluous data.

By using the virtual technology, the core components of the big data platform including Hadoop, Spark, HDFS, HBase, Zookeeper and BigData Software Platform within Liferay [5] IDE (Eclipse [6]) are all integrated into this platform. Spark is built on the top of Hadoop system, and is managed by YARN for the resource management. Specifically, this big data multi-VM platform can be simply and easily deployed by extracting its VM image file into VirtualBox under the Windows operating system. Therefore, the big data multi-VM platform could be a very helpful software platform for big data training, deployment and testing of Spark development and Map/Reduce programming, especially for beginners.

In addition, the big data developers can also integrate the backend computing by implementing the specific portlets such as Job Submission, Job Status and other application portlets for their own purposes. For the big data applications of portal development and Map/Reduce programming, this big data multi-VM platform provides developers to reduce application complexity, development time, and improving application performance. It also provides the well-designed platform in the fields of big data computing and e-Commerce for Java developers to design, implement, configure and deploy. In order to satisfy the presentation tier, the business tier and the database tier at the same time, this platform adopts the enterprise architecture to make it robust, stable and advanced by utilizing the enterprise portal engine and Java EE application server.

2 Platform Architecture

As shown in Fig. 1, this platform utilizes the Oracle VirtualBox to construct multi-node Spark and Hadoop systems including the development tier, the middleware tier and the system tier. The software adopted in each tier is listed in Table 1. In this architecture, the development tier contains four modules, which are Liferay Portal development, Hadoop Map/Reduce Samples, big data App and Hadoop Library API Generator. The middleware tier contains the multi-node Hadoop Cluster for developers to run their Map/Reduce source code during the development period. The developers can also implement Scala or Java to access Spark cluster within this tier. Users do not have to

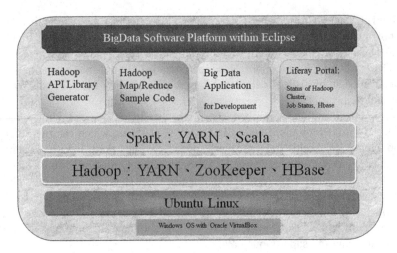

Fig. 1. Architecture of Big Data (1+3) multi-VM platform

Table 1. Software architecture of Big Data (1+3) Multi-Node VM Platform

BigData software platform	• Liferay IDE (Eclipse) 2.2.4 GA5 • Liferay Portal 6.2.5 GA6 • Hadoop Library 2.6.0 • Sample Code of Hadoop 2.x Map/Reduce
Spark and Scala	• spark-1.6.0-bin-hadoop2.6.tgz • scala-2.10.0.gz
Hadoop 2.6.0 (CDH 5.5.1)	• hadoop-2.6.0-cdh5.5.1.tar.gz • zookeeper-3.4.5-cdh5.5.1.tar.gz • hbase-1.0.0-cdh5.5.1.tar.gz
Java runtime machine	JDK 8.0 (jdk-8u66-linux-x64.tar.gz)
Operating system	Ubuntu 16.04.3 LTS (Desktop Edition; 64 bit)

worry about the complexity of building the Spark and Hadoop Cluster. Instead, they can just concentrate on the big data development for their specific applications [7]. Furthermore, by many times and long time running tests, the provided big-data multi-VM platform has been approved to be a stable and robust version.

To deploy the big data multi-VM platform, firstly, users can install VirtualBox software on a Windows-based PC. Then, the multi-node VM image files which contain NameNode, DataNode1, DataNode2 and DataNode3 can be loaded in a one-by-one way and the VirtualBox will automatically build the corresponding Virtual Machine, the multi-node big data VM platform, as shown in Fig. 2. Meanwhile, all the relevant software and tools are well established, in which the host name and fixed IP addresses are shown in Table 2.

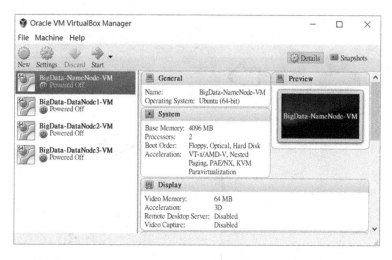

Fig. 2. Big Data multi-VM platform with Spark and Hadoop cluster

Table 2. Hosts and IP of big data multi-VM platform

VirtualBox (VM)	Host	IP
BigData-NameNode-VM	master	192.168.56.100
BigData-DataNode1-VM	slave1	192.168.56.101
BigData-DataNode2-VM	slave2	192.168.56.102
BigData-DataNode3-VM	slave3	192.168.56.103

Basically, these multi-node Virtual Machines utilize the highly stable Ubuntu Linux operation system, on which Hadoop, Spark, Hadoop database/HBase, Scala development and MapReduce programming tools, Java development tools/Eclipse, ZooKeeper/ distributed application coordination tool, have been pre-installed and pre-configured, so users can immediately start using this platform for their own applications. In general, the overall time to import Big Data Multi-VM Platform can be completed within 45 min; hence it really reduces the difficulty for users to construct the Spark and Hadoop system.

The developers can perform programming, submitting jobs, analyzing data in this big data VM platform just like in a physical Spark and Hadoop cluster. Besides, this platform also provides important sample codes and readme instruction with respect to Hadoop Map/Reduce programming and Liferay portal development for users to learn and modify the source codes. This platform is not only for development but also for education and training purposes. Hence, this platform is a good initiation step for the beginners with science or engineering backgrounds to move a step into the big data world. With this Big Data multi-VM Platform, both the powerful four-node Spark and Hadoop systems can be at hand.

As shown in Fig. 3, the usage procedure of Big Data (1+3) Multi-Node VM Platform is shown below.

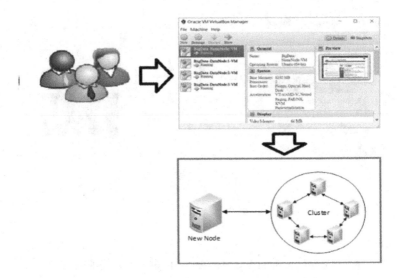

Fig. 3. Usage procedure of Big Data multi-VM platform

1. Startup Virtual Machines (1 NameNode and 3 DataNodes)
2. Startup Spark and Hadoop cluster
3. Start the BigData Software Platform by Eclipse IDE
4. Design, implement and test the deep learning or big data applications
5. Send the executable applications into physical Spark and Hadoop cluster

One thing needs to address is that the common method to build the Big Data Multi-VM Platform is to create a single VM of NameNode with Spark and Hadoop cluster, and then to clone this single VM into DataNode1, DataNode2 and DataNode3, respectively. This procedure will work, but problems may occur because of the existed dirty data within the DataNode1, DataNode2 and DataNode3. For example, the Hbase service may be unstable without finding the actual root cause. In order to avoid the mentioned problems, we designed an alternative method to build the stable multi-node big data VM platform. The procedure is described as following steps:

(1) To create a pure VM with Ubuntu Linux
(2) To clone this VM into other 3 VMs
(3) To configure the fixed IP and create SSH in NameNode, and then send the SSH key to DataNode1, DataNode2 and DataNode3
(4) Manually and carefully to build up Spark and Hadoop cluster with ZooKeeper and Hbase

Finally, a stable Big Data Multi-VM Platform can be obtained. As shown in Table 3, this new multi-VM platform is also light-weighted than the old one.

Table 3. Stable version of (1+3) VM Hadoop

VM	Old file size	New file size
BigData-NameNode-VM	10.9 GB	8.7 GB
BigData-DataNode1-VM	10.4 GB	3.9 GB
BigData-DataNode2-VM	10.3 GB	3.8 GB
BigData-DataNode3-VM	10.4 GB	3.8 GB

3 AI Application and Benchmark

3.1 AI Application

Deep learning is the area of artificial intelligence where the real magic is happening right now. Traditional computers, while being very fast, have not been very smart – they have no ability to learn from their mistakes and have to be given precise instructions in order to carry out any task. Deep learning involves building artificial neural networks which attempt to mimic the way human brains sort and process information. The "deep" in deep learning signifies the use of many layers of neural networks all stacked on top of each other. This data processing configuration is known as a deep neural network, and its complexity means it is able to process data to a more thorough and refined degree than other AI technologies which have come before it. Recently, deep learning applications have shown impressive results across a wide variety of domains [8]. However, training neural networks is relatively time-consuming, even on a single GPU-equipped machine. Fortunately, setting up distributed environments can be an approach to greatly help us accelerate the training process and handle larger size of datasets. The big data multi-VM platform provides the rapid prototyping of distributed deep learning by using a cutting-edge framework TensorFlowOnSpark (TFoS) [9] for AI developers.

TensorFlowOnSpark integrates Apache Hadoop and Apache Spark with the deep learning framework, TensorFlow [10]. Most of the existing distributed deep learning frameworks need to set up an additional cluster for deep learning separately, while TensorFlowOnSpark does not. As shown in Fig. 4, the separated clusters not only require copying data back and forth between the clusters but also encounter the issues of unwanted system complexity and end-to-end learning latency. In contrast, as shown in Fig. 5, TensorFlowOnSpark allows distributed deep learning execution on the identical Spark clusters, which achieves faster learning.

Data ingestion is the first step to utilize the power of Hadoop. Choosing a proper method of data ingestion can accelerate training processes significantly. In this study, we benchmark two types of data ingestion which TensorFlowOnSpark offers: InputMode. SPARK and InputMode.TENSORFLOW with MNIST handwritten digit dataset [11] on the big data multi-VM platform and the physical server. InputMode.SPARK is provided as a way to send Spark RDDs into the TensorFlow nodes, it enables easier integrations between Spark and TensorFlow once you have had an existing data pipeline generating data via Spark; on the other hand, InputMode.TENSORFLOW takes advantage of TensorFlow's QueueRunners mechanism to read data directly from HDFS files, and Spark is not involved in accessing data. Besides, for each type of data ingestion, we

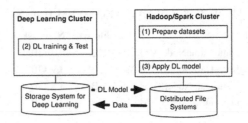

Fig. 4. ML pipeline with multiple programs on separated clusters

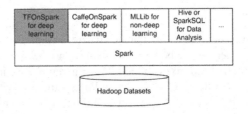

Fig. 5. TFoS for deep learning on spark clusters

benchmark the data ingestion of both TFRecords and CSV input formats, which are standard TensorFlow format and comma-separated value format respectively.

The results are illustrated in Fig. 6, showing InputMode.SPARK can get 3.3 times faster than InputMode.TENSORFLOW on the big data VM platform and even achieve 6.1 times faster on the physical server. Furthermore, whatever the data ingestion we tested in our experiments, TFRecords always get better performance than CSV does.

The following benchmark we hold the combination of InputMode.SPARK and TFRecords to delve deeper into the effect of tuning the number of executors and the number of data nodes on the big data VM platform. The architecture of TensorFlowOnSpark uses only one core/task per executor for easier debugging and Spark/Yarn configuration settings. Nonetheless, it is not a problem to set multiple executors per node as long as the physical server is providing enough memory and CPU

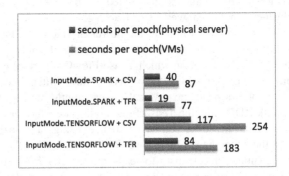

Fig. 6. Comparative experiment results of distributed deep learning

resources. Accordingly, we benchmark 5 cases within acceptable resource limits as shown in Table 4. Note that at least two executors must be set owing to one of them must be a parameter server (ps) to maintain and update the model weights while the other performs the main computations such as reading data and computing the gradient. In general, if the number of executors is greater than two, we tend to configure only one ps task and the others are all set to do parallel computations.

Table 4. The experimental setup of VMs

Case	Platform	Nodes	Number of executors	Cores per data node	Images per second
1	(1+1) VM Hadoop	1 name node + 1 data node	2	2	923
2	(1+1) VM Hadoop	1 name node + 1 data node	3	3	1224
3	(1+2) VM Hadoop	1 name node + 2 data nodes	2	1	845
4	(1+2) VM Hadoop	1 name node + 2 data nodes	4	2	1091
5	(1+3) VM Hadoop	1 name node + 3 data nodes	3	1	909

Under the same number of data nodes, comparing case 1 to case 2 or case 3 to case 4, setting a greater number of executors gets better performance. As a whole, we can allow each worker to process more records by adding more executors to a Spark job. This will accelerate the training process. In addition, comparing case 2 to case 5 or case 1 to case 3, when we set the same number of executors, adding more data nodes negatively affect performance. Overall, performance degradation on VM clusters may be due to the overhead of the virtualization. For example, if we are running 4 VMs on a single physical machine, it means that 4 OSs, 4 namenode/datanode services and 4 of any other VM services are running simultaneously. Obviously, the overhead system loading of 4 VMs is much heavier than that of 2 VMs thus to reduce its computing capacity.

3.2 Benchmark

Finally, this study performs the benchmark in different types of deep learning systems including the multi-node big data VM platform, physical standalone system and the physical small-cluster system. In terms of hardware consideration, we conduct experiments as follows:

- CPU only: single-node, single-CPU (non-distributed)
- GPU only: single-node, single-GPU (non-distributed)
- SPARK+CPU: 3 data nodes, single-CPU-per-node
- SPARK+GPU: 3 data nodes, single-GPU-per-node
- VMs+vCPUs: 3 data nodes VMs on a single physical node

The experimental hardware configuration on each physical node is given as above:

- CPU(s): 4
- RAM: 24 GB
- Graphics card: Quadro K620 (see Table 5 for details)

Table 5. The specifications of Nvidia Quadro K620

Specification	
CUDA Cores	384
Memory size	2 GB
Bus width	128-bit
Bandwidth	29 GBps

In general, fast training on large-scale datasets is a crucial factor for a distributed computing system. In this experiment, the dataset we train to perform the benchmark is Dogs vs. Cats dataset from Kaggle [12], which is a larger-scale dataset compared to MNIST. The results of testing different types of deep learning systems are shown in Fig. 7.

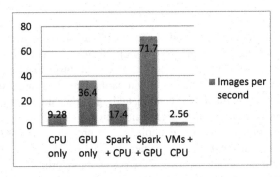

Fig. 7. Deep learning's training performance on physical servers and VMs.

In non-distributed systems for CPU-only and GPU-only cases, an entry-level graphic card achieves nearly 4x speedup. On top of that, the distributed deep learning framework of TensorFlowOnSpark runs even two times faster than non-distributed systems. Note that we only configured 3 data nodes and one executor per node in this example. Therefore, 2 data nodes execute computations and 1 data node executes weights update, and that is the main reason why 3 data nodes setup only achieves 2x speedup. Additionally, the case for applying distributed computing with GPUs turns out to get almost 8 times faster than the non-distributed CPU system and even 28 times faster than the VMs distributed environment.

Next, the computational time using the same distributed computing way in the physical small-cluster system is 6.8 times faster than in the VM distributed system. The main reason is that the big data VM platform is the guest operating system on Windows OS, therefore it shares the same resources of CPU, memory and hard disk with the Windows operating system to reduce its computational capacity.

In brief, InputMode.SPARK data ingestion and TFRecords input format get the best performance under the framework of TensorFlowOnSpark. By tuning the number of executors properly can decrease total training time. Although the computational efficiency in Big Data multi-VM Platform system is not as good as the physical small-cluster system, users are encouraged to use the big data multi-VM platform for distributed deep learning programming, compilation, testing and running. Therefore, the big data VM platform is an ideal platform for the preparing processing of development of big data application.

3.3 Demonstration

During the paper presentation, the first author of this paper would like to give a demonstration for the proposed Big Data Multi-VM Platform system to show its capability and applicability.

4 Concluding Remarks

The present study is focused on the development of big data multi-VM platform, and the benchmark in different types of deep learning systems including the multi-node big data VM platform, physical standalone system and the physical small-cluster system. The conclusion and suggestion are provided below:

A. This study utilizes the virtual machine technology by VirtualBox to construct big data multi-VM platform and obtain its four VM image files for quick installation. During the design and implementation phase, the system resource of physical multi-node Spark and Hadoop Cluster can be occupied and shared by multiple developers. This would cause the major impact when other users running their jobs in the system. Therefore, the big data multi-VM platform provides a well-designed personal development platform for developers to design, implement and test before running their deep learning applications in physical multi-node Spark and Hadoop Cluster. This can save more system resource and waiting time for users. This multi-VM platform provides an environment for users to use just like a real environment in the physical Spark and Hadoop cluster. Hence, users can quickly access the Spark and Hadoop Map/Reduce programming and portal development to speed up building their big data solutions.

B. The big data multi-VM platform provides the rapid prototyping of distributed deep learning by using a cutting-edge framework TensorFlowOnSpark(TFoS for AI developers. In this study, we benchmark two types of data ingestion which TensorFlowOnSpark offers: InputMode.SPARK and InputMode.TENSORFLOW with MNIST handwritten digit dataset on the big data multi-VM platform and the physical server.

C. InputMode.SPARK data ingestion and TFRecords input format get the best performance under the framework of TensorFlowOnSpark. By tuning the number of executors properly can decrease total training time. Although the computational efficiency in Big Data multi-VM Platform system is not as good as the physical small-cluster system, users do be able to use the big data multi-VM platform for distributed deep learning programming, compilation, testing and running. Therefore, the big data VM platform is an ideal platform for the development of big data application.

D. To speed up the development of big data applications for worldwide users, the proposed big data VM platform is ready for download [13]. We hope it can benefit all the users in big data computing community.

5 Future Work

Due to the use of most Spark and Hadoop systems is still in the command-line interface, it may be inconvenient for beginners to use. Therefore, the future work will be focused on the big data portal development by using Liferay Portal, which provides users friendly interface to access Spark and Hadoop systems in multiple Virtual Machines. This would improve the working efficiency for users in Spark and Hadoop management and applications.

When Spark and Hadoop cluster starts up several times, the file size of each VM image file increases rapidly. It's still hard to identify root causes. To temporarily deal with this problem, the current solution is to export each of VM image file immediately when the deep learning application software is installed and configured completely. Further studies are needed to address this issue.

In addition, Spark possesses the ability of in-memory computing and has been shown to be potentially 10–100 times faster than Hadoop. The future work of research will be focused on deep learning enabling and streaming applications by using Spark in the big data multi-VM platform. Such as the techniques of GPU access in VMs would be worthy to develop in the future. This would allow users to have more choice for their big data development and applications.

Acknowledgement. Financial support from the Ministry of Science and Technology, Taiwan, under grants MOST 105-2634-E-492-001 is highly appreciated. We are also grateful to the National Center for High-Performance Computing for computer time and facilities.

References

1. Dean, J., Ghemawat, S.: MapReduce: simplified data processing on large clusters. In: 6th Symposium on Operating Systems Design & Implementation (OSDI 2004), 6–8 December 2004, San Francisco (2004)
2. http://hadoop.apache.org/

3. Apache Spark. http://spark.apache.org/
4. VirtualBox. https://www.VirtualBox.org/
5. Liferay Portal. https://www.liferay.com
6. Eclipse. https://eclipse.org/ide/
7. Wu, C.-H., Tsai, W.-F., Lin, F., Chang, W.-Y., Lin, S.-C., Yang, C.-T.: Big Data development platform for engineering applications. In: Proceedings of 2016 IEEE International Conference on Big Data (IEEE BigData), pp. 2699–2702 (2016)
8. The Amazing Ways Google Uses Deep Learning AI. https://www.forbes.com/sites/bernardmarr/2017/08/08/the-amazing-ways-how-google-uses-deep-learning-ai/#31e431d13204
9. TensorFlowOnSpark. https://github.com/yahoo/TensorFlowOnSpark
10. Tensorflow. https://www.tensorflow.org
11. MNIST Dataset. http://yann.lecun.com/exdb/mnist/
12. Dogs vs. Cats Dataset. https://www.kaggle.com/c/dogs-vs-cats
13. FTP Site (IP = 140.110.20.15; Port = 21; Anonymous)

Developing Cost-Effective Data Rescue Schemes to Tackle Disk Failures in Data Centers

Zhi Qiao[1(✉)], Jacob Hochstetler[1(✉)], Shuwen Liang[1(✉)], Song Fu[1(✉)],
Hsing-bung Chen[2(✉)], and Bradley Settlemyer[2(✉)]

[1] Department of Computer Science and Engineering, University of North Texas,
Denton, USA
{ZhiQiao,JacobHochstetler,ShuwenLiang}@my.unt.edu, SongFu@unt.edu
[2] HPC-DES Group, Los Alamos National Laboratory, New Mexico, USA
{hbchen,bws}@lanl.gov

Abstract. Ensuring the reliability of large-scale storage systems remains a challenge, especially when there are millions of disk drives deployed. Post-failure disk rebuild takes much longer time nowadays due to the ever-increasing disk capacity, which also increases the risk of service unavailability and even data loss. In this paper, we present a proactive data protection (PDP) framework in the ZFS file system to rescue data from disks before actual failure onset. By reducing the risk of data loss and mitigating the prolonged disk rebuilds caused by disk failures, PDP is designed to enhance the overall storage reliability. We extensively evaluate the recovery performance of ZFS with diverse configurations, and further explore disk failure prediction techniques to develop a proactive data protection mechanism in ZFS. We further compare the performance of different data protection strategies, including post-failure disk recovery, proactive disk cloning, and proactive data recovery. We propose an analytic model that uses storage utilization and contextual system information to select the best data protection strategy to achieve cost-effective and enhanced storage reliability.

Keywords: Storage reliability · Data protection · Failure prediction

1 Introduction

Nowadays, big data applications require storage systems to possess exabytes of capacity, provided by millions of hard disk drives. At such a scale, disk failures become the norm. Redundant array of inexpensive disks (RAID) is a common practice in most large-scale storage systems for redundancy and performance [1]. In the face of disk failures, a RAID system uses parity data on the remaining, working disk drives to compute and recover the lost data on a spare drive. However, RAID recovery is a time-consuming process which demands considerable computing resources and stalls user applications due by bringing the RAID system offline

© Springer International Publishing AG, part of Springer Nature 2018
F. Y. L. Chin et al. (Eds.): BIGDATA 2018, LNCS 10968, pp. 194–208, 2018.
https://doi.org/10.1007/978-3-319-94301-5_15

for repair or consuming a large portion of the I/O bandwidth. In recent years, the dramatic growth of disk capacity far outpaces the improvement of disk speed, resulting in an even longer RAID recovery time. It takes days and even weeks, to recover an enterprise-grade RAID group composed of helium-filled large-capacity hard drives. Furthermore, the RAID recovery process places additional stress on the remaining disk drives due to the intensive read and write activities. During the day or week-long RAID recovery, it becomes increasingly likely that other disk drives(s) in the same RAID group may become failed. Such multiple disk failures can cause data loss and high monetary cost. The prolonged disk rebuilds also compromise the overall system performance, as data accesses to the affected RAID system are delayed for a long period of time.

The existing methods and products are mostly reactive, providing disk or storage remediation after failures occur. Reactive data protection schemes suffer from high recovery overhead which affects storage availability and system performance. Proactive data protection (PDP), on the other hand, can explore the lead time prior to disk failures to overlap data rescue with regular storage operations. From the system's perspective, the disk drives and storage system continuously service I/O requests without obvious disruptions. Failure prediction is an enabling technology for proactive data protection, as it allows data rescue mechanisms to be performed before failures truly happen. Disk manufacturers embedded SMART (Self-Monitoring Analysis and Reporting Technology) monitoring in their products, which reports the health status of a disk. But they set the thresholds of SMART attributes for disk failure detection as low as possible in order to minimize the false alarm rate. As a result, disk drives fail way before these thresholds are reached in the field, which makes the proactive RAID controllers inefficient to protect data.

Studies leveraging advanced statistic models and machine learning technology show promising results of predicting disk failures ahead of their actual occurrences. For early attempts such as [2,3], the effectiveness of failure prediction was questioned, for example in [4]. Recently, many researchers used advanced machine learning algorithms and obtained accurate prediction results. Mahdisoltani et al. [5] evaluated the effectiveness of a set of machine learning techniques and discussed the proactive error prediction method that aims to improve storage reliability. Their work inspires us to incorporate failure prediction in file systems to develop a cost-effective data protection solution with important contextual information.

As an open-source high performance file system, ZFS has been used by data centers in their production storage systems, as well as by industry vendors in their software-defined storage products. ZFS implements software RAID and efficient data compression. It was designed from scratch to address fault management issues in storage systems. ZFS can detect and repair silent disk corruptions [6], and rebuild software RAID from disk failures. The performance of RAID recovery in ZFS is affected by many factors, including:

- storage pool utilization,
- the number of virtual disks in a RAID group,
- software RAID configuration,
- the amount of corrupted data that needs to be recovered.

Hardware configuration of the storage system also affects the performance of ZFS recovery through

- the use of flash drives,
- the amount of DRAM available,
- CPU cycles available.

In this paper, we propose a proof-of-concept, proactive data protection scheme that explores failure prediction in file systems to enhance the storage reliability and reduce the risk of data loss caused by disk failures. By performing counter measures prior to disk failures, PDP mitigates the performance impact caused by lengthy disk rebuilds. Due to the popularity in both the research and industrial communities, we choose the ZFS file system to implement a prototype proactive data protection system. This combination of file-system-level control and data protection with disk-drive-level reliability monitoring and failure prediction provides a pragmatic and holistic approach to build highly reliable storage systems. The disk-drive failure prediction technologies enable ZFS to proactively move or re-compute data from a failing disk to an available and healthy drive. Since there are many factors that affect ZFS' recovery performance, we first present an analytic model (in Sect. 3) for evaluating the I/O and recovery performance of ZFS using different system configurations. Due to the increasing adoption of flash drives in storage systems, we conduct our experiments in a heterogeneous storage environment that consists of both magnetic hard disk drives (HDD) and solid-state drives (SSD). We then propose three PDP strategies (in Sect. 4) designed for ZFS. Our performance evaluation of these strategies shows that ZFS can make an optimal decision with the help of contextual file-system information. That is, ZFS can determine the best PDP strategy based on the current system state. The major contributions of our paper are as follows.

- We extensively evaluate the performance of software RAID and observe that the performance of I/O and data recovery in ZFS degrades as the storage utilization increases, which influences the selection of data protection strategies.
- As far as we know, we are the first to integrate proactive data protection with file systems and leverage the strengths of both to enhance storage reliability. The contextual file-system information enables the maintenance (such as disk drives early retirement and proactive data migration) to be scheduled at off-peak time, thus preserves the valuable I/O and network bandwidth for user's request.
- We provide a quantitative analysis of each proactive data protection strategy and design a method to select the optimal strategy by exploring the run-time system status and cost functions to best protect storage data.

2 Background

2.1 RAID, Striping, and Levels

RAID (Redundant Array of Independent Disks, originally Redundant Array of Inexpensive Disks) is a virtualized storage subsystem that combines multiple physical disk drives into one or more logical units to improve the performance and fault tolerance of storage systems. Over the years of development, several *RAID levels* have evolved. These include the standard schemes such as mirroring data and striping data, as well as other variations such as nested levels and proprietary schemes. Depending on the specific RAID level, an array of multiple disks can be configured to achieve a balance between performance, reliability, and capacity. Many RAID levels employ an error protection scheme called *parity*, which uses a relatively small amount of data recover more user data. *Striping* is the underlying concept of all RAID levels other than RAID 1 (mirroring). Striping is a mechanism to split up disk partitions into stripes of contiguous sequences of disk blocks. A RAID Stripe usually consists of multiple data blocks and one or more parity blocks. Disk striping without any data redundancy (or parity) is RAID 0 and is used for increasing the I/O performance. Different RAID levels organize the stripes and parity data differently. RAID 5 uses a distributed parity block with a disk stripe, while RAID 6 employs two parity blocks distributed across all disks in the stripe. In terms of ZFS, *Stripe* is equivalent to RAID 0, whereas *Mirror* corresponds to RAID 1 and *RAID-Z* and *RAIDZ-2* are equivalent to the standard RAID 5 and RAID 6, respectively.

2.2   RAID Rebuild in ZFS

In ZFS the RAID rebuild process is called resilvering. This comes from the word used for the actual repairing of physical mirrors. Resilvering occurs when a disk needs to be replaced due to either failure or data corruption. If a disk fails unexpectedly, the ZFS rebuild algorithm reassembles data on a new disk using either mirrored or parity data. The resilvering process can take an extended period of time, depending on the size of the drive and the amount of data that needs to be recovered. During resilvering, the performance of a RAID array is degraded because (1) the system resources are usually prioritized for data recovery, and (2) the rest of the disks in the group cannot provide optimal data redundancy for the configured RAID level. The resilvering process causes extra wear and accelerates the failure of the remaining healthy disks. Two or more disk failures can occur when I/O workloads stress individual disks too much. Therefore, the total number of disk failures might exceed the maximum fault tolerance for the RAID level, which causes resilvering to fail. The RAID array then becomes unavailable once such nested failures occur. If no additional backup exists, data on that array will become lost.

ZFS provides two different redundancy strategies, i.e., data mirroring and parity. Data mirroring used in *mirror* is a relatively simply method for rebuilding a RAID group. Since data is simultaneously stored on each component disk in the array, the content of any disk is identical to the rest. In the case of a disk failure, ZFS copies blocks of data from the remaining healthy disks to a spare disk. Although this type of redundancy provides fast RAID rebuild, it suffers from a low space utilization. The cost of scaling such an array increases linearly as a full drive is needed for every copy of the data. On the other hand, parity is more computationally expensive, but has a lower disk cost since it does not need a full set of duplicated disks to operate. *Raidz* uses simple logical operations, such as *exclusive or* (XOR), across the stripe to compute parity P

$$P = \bigoplus_i D_i = D_0 \oplus D_1 \oplus D_2 \oplus \cdots \oplus D_{n-1} \tag{1}$$

where D_i is a data block within a stripe.

If one disk fails, ZFS performs XOR operations on the data and parity blocks from the remaining healthy stripes to recover the original data. However, XOR operations do not specify the exact location of each data block in the stripe. If two disks fail in a *raidz2* array, applying XOR twice would be useless, since ZFS cannot determine which data block belongs to which failed disk. To handle this problem, *raidz2* utilizes the Reed-Solomon coding method and *Galois field* $GF(m)$ to generate the second parity Q and embeds the ownership information in the block.

$$Q = \bigoplus_i g^i D_i = g^0 D_0 \oplus g^1 D_1 \oplus \cdots \oplus g^{n-1} D_{n-1} \tag{2}$$

Calculating parities and data recovery (i.e., resilvering) based on *Galois field* $GF(m)$ are more compute-intensive than performing XOR in *raidz*. On the other hand, *raidz2* achieves a better redundancy.

3   Performance Modeling of I/O and Resilvering in ZFS

ZFS has been widely used in production storage systems, due to its support for high storage capacity, efficient data compression, integrated volume management and reliability management features. Data resilvering in ZFS is reactive, that is, data and parity blocks are read, regenerated, and stored after failures are detected. Although there are a few works that evaluate the performance of ZFS [7–9], they focus on certain features such as data compression or the read/write speed as a file system. Little work has been conducted to understand the performance of the new fault management techniques in ZFS. In this section, we evaluate the performance of ZFS' software RAID and the cost of the resilvering process. These results help us obtain a deeper understanding of ZFS' fault management mechanisms influencing the design of our proactive data protection scheme.

3.1   Test Platform Configuration

Table 1 shows the parameters and their values used in our experiments. The servers in the test platform were equipped with eight Intel Xeon cores (3 GHz), 32 GB DRAM, Ubuntu 16.04 LTS, and ZFS version 0.6.5.11. The disk drives model we used in the tests are Seagate BarraCuda ST2000DMs (magnetic HDD) and Intel DC3520s (data-center class SSD). We use the Bonnie++ benchmark suite to test ZFS' I/O performance. Specifically, we use the three subtests 'sequential

Table 1. Test platform configuration and experiment setting.

System configuration	Setting
Disk media	HDD (magnetic spinning), SSD (solid-state flash)
RAID (ZFS) level	0 (*stripe*), 1 (*mirror*), 5 (*raidz*), 6 (*raidz2*)
RAID stripe size	2–6
DRAM size for ZFS	8, 16, 32 GB
Storage utilization	0%–90%

output', 'sequential input', and 'rewrite' to evaluate the file system functions *write*, *read*, and *modify*. As ZFS is memory intensive, we vary the DRAM size allocated to ZFS to characterize its performance. To evaluate the resilvering performance of ZFS, we create a disk logical error to activate the resilvering process, then we measure the turnaround time of resilvering and the amount of data that ZFS recovers. The disk logical failure is created by injecting an error into ZFS' reserved area on a target disk to disable the communication channel between ZFS and the target disk.

3.2   ZFS Performance Characterization

We run each benchmark five times and compute the average of results. Each of the five runs is within 5% of the average. Thus we believe the experimental results are relatively stable. We present the average values of the results in the following figures for a clear presentation and interpretation. We expect to address the following important questions in our experiments.

How does strip width affect ZFS' performance?
As discussed in Sect. 2.1, the striping process distributes file chunks across multiple disks, which means a file access request is served by all disks within the stripe. Our experimental results as presented in Fig. 1 show that for the software RAID in ZFS, **increased stripe width improves I/O throughput and the improvement is super-linear**. The x-axis in the figure is the stripe width of RAID array. *4+1* denotes that a stripe consists of four data blocks and one parity block. When the stripe width is increased from 2 to 5 in a *raidz* array, or from 3 to 5 in a *raidz2* array, the overall I/O throughput increases linearly. Note that a RAID array using SSDs does not always outperform an HDD-based RAID array. For sequential I/O, SSDs arrays only outperforms the HDD-based array in the largest stripe size (i.e., *4+1* and *3+2* on *raidz* and *raidz2*, respectively).

Unlike I/O workload which can be serviced in parallel, resilvering is more computation intensive that reconstructs data sequentially. Although employing more disks in the rebuild process can improve the aggregated I/O throughput, the wider stripe increases the amount of data and parity used during the resilvering calculation which offsets the throughput gain. Figure 2a shows the **recovery speed does not monotonically decrease as the stripe expands**. This finding inspires us to

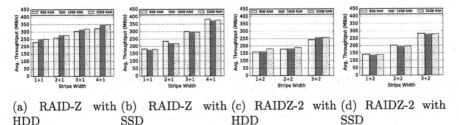

(a) RAID-Z with HDD   (b) RAID-Z with SSD   (c) RAIDZ-2 with HDD   (d) RAIDZ-2 with SSD

Fig. 1. I/O throughput under different RAID level, stripe width, and DRAM size.

explore the search space of different hardware and environment settings to achieve the best recovery performance.

How does the size of DRAM affect ZFS' performance?

ZFS requires a large amount of DRAM for file caching and metadata management. A common practice for system configuration is to provide 5 GB of DRAM for each terabyte of storage. As shown in Fig. 1, we test three DRAM sizes: 8, 16, and 32 GB for each stripe width and RAID level. From the figure we can see that larger DRAM size leads to about 10% of improvement of sequential I/O performance in HDD-based RAID array. But SSD-based array does not show significant benefit from using more DRAM. However, we cannot simply assume DRAM size is not important for ZFS, since our synthetic benchmark only consider sequential I/O performance. Real-world workload consist of both random and sequential I/Os. The random data access pattern in ZFS will gain huge benefit from large DRAM size.

The results from the RAID recovery experiments show that **adding more DRAM does not always improve ZFS' recovery performance**. Figure 2b shows that using more DRAM for an HDD-based array improves the recovery speed by 3.8% on average, while the performance on an SSD-based array drops by 5.6%. In order to explain this phenomenon, we decompose the resilvering process of ZFS into three steps: (1) reading data and parity blocks, (2) computing the data that is lost, and (3) writing the recovered data to a spare. The performance of the second stage (data recovery computation) significantly influences the overall resilvering speed. We can compare the resilvering process with the *rewrite* operations that Bonnie++ performs, which also consists of reading, modifying, and writing blocks of data. From Fig. 2c, we can see that the **recovery is a time-consuming process and the CPU performance is a dominant factor**. Although larger DRAM accommodates more metadata and file caching for regular I/O operations, the resilvering process rarely uses cached files. Hence, the DRAM size does not significantly affect the recovery performance of ZFS.

Does the storage utilization affect ZFS' performance?

ZFS employs *copy-on-write* (COW), that is a new copy is created only when the data is modified. It efficiently shares duplicate data to avoid unnecessary resource consumption. However, COW causes more fragmentations as the utilization of the

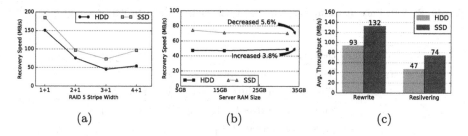

Fig. 2. The I/O and recovery performance of ZFS.

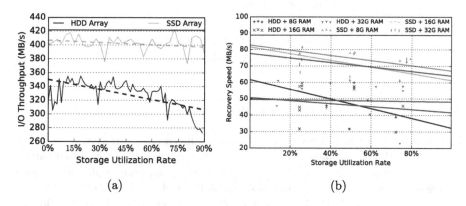

Fig. 3. Storage utilization influences ZFS' performance.

storage increases. Moreover, when the storage usage increases to approximately 80%, ZFS switches to a space-conserving (rather than speed-oriented) mode to preserve working space on the volume. In our experiment, we profile ZFS and measure its throughput as we ramp up zpool's utilization from 0% to 90%. Figure 3a presents the experimental results. The solid curve shows the average throughput of ZFS under the RAIDZ configuration, and the dashed curve is the corresponding trendline using a linear regression. From the figure, we can see that **as the utilization of zpool increases, the throughput of ZFS decreases significantly in the RAID array using HDDs**. As a zpool becomes fully used, ZFS' throughput drops by up to 25%.

Similar to the preceding results, **the recovery throughput degrades as the zpool utilization increases**. Figure 3b shows the recovery throughput under different zpool utilization for both HDDs and SSDs based array. The least-squares linear regression model fits the results the best. We observe that the recovery performance decreases as the zpool utilization increases. When the storage array is close to a full utilization, the recovery speed is about 37% slower than the average throughput, and 47% slower than the peak speed. The first and last steps of the resilvering process consist of reads and writes. Therefore the performance degradation due to increased zpool utilization plays a major role for the reduced recovery speed.

Table 2. Variables used in the analysis of proactive data protection and strategy selection

Variables	Description
t_l	Lead time of a failure prediction
t_i	Time when a failure prediction is performed
t_f	Time when a disk failure is predicted to happen
t_v	Validation period of a prediction
T	Duration of the data rescue process
S	Data rescue speed
A	Amount of data to be rescued
W	Wasted time due to failure misprediction
p	Precision of failure prediction

4   Proactive Data Protection

Disk access speed has been outpaced by the increasing capacity. The stripe width of RAID arrays have grown to fill the gap between disk speed and capacity. Unfortunately, the probability of having double and even triple failures also increases as the disk recovery time is significantly prolonged. Although ZFS supports RAID 6 and RAID 7 (triple parity) to handle disk failures, the longer performance degradation and even unavailability of disk arrays compromise the overall system performance and users' satisfaction. Complementary to post-failure disk rebuilds, data can be rescued proactively prior to disk failures. This is enabled by disk failure prediction techniques [10–12] which forecast when failures will happen on which drives with promising accuracy. We aim to incorporate disk failure prediction methods in ZFS so that ZFS becomes capable of replacing a failing drive before the failure actually happens. Data on the failing drive (in contrast to failed drive) can be moved to a spare drive without the expensive disk rebuild process, thereby avoiding service disruption of the storage system. In addition, proactive data protection can be scheduled to perform during off-peak hours, which can further improve storage availability and achieve service-level objectives (SLO). Disk failure prediction remains an active research topic. In this work, we leverage the existing techniques of prediction disk failures and explore them in ZFS to develop proactive data protection. Research on the prediction methods is not the focus of this paper. Table 2 lists the variables that we use in the following discussion and analysis.

4.1   Proactive Strategies for Handling Disk Failures

A disk failure prediction model (FPM) uses monitored, real-time status data of disk drives to compute the probability at which the drives will fail in the future. At time t_i, if FPM predicts that a disk is going to fail, it reports the predicted failure occurrence which will happen at time t_f. We use lead time t_l, i.e., the length of time between the point when FPM makes prediction and the predicted failure occurrence time, to represent the *urgency* of a failure. We can calculate the lead time using $t_l = t_f - t_i$. Proactive data rescue from the failing disk to a spare or available disk also takes time. We use T to denote the time that a proactive action uses to rescue data. If $t_l > T$, then proactive data rescue can ensure the safety of the data on the failing drive(s). In this paper, we discuss several proactive strategies to handle disk failures. To proactively rescue data on a failing disk, we need to calculate the estimated recovery time T for each proactive strategy and compare it with the lead time t_l.

1. Proactive **Disk** **C**loning (P-DISCO)
2. Proactive **A**ctive-**da**ta **R**ecovery (P-ADAR)
3. Proactive **A**ctive-**da**ta **C**loning (P-DACO)

The *Proactive Disk Cloning*, or P-DISCO, migrates data on a predicted, failing drive to a hot spare using disk cloning. In the conventional RAID rebuilds, data reconstruction involves massive data movement and intensive parity computation using data and parity blocks from all of the other disks in an array. CPU and interconnect become the performance bottleneck as they determine the recovery throughput. Disk cloning, on the other hand, only migrates the data from the failing disk to a spare one without involving other disks or calculating parities. Theoretically, proactive disk cloning can achieve a much higher throughput. The additional workload might accelerate the death of the already failing drives, but other disk drives in the same RAID array stays intact. Moreover, the recovery speed is independent of the disk space utilization. In contrast, the performance of ZFS' resilvering process is affected by disk utilization as shown in Sect. 3.2. Figure 4a compares the two recovery strategies, i.e., ZFS' resilvering and proactive disk cloning, for a fully used drive. From this figure, we can see P-DISCO completes data rescue in 17.59 h for an 8TB HDD, and in 4.17 h for a 3.2 TB SSD. Most importantly, data rescue happens before a disk failure happens, thereby preventing a storage system from operating in a degraded state and nested disk failures. In contrast, after the disk fails, the default ZFS data recovery process takes over two times longer period of time to rebuild the same disk. During this period, the data reconstruction process competes with regular I/O workloads for system resources, which prolongs the recovery time and the vulnerable period spent in a degraded state. The volume manager masks data locations to make the recovery process transparent to the system. Without knowing which disk sectors contain useful data, disk cloning has to copy everything from every sector to the spare drive. For a nearly empty disk, P-DISCO may not be time and resource efficient.

$$T_{P-DISCO} = \frac{A_{disk\ capacity}}{S_{cloning}} \tag{3}$$

ZFS combines a file system and volume manager. It keeps tracking "active data" and provides the opportunity to recover only those tagged data areas when

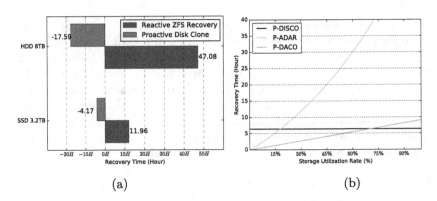

Fig. 4. Effects of storage utilization on ZFS' performance.

a disk fails. This results in an efficient data rescue strategy. *Proactive Active-data Recovery*, or P-ADAR, enables ZFS to proactively rescue only active-data instead of the entire disk to the spares. Although resilvering is computationally intensive, recovering the minimum amount of necessary data can save time. The amount of active data on a disk is no more than the disk's overall capacity, that is $A_{active\ data} < A_{disk\ capacity}$. Additionally, ZFS' resilvering is safe to interrupt. If power outage or reboot occurs during data rescue, the resilvering process resumes at the exact location where it is interrupted without human intervention. However, the resilvering process is more complex than cloning, which involves a full disk scan to gather active data locations, followed by compute-intensive data reconstruction. Therefore, $S_{resilvering} < S_{cloning}$. The actual recovery time T equals

$$T_{P-ADAR} = \frac{A_{active\ data}}{S_{resilvering}} \tag{4}$$

To further speed data recovery, we leverage computation-light cloning to rescue the necessary data tagged by ZFS. The proposed *Proactive Active-data Cloning*, or P-DACO, only moves the active data during recovery, and avoids parity computation. It combines the advantages of the preceding two strategies. Thus the recovery time equals to

$$T_{P-DACO} = \frac{A_{active\ data}}{S_{cloning}} \tag{5}$$

To obtain the value of S and A, a set of daemons frequently check disk's SMART (Self-Monitoring Analysis and Reporting Technology) data and zpools' utilization. A daemon process also runs a set of micro-benchmarks to determine the I/O performance and data recovery speed S at runtime. The amount of data to be rescued A is determined when the proactive action starts to rescue data. Meanwhile, the SMART data of each disk in the system is transferred to FPM for failure prediction and disk health analysis.

4.2   Selecting the Best Data Rescue Strategy

From the preceding discussion, we can see each proactive data rescue strategy is only suitable for some specific situations. How to select the appropriate strategy to handle disk failures is critical. Figure 4b illustrates the performance of each proactive data rescue strategy to recover data from an 8TB enterprise-grade HDD with different disk utilization. From the figure, we find that P-ADAR is more efficient when the zpool is lightly used (i.e., below 15%). As more active data is stored, P-DISCO becomes a better choice. P-DACO is a promising strategy that yields better results than P-ADAR and P-DISCO most of the time. However, P-DACO cannot tolerate interrupts during data rescue, which needs trade-off between dependability and performance.

To select the best data rescue strategy systematically and accurately to handle disk failures, we design two constraints for the selection process: *urgency* and *dependability*. Specifically, the urgency indicates how fast a data rescue strategy can complete data recovery, while the dependability measures to what extent the data rescue strategy is tolerant to interruptions. For example, to minimize storage downtime caused by disk failures, we can prioritize *urgency* over *dependability*. In this case, if failure predictions provide enough lead time, where $t_l > min(T_{P-DISCO}, T_{P-ADAR}, T_{P-DACO})$, then we can select the strategy with shortest data rescue time T. For a storage system that prioritizes *dependability* over *urgency*, we adopt a simpler approach to narrow down the strategy selection set, since *active-data resilvering* is the only one that is safe to interruptions. Therefore, if FPM provides enough lead time for P-ADAR, we choose *active-data resilvering* as the proactive data rescue strategy to handle disk failures. Otherwise, the default reactive data rescue, or R-ADAR, is employed.

4.3   Analysis of Data Rescue Cost

In practice, each FPM only performs well for certain model and type of disk drives. For drives of a different model or type, it may generate many false alarms that incur unnecessary data rescues, and false negatives which need the expensive disk rebuild. In this section, we analyze the effectiveness and cost of proactive data protection strategies.

When FPM predicts a failure to occur at time t_f, it also provides a valid period t_v, which indicates the failure occurrence is likely within $t_f \pm \frac{1}{2}t_v$. We can calculate the adjusted lead time as $t_l' = t_f - \frac{1}{2}t_v - t_i$. If the disk is still healthy beyond $t_f + \frac{1}{2}t_v$, we say the prediction is a false positive. For proactive data protection, the cost of a false positive is a replacement of a healthy disk, in addition to the wasted resources used during data rescue. If the failure actually happens before $t_f - \frac{1}{2}t_v$, or FPM does not report that failure, we say the prediction is a false negative. In our discussion, we assume the worst case of false negative is that there is no proactive action to rescue data. Since proactive data protection should be complemented by reactive data rescue (R-ADAR) to address mis-predictions, the cost of false negatives becomes the cost from performing R-ADAR. The default R-ADAR and proactive strategy P-ADAR take the same time to do data rescue ($T_{P-ADAR} = T_{R-ADAR}$). The difference is their start time. At time t_i, when FPM predicts that a failure will occur by time t_f, P-ADAR starts data rescue and complete it by $t'_l - T_{P-ADAR}$. This is equivalent to $t_f - \frac{1}{2}t_v - T_{P-ADAR}$. For R-ADAR, the rescue process starts only after the actual disk failure occurrence at t'_f, and completes by $t'_f + T_{R-ADAR}$. Assume the probability that $t_f = t'_f$ is p. Once FPM provides enough lead time for proactive data rescue, i.e., $t'_l > T$, we say that the proactive data rescue strategy can be completed before the reactive action starts. With a probability of $(1 - p)$ that FPM makes a wrong prediction, the cost is no higher than that of the default reactive strategy. Therefore, proactive data rescue improves the overall storage reliability by reducing the risk of data loss risk and the cost of data recovery.

We note that proactive data rescue strategies are not to replace the conventional resilvering process in ZFS. They are complementary to resilvering, aiming to enhance the storage reliability further. When failure predictions are sufficiently accurate, that is the majority of disk failures can be handled by proactive data rescue strategies prior to failure occurrences, the data recovery time is significantly shortened and the storage availability is improved. The probability of nested/multiple disk failures can also be reduced, which mitigates the risk of data loss.

5   Related Works

To improve storage reliability, some researches utilizes erasure coding based redundancy scheme for cost-effective error tolerant. Although we have seen its application in Windows Azure storage [13] and NEC HYDRAstor [14], the usage of erasure coding in primary storage, which requires high throughput and low latency, has not been widely adopted yet. To address such issue, our previous work [15,16] leverages the parallelism to improve Jerasure's [17] coding performance for storage systems. That been said, erasure code is still a reactive solution to improve the storage reliability.

Other researches focus on RAID related technologies. For example, early works for tolerating multiple failures in an RAID array include [18–20]. Researchers have also investigated remediation mechanisms to mitigate performance degradation caused by RAID recovery. For example, in [21], the authors presented a new RAID organization called multi-partition RAID to reduce the performance degradation during RAID rebuilds. In [22], the workload that targeting degraded RAID sets were outsourcing to surrogate RAID sets, hence improving the overall availability of a storage system. Parity declustering [23,24] recently gained attention as it could reduce the reconstruction time of RAID 5 and 6. However, as we mentioned in Sect. 1, most of the problems these researches tries to addressed can be eliminated if we can proactively rescue data prior to disk drives failure.

6 Conclusions

Storage reliability imposes a major challenge to big data systems and applications. In this paper, we characterize storage performance under disk failures with a variety of ZFS configurations. We propose a proactive data protection scheme that leverages promising disk failure prediction techniques and rescues data prior to disk failure occurs. We explore the findings from our experiments to design an analytic model that aims to find the optimal data rescue strategy. Our analytic model uses zpool utilization and the configuration of a storage system to select the best strategy that minimizes the data rescue cost and maximize storage availability.

Acknowledgment. This work was supported in part by an LANL grant and an NSF grant CCF-1563750. Los Alamos National Laboratory is supported by the U.S. Department of Energy contract DE-AC52-06NA25396. This publication has been assigned a LANL identifier LA-UR-17-27639.

References

1. Gibson, G.A., Patterson, D.A.: Designing disk arrays for high data reliability. J. Parallel Distrib. Comput. **17**(1–2), 4–27 (1993)
2. Murray, J.F., Hughes, G.F., Kreutz-Delgado, K.: Hard drive failure prediction using non-parametric statistical methods. In: Proceedings of the ICANN/ICONIP (2003)
3. Murray, J.F., Hughes, G.F., Kreutz-Delgado, K.: Machine learning methods for predicting failures in hard drives: a multiple-instance application. J. Mach. Learn. Res. **6**, 783–816 (2005)
4. Pinheiro, E., Weber, W.D., Barroso, L.A.: Failure trends in a large disk drive population. In: Proceedings of the 8th USENIX Conference on File and Storage Technologies (2007)
5. Mahdisoltani, F., Stefanovici, I.A., Schroeder, B.: Proactive error prediction to improve storage system reliability. In: USENIX Annual Technical Conference (2017)
6. Bonwick, J., Ahrens, M., Henson, V., Maybee, M., Shellenbaum, M.: The zettabyte file system. In: Proceedings of the 2nd USENIX Conference on File and Storage Technologies, vol. 215 (2003)
7. Heger, D.A.: Workload dependent performance evaluation of the Btrfs and ZFS filesystems. In: Proceedings of the International Conference of CMG (2009)
8. Phromchana, V., Nupairoj, N., Piromsopa, K.: Performance evaluation of ZFS and LVM (with ext4) for scalable storage system. In: 2011 Eighth International Joint Conference on Computer Science and Software Engineering (JCSSE), pp. 250–253. IEEE (2011)
9. Mohr, R., Peltz Jr., P.: Benchmarking SSD-based lustre file system configurations. In: Proceedings of the 2014 Annual Conference on Extreme Science and Engineering Discovery Environment. ACM (2014). Article no. 32
10. Goldszmidt, M.: Finding soon-to-fail disks in a haystack. In: Proceedings of the HotStorage (2012)
11. Huang, S., Fu, S., Zhang, Q., Shi, W.: Characterizing disk failures with quantified disk degradation signatures: an early experience. In: IEEE International Symposium on Workload Characterization (IISWC), pp. 150–159. IEEE (2015)

12. Botezatu, M.M., Giurgiu, I., Bogojeska, J., Wiesmann, D.: Predicting disk replacement towards reliable data centers. In: Proceedings of the 22nd International Conference on Knowledge Discovery and Data Mining ACM SIGKDD, pp. 39–48. ACM (2016)

13. Huang, C., Simitci, H., Xu, Y., Ogus, A., Calder, B., Gopalan, P., Li, J., Yekhanin, S., et al.: Erasure coding in windows azure storage. In: USENIX ATC, Boston, MA, pp. 15–26 (2012)

14. Dubnicki, C., Gryz, L., Heldt, L., Kaczmarczyk, M., Kilian, W., Strzelczak, P., Szczepkowski, J., Ungureanu, C., Welnicki, M.: HYDRAstor: a scalable secondary storage. In: FAST 2009, pp. 197–210 (2009)

15. Chen, H.B., Fu, S.: Improving coding performance and energy efficiency of erasure coding process for storage systems-a parallel and scalable approach. In: 2016 IEEE 9th International Conference on Cloud Computing (CLOUD), pp. 933–936. IEEE (2016)

16. Chen, H.B., Fu, S.: Parallel erasure coding: exploring task parallelism in erasure coding for enhanced bandwidth and energy efficiency. In: 2016 IEEE International Conference on Networking, Architecture and Storage (NAS), pp. 1–4. IEEE (2016)

17. Plank, J.S., Simmerman, S., Schuman, C.D.: Jerasure: a library in c/c++ facilitating erasure coding for storage applications-version 1.2. University of Tennessee, Technical report CS-08-627 23 (2008)

18. Blaum, M., Brady, J., Bruck, J., Menon, J.: EVENODD: an efficient scheme for tolerating double disk failures in raid architectures. IEEE Trans. Comput. 44(2), 192–202 (1995)

19. Alvarez, G.A., Burkhard, W.A., Cristian, F.: Tolerating multiple failures in raid architectures with optimal storage and uniform declustering. ACM SIGARCH Comput. Archit. News 25, 62–72 (1997)

20. Corbett, P., English, B., Goel, A., Grcanac, T., Kleiman, S., Leong, J., Sankar, S.: Row-diagonal parity for double disk failure correction. In: Proceedings of the 8th USENIX Conference on File and Storage Technologies (2004)

21. Tsai, W.J., Lee, S.Y.: Multi-partition raid: a new method for improving performance of disk arrays under failure. Comput. J. 40(1), 30–42 (1997)

22. Wu, S., Jiang, H., Feng, D., Tian, L., Mao, B.: Improving availability of raid-structured storage systems by workload outsourcing. IEEE Trans. Comput. 60(1), 64–79 (2011)

23. Holland, M., Gibson, G.A.: Parity declustering for continuous operation in redundant disk arrays, vol. 27. ACM (1992)

24. Chau, S.C., Fu, A.W.C.: A gracefully degradable declustered raid architecture. Cluster Comput. 5(1), 97–105 (2002)

On Scalability of Distributed Machine Learning with Big Data on Apache Spark

Ameen Abdel Hai$^{(\boxtimes)}$ and Babak Forouraghi$^{(\boxtimes)}$

Department of Computer Science, Saint Joseph's University,
Philadelphia, PA 19131, USA
{aa671849,bforoura}@sju.edu

Abstract. Performance of traditional machine learning systems does not scale up while working in the world of Big Data with training sets that can easily contain petabytes of data. Thus, new technologies and approaches are needed that can efficiently perform complex and time-consuming data analytics without having to rely on expensive super machines.

This paper discusses how a distributed machine learning system can be created to efficiently perform Big Data machine learning using classification algorithms. Specifically, it is shown how the Machine Learning Library (MLlib) of Apache Spark on Databricks can be utilized with several instances residing on Elastic Compute Cloud (EC2) of Amazon Web Services (AWS). In addition to performing predictive analytics on different numbers of executors, both in-memory processing and on-table scans were used to utilize the computing efficiency and flexibility of Spark. The conducted experiments, which were run multiple times on several instances and executors, demonstrate how to parallelize executions as well as to perform in-memory processing in order to drastically improve a learning system's performance. To highlight the advantages of the proposed system, two very large data sets and three different supervised classification algorithms were used in each experiment.

Keywords: Big data · Machine Learning · Apache spark · Timing analysis
Accuracy prediction · Data analysis · MLlib · Databricks

1 Introduction

The Internet has widely become the universal source of information for a considerable number of users. Astounding growth rates of online systems such as social media, mobile e-commerce applications, and many others continues to generate large amounts of data on a daily basis, and that has led to the phenomenon of Big Data [15]. Gathering enormous datasets of various types and structures can help enterprises perform advanced machine learning and predictive analytics such as fraud detection, risk analysis, economic and weather forecasts as well as advanced biometrics, which require very large training sets [1, 9–11, 13–15]. Thus, the world of Big Data has become of significant importance in our daily lives.

Unlike traditional structured data sets, Big Data is mostly unstructured (schema-less), and systems analyzing them utilize unconventional data manipulation techniques

© Springer International Publishing AG, part of Springer Nature 2018
F. Y. L. Chin et al. (Eds.): BIGDATA 2018, LNCS 10968, pp. 209–219, 2018.
https://doi.org/10.1007/978-3-319-94301-5_16

in order to efficiently perform complex operations. These operations should ideally be performed in-memory, rather than on-disk, with minimal computational latency and without the use of expensive hardware [1, 12]. Therefore, the performance and real-time training speed of machine learning systems on big data sets is of primary concern.

The Apache Hadoop is a distributed computational framework, which allows processing of large datasets across clusters of computers using the MapReduce programming model, and it has been used in advanced machine learning applications [15, 16, 18]. However, Hadoop exhibits high latency, which is directly due to its disk-persistent HDFS write operations as well as the general limitations of MapReduce such as the overhead of map jobs, high latency of storing intermediate computational results on disk and fault-intolerance [10, 16].

Apache Spark is a powerful and efficient in-memory framework for distributed processing, and it is fast becoming the mainstream distributed engine for performing advanced data analytics on big data sets [8, 10, 17]. A key feature of Spark is that its distributed computational platform is based on the Resilient Distributed Dataset (RDD) architecture, which caches intermediate results in order to perform in-memory processing [8, 11]. Resilient data sets are then split into a number of partitions and distributed across a cluster of workers/executors to prevent fault intolerance [1].

Spark's Machine Learning Library (MLlib) drastically reduces time-consuming analysis of mid-sized to very large datasets, and it enhances machine learning in general by its scalability and excellent ability for iterative processing [1, 3]. Further, MLlib supports combination of multiple algorithms in a single pipeline (workflow) to distribute stages across multiple executors as well as integration of various streaming and graphics tools [11]. Spark is flexible in that it supports a variety of programming languages for data scientists and engineers to develop workflows in Python, Scala, R, and Java [1].

To-date, there have been a few attempts to demonstrate the computational power of Spark on learning highly complex models using large data sets of high dimensionality [1, 4, 8, 13, 14]. However, the experiments reported in these works fail to discuss the scalability issue as it specifically relates to the speed-up of a supervised machine learning system across distributed Spark clusters of varied sizes. The aim of this paper, therefore, is to introduce a robust and efficient big data machine learning system, which can perform real-time, distributed learning of large and complex data sets in a computationally efficient manner. The implemented system utilizes Apache Spark in order to take advantage of its low latency and powerful parallel processing framework. To highlight accuracy and efficiency of deep learning on Spark, two large real-world datasets were analyzed using several classification algorithms. To measure the predictive accuracy of the learners on each dataset, several statistically independent experiments were conducted on Spark clusters of 1 to 5 workers/executors. Finally, to assess efficiency of Spark's in-memory persistence mode, experiments were performed using both in-memory and on-disk processing.

The remainder of this paper is organized as follows. Section 2 provides a brief description of Spark's key architectural features and its machine learning library MLlib. Section 3 reports on the outcomes of performing deep learning across several Spark clusters of varying sizes.

2 Apache Spark

Apache Spark was designed to work in conjunction with Apache Hadoop in order to address some of its limitations and downsides such as high latency [1]. Hence, Hadoop lacks the ability to perform Streaming/Real-time analytics since streaming requires persistent in-memory data processing capabilities. Fault-intolerance is also a significant downside of Hadoop, specifically, as regards insufficient memory capacity [4].

Spark overcomes Hadoop's design inefficiencies by utilizing the Resilient Distributed Dataset (RDD) architecture to perform in-memory parallel computation, and it is fault tolerant [1, 10]. An RDD is a collection of elements partitioned across nodes of the cluster that can be operated on in parallel [3]. Optionally, users might persist an RDD in memory, allowing it to be reused efficiently across parallel operations. In cases of failure, RDDs are automatically recovered from the executor failures. RDDs can be referenced to a dataset in external data storage system such as Amazon S3, HDFS, HBase, or any other supported Hadoop input format or by transforming a collection of objects in the program to RDD. Furthermore, RDDs support popular types of transformations such as maps, filters and aggregations [3].

Apache Spark jobs run as independent sets of processes on a cluster, coordinated by the object of SparkContext in the driver program [1]. Specifically, SparkContext connects to either Spark's standalone cluster manager, or YARN (resource manager) on Hadoop to allocate resources across applications. Upon establishing a successful connection between SparkContext and cluster manager, Spark demands workers/executors to compute and store data for the driver. Figure 1 depicts a typical Spark cluster consisting of a master node or driver that runs the entire process with a minimum of one worker or executor [10].

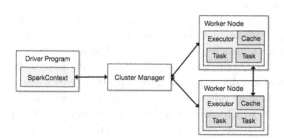

Fig. 1. A spark cluster

Spark's scalable Machine Learning Library (MLlib) consists of the most common learning algorithms and utilities, including classification, regression, clustering, collaborative filtering, and dimensionality reduction [3, 10]. MLlib revolves around the concept of pipelines, which allows users to combine multiple algorithms represented as an array of stages in a single pipeline to distribute stages to clusters of machines. Additionally, MLlib utilities support a variety of open source machine learning libraries and data formats such as LIBSVM datasets [1], which aids to perform faster and more efficient analysis since, it is a numeric dataset that excludes headers.

Having discussed the pertinent key features of Apache Spark in this section, Sect. 3 provides specific details as to how to learn from big data sets using Spark clusters of varying sizes and also how to implement classification evaluators for learning from big training data sets and making predictions on unknown test cases.

3 The Experiments

The main goal of the experiments conducted in this work was to assess scalability of big data machine learning on several configurations of Spark clusters. Specifically, the experiments were designed to directly measure relevant performance indices using two very large datasets and three machine learning algorithms. The following sections provide specific details of the utilized classification algorithms, the hardware configuration of the master and slave nodes in Spark clusters, and the obtained results for experiments conducted on each of the two big datasets.

3.1 Machine Learning Algorithms

The process of learning from big data and performing predictive analytics on Apache Spark comprises of two phases.

In Phase 1, datasets are randomly shuffled and split to perform training and testing; for instance, 70% of data may be used for training while the remaining 30% is set aside to evaluate the performance of the generated model. To take advantage of Spark's low latency, both training and testing datasets were made persistent in-memory to perform iterative operations during the learning phase. In addition, to further assess the advantages of in-memory RDD operations, the on-disk persistent model was used in each conducted experiment.

Phase 2 revolves around the concept of machine learning pipeline, which is chosen to combine multiple stages in a single pipeline, represented as an array of stages to scale and distribute to clusters of varying sizes. The first stage uses index values to perform evaluation of classification models, and datasets have to be enumerated and any arbitrary form of text data is not accepted. This is done by using the Vector-Indexer that indexes classification values to be predicted (*i.e.*, features) [5, 10]. The second stage is the classifier, which specifies which machine learning classification algorithm to use. For the experiments conducted in this work, it was decided to utilize three algorithms: Decision Tree, Logistic Regression, and Naïve Bayes.

A Decision Tree (DT) classifier is a machine learning algorithm that generates a predictive model to classify a dataset into target classes based on the training data. Each internal node in an induced DT represents a test on some property (feature) and leaf nodes represent classifications. DT performs well with large datasets, and the generated tree can help users easily visualize the diversion/flow of the data [9]. DT is slow because of representation structure of *replication and fragmentation,* and its ability to deal with missing values; hence, DT is known as Lazy Decision Tree [7].

Logistic Regression is a popular method to predict categorical responses. It is a special case of generalized linear models that predict the probability of outcomes [2]. This method can be used to predict binary (binomial) as well as multiclass (multinomial) outcomes. Multinomial logistic regression was used in the experiments reported here because of the many categorical groups of outcomes present in the training datasets.

Naïve Bayes is one of the most commonly used classifiers, that is based on Bayes Theorem to classify datasets by assuming that the presence of a particular feature in a class is unrelated to the presence of other features in the dataset [9].

Spark's MLlib provides Classification Evaluators with a suit of available metrics to evaluate the performance of its various classifier models [5]. The accuracy metrics used in this work to evaluate the actual versus the predicted outcomes are summarized in Fig. 2.

	Predicted = TRUE	Predicted = FALSE
Actual = TRUE	TP (True Positive)	FN (False Negative)
Actual = FALSE	FP (False Positive)	TP (True Negative)

Fig. 2. The confusion matrix

Spark's Multiclass Classification Evaluator was selected due to the fact that the two datasets utilized in this work were multinomial. The Positive Predictive Value (PPV) metric (see Eq. 1) was used to evaluate the accuracy of each classifier model since PPV considers the type of error, and hence it is the ideal metric to consider both the classification and misclassification rates to evaluate the pipelined transformations made by each classifier [5].

$$PPV = \frac{TP}{TP + FP} \tag{1}$$

3.2 Cluster Configurations

In order to assess Spark's efficiency, memory optimized EC2 (Elastic Compute 2) instances of Amazon Web Services (AWS) were created to distribute Spark jobs to cluster's executors, along with a general-purpose driver node on the cloud-based platform of Databricks, which is a unified analytics platform used to run Spark application on supported could platforms. In the experiments reported in the next section, memory-optimized instances were not utilized for the master node since computations do not reside on the driver, and hence, heavy instances are not required or necessary.

The Spark hardware instances used in all conducted experiments were identical in order to accurately and impartially perform timing analysis of both in-memory and on-disk scans of one to five executors including a driver. Specific cluster configurations are depicted in Fig. 3.

Driver Instances Type	General Purpose of m4.large
Worker Instances Type	Memory Optimized of r3.2xlarge
Worker Instances Memory/Core	61GB memory, 8 Cores
Driver Instances Memory/Core	8GB memory, 2 Cores
Operating System	Ubuntu 16.04
EBS Volume	General Purpose Solid State Drive

Fig. 3. Spark cluster configurations

3.3   Experiment 1

The dataset used in the first experiment is H-1B visa petitions, which includes over three million rows of data collected in 2011–2016 by the U.S. Department of Immigration Services [1, 12]. The dataset's unnecessary columns that had no relevance on the performance of the model (*e.g.* unique identifiers, dates, etc.) were removed; therefore, a total of seven features were selected along with the classification outcome *Case Status* as shown in Fig. 4.

Variable Name	Type
Company	~220K Categorical Classes/Features
SOC	~2K Categorical Classes/Features
Job Title	~60K Categorical Classes/Features
Full Time (0, 1)	Binomial Categorical Class/Feature
Wage (grouped to 5 groups)	5 Categorical Classes/Features
Year	6 Real-valued features
Worksite (grouped by state)	50 Categorical Classes/Features
Case Status	8 Classification labels

Fig. 4. H-1B visa petitions features

Timing analyses to learn from the training data and evaluate the generated model's classification accuracy required that classification labels and categorical data be enumerated (*i.e.,* indexed) and combined in a single vector. Thus, data cleaning was performed to format the dataset to LIBSVM and process it via Machine Learning Utilities (MLUtils) and MLlib Pipelines.

After converting the dataset into the required format, the three leaning models (see Sect. 3.1) were generated using clusters of one to five executors to assess Spark's performance and scalability. The results of timing analyses are depicted in Fig. 5, where each measurement is the average of five statistically independent runs for both in-memory and on-disk operations.

As shown in Fig. 5, there was a drastic difference between in-memory and on-disk operations, which can be directly attributed to Spark's low latency. Further, the most significant speedups occurred by increasing the number of executors from one to four. To determine whether further speedups would be possible, the Auto Scaling [1] feature,

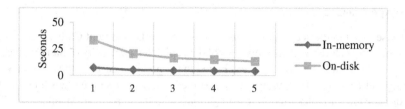

(a) Logistic Regression

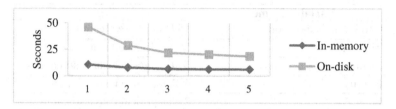

(b) Decision Tree

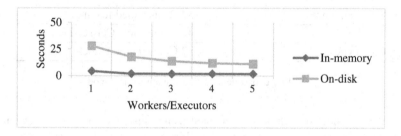

(c) Naïve Bayes

Fig. 5. H-1B visa data

which monitors the status of distributed jobs and automatically adjusts the number of executors on demand to improve performance, was enabled. However, no further speedups could be attained due to the size of the H-1B dataset.

Finally, the obtained classification accuracies for the three algorithms are shown in Fig. 6.

Learning Model	Classification Accuracy
Decision Tree	87%
Logistic Regression (multinomial) Maximum iterations = 3	87%
Naïve Bayes (multinomial) Smoothing = 1.0	56%

Fig. 6. Positive Predictive Value (PPV)

3.4   Experiment 2

The dataset used in the second experiment was the Fire Department Calls for Service of San Francisco, including 4.6 million rows of data collected by the city of San Francisco in 2015–2018 [6]. Similar to what was explained in the previous section, the necessary data cleaning and LIBSVM formatting was performed, and Fig. 7 summarizes the nine features or decision variables used in learning along with the classification label *Call Type*.

Variable Name	Type
Call Final Disposition	15 Real-valued features
Zip code	53 Real-valued features
Station Area	53 Categorical Classes/Features
Priority	6 Categorical Class/Feature
Final Priority	2 Categorical Classes/Features
ALS Unit	Binomial Real-valued features
Number of Alarms	5 Categorical Classes/Features
Unit Sequence	84 Real-valued features
Neighborhoods	42 Categorical Classes/Features
Call Type	32 Classification labels

Fig. 7. Fire department of San Francisco data model

The three leaning models were generated using clusters of one to five executors. The timing analyses results are depicted in Fig. 8, where each measurement is the average of five statistically independent runs for both in-memory and on-disk operations.

Considering the larger dimensionality of the dataset used in this experiment, the learning task for each algorithm required more time. Clearly, distributing very large datasets among clusters with multiple executors is more efficient for big data, which naturally require more time for processing (*e.g.* 1 min or more).

As depicted in Fig. 8, there were generally significant speedups measured for larger clusters of executors for in-memory but especially for on-disk operations. Logistic regression was the slowest algorithms due to the fact that in order to obtain reasonably high classification accuracy the number of iterations had to be increased to 20. Figure 9 summarizes the classification accuracies obtained from the three classifiers.

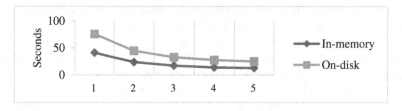

(a)   Logistic Regression

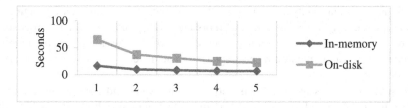

(b)   Decision Tree

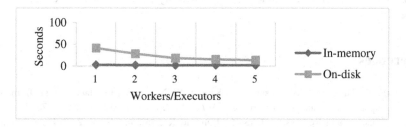

(c)   Naïve Bayes

Fig. 8. San Francisco fire department data

Learning Model	Classification Accuracy
Decision Tree	73%
Logistic Regression (multinomial) Maximum iterations = 20	69%
Naïve Bayes (multinomial) Smoothing = 500.0	52%

Fig. 9. Positive Predictive Value (PPV)

4   Conclusions

In this paper, it was demonstrated how a distributed system can be created to efficiently perform Big Data machine learning using various classification algorithms and very large datasets. Specifically, it was shown that Apache Spark's Machine Learning Library

(MLlib) on Databricks can be utilized using several instances (executors) residing on the Elastic Compute Cloud (EC2) of Amazon Web Services (AWS).

In addition to performing predictive analytics on different numbers of executors, both in-memory and on-table scans were utilized to assess the scalability and computing efficiency of Spark. The conducted experiments, which were run multiple times on several instances and executors, demonstrated how to parallelize executions as well as to perform in-memory processing to drastically improve a learning system's performance. To highlight the advantages of the proposed system, two very large data sets and three different supervised classification algorithms were used in each experiment. The obtained timing analyses confirmed that significant speedups can be attained by manually increasing the number of executors across the cluster. Further, it was shown that by using Spark's auto scaling feature, which increases the configuration of instances on demand, maximum speedups can be achieved depending upon the size of the training data.

In summary, the size of datasets chosen in this work did not warrant the use of larger clusters of executor/worker nodes since maximum speedups were achieved by using clusters of size three and four. However, in terms of future directions of this research, it will be beneficial to assess the scalability issue on much larger datasets and clusters.

References

1. Gupta, A., Thakur, H., Shrivastava, R., Kumar, P., Nag, S.: A big data analysis framework using apache spark and deep learning (2017). https://doi.org/10.1109/icdmw.2017.9
2. Classification and Regression: Classification and Regression - Spark 2.2.0 Documentation. https://spark.apache.org/docs/2.2.0/ml-classification-regression.html. Accessed 13 Mar 2018
3. Harnie, D., et al.: 2015 15th IEEE/ACM International Symposium on Cluster, Cloud and Grid Computing, pp. 871–879. IEEE (2015). http://ieeexplore.ieee.org/document/7152571/
4. Miji, D., Varga, E., Member, S.: Machine Learning Driven Responsible Gaming Framework with Apache Spark, pp. 31–34 (2017)
5. Evaluation Metrics - RDD-based API. Evaluation Metrics - RDD-based API - Spark 2.2.0 Documentation. https://spark.apache.org/docs/2.2.0/mllib-evaluation-metrics.html. Accessed 13 Mar 2018
6. Fire Department Calls for Service. Open Data of San Francisco. https://data.sfgov.org/Public-Safety/Fire-Department-Calls-for-Service/nuek-vuh3. Accessed 7 Feb 2018
7. Friedman, J.H.: Lazy decision trees. AAAI **34**, 167–180 (1997)
8. Berral-garcía, J.L.: A quick view on current techniques and machine learning algorithms for big data analytics. In: 18th International Conference on Transparent Optical Networks (ICTON), pp. 1–4 (2016)
9. Vimalkumar, K., Radhika, N.: A big data framework for intrusion detection, pp. 198–204 (2017)
10. Wang, K., Fu, J., Wang, K.: SPARK – a big data processing platform for machine learning. In: 2016 International Conference on Industrial Informatics - Computing Technology, Intelligent Technology, Industrial Information Integration, pp. 48–51 (2016)
11. Capuccini, M., Carlsson, L., Norinder, U., Spjuth, O.: Proceedings of 2015 2nd IEEE/ACM International Symposium on Big Data Computing, BDC 2015. Institute of Electrical and Electronics Engineers Inc., pp. 61–67 (2016)

12. Naribole, S.: H-1B Visa Petitions 2011–2016. In: H-1B Visa Petitions 2011–2016 | Kaggle (2017). https://www.kaggle.com/nsharan/h-1b-visa/version/2. Accessed 15 Mar 2018
13. Alfred, R.: [Plenary Speaker] The Rise of Machine Learning for Big Data Analytics, 2016 (2016)
14. Haupt, S.E., Kosovic, B.: Big data and machine learning for applied weather forecasts: Forecasting solar power for utility operations. In: Proceedings of 2015 IEEE Symposium Series on Computational Intelligence, SSCI 2015, pp. 496–501 (2016)
15. Mall, S., Rana, S.: Overview of big data and hadoop. Imperial J. Interdisc. Res. 2(5), 1399–1406 (2016)
16. Biku, T., Rao, N., Akepogu, A.: Hadoop based feature selection and decision making models on big data. Indian J. Sci. Technol. 9(10) (2016). https://doi.org/10.17485/ijst/2016/v9i10/88905
17. Liu, T., Fang, Z., Zhao, C., Zhou, Y.: Proceedings of 2016 IEEE/ACIS 15th International Conference on Computer and Information Science, ICIS 2016. Institute of Electrical and Electronics Engineers Inc. (2016)
18. Eluri, V., Ramesh, M., Al-Jabri, A., Jane, M.: A comparative study of various clustering techniques on big data sets using Apache Mahout. In: International Conference on Big Data and Smart City (ICBDSC), pp. 1–4 (2016). https://doi.org/10.1109/icbdsc.2016.7460397

Development of a Big Data Platform for Analysis of Road Driving Environment Using Vehicle Sensing Data and Public Data

Intaek Jung and Kyusoo Chong[(✉)]

Department of Future Technology and Convergence Research,
Korea Institute of Civil Engineering and Building Technology,
Goyang, Gyeonggi-do, Republic of Korea
{jungintaek,ksc}@kict.re.kr

Abstract. The driving environment on the road can be rapidly changed due to various event factors such as bad weather, traffic accident, congestion, etc. If drivers fail to recognize these dangerous situations in advance, they can lead to major traffic accidents. For this purpose, it is very important to provide real-time driving environment information to drivers. In order to provide real-time driving environment information, data collection devices are required. Current data collection devices are types of fixed collection systems that collect driving environment data at specific points or intervals. This fixed collection system is limited in time and space, and if it collects all the roads nationwide, there is a huge installation cost. In order to overcome the limitations of the fixed data collection system, this study utilize vehicle sensing data collected from individual vehicle sensors as a mobile collector. Since the vehicle sensing data collected from individual vehicles nationwide correspond to the spatial big data, an analysis system for processing big data is needed. Since this analysis system should utilize various collection data such as public data as well as vehicle sensing data, it should be developed in a platform form considering data scalability. Therefore, this study developed a big data platform for collecting, storing, processing, analyzing and visualizing various kinds of big data such as vehicle sensing data and public data. The development platform consists of H/W and S/W, and it is applied for providing real time driving environment information and analyzing/forecasting driving environment information including road surface freezing, road rainfall/snowfall, incident situation, traffic congestion, etc.

Keywords: Big data · Platform · Road driving environment
Vehicle sensing data · Public data

1 Introduction

Due to the recent development of information and communication technology, digital data generated from various collectors such as web, mobile, IOT, various smart devices, and social media is increasing exponentially. According to the International Data Corporation (IDC) report, the global digital data generated in 2020 will amount to

© Springer International Publishing AG, part of Springer Nature 2018
F. Y. L. Chin et al. (Eds.): BIGDATA 2018, LNCS 10968, pp. 220–234, 2018.
https://doi.org/10.1007/978-3-319-94301-5_17

44 trillion gigabytes [1]. Now is the time for a big data age to deal with large volumes of data beyond the data collection and storage technologies of existing systems. Since the types of collected data are various, a system capable of collecting and storing various types of big data is needed.

The driving environment on the road can be rapidly changed due to various event factors such as bad weather, traffic accident, congestion, etc. If drivers fail to recognize these dangerous situations in advance, they can lead to major traffic accidents. For this purpose, it is very important to provide real-time driving environment information to drivers. In order to provide real-time driving environment information, data collection devices are required. Current data collection devices are types of fixed collection systems that collect driving environment data at specific points or intervals. This fixed collection system is limited in time and space, and if it collects all the roads nationwide, there is a huge installation cost. In order to overcome the limitations of the fixed data collection system, this study utilize vehicle sensing data collected from individual vehicle sensors as a mobile collector. Since the vehicle sensing data collected from individual vehicles nationwide correspond to the spatial big data, an analysis system for processing big data is needed. Since this analysis system should utilize various collection data such as public data as well as vehicle sensing data, it should be developed in a platform form considering data scalability. The government has been pushing public data to use openly by the private sector since 2013. According to a public data portal, the number of public data releases increased from 5,272 in 2013 to 23,084 in 2017, about 4 times more than in 2013. The number of public data applications also increased from 13,923 in 2013 to 3,505,731 in 2017, about 252 times more than in 2013. This shows that the use of public data is continuously increasing [2].

Therefore, this study aims to build a big data platform to provide various driving environment information by using various big data such as vehicle sensing data and public data. We develop H/W and S/W for collection, storage, processing, analysis and visualization of various big data. Through the web GIS service platform under development, we aim to provide real time driving environment information and analyzing/forecasting driving environment information including road surface freezing, road rain-fall/snowfall, incident situation, traffic congestion, etc.

2 Literature Review

2.1 H/W and S/W for Building Big Data Platform

Recently, as shown in Fig. 1, H/W and S/W technologies have been developed to efficiently process big data, which is increasing exponentially. First, major technology development trends of Big Data Platform H/W include processor core of high performance and low power, process for large data processing and memory integrated computing technology, high performance and low power next generation memory device (STT-MRAM, ReRAM, PCRAM etc.), commercialization of artificial intelligence computers.

Next, major technology development trends of Big Data Platform S/W are as follows. Big data collection technologies include Crawling, Open API, File Transfer

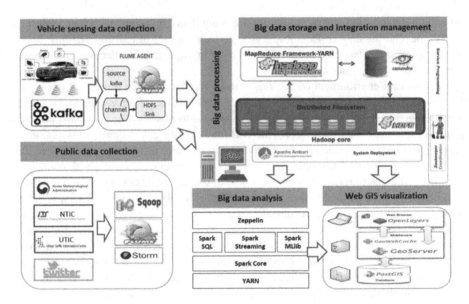

Fig. 1. Development structure of driving environment analysis platform

Protocol (FTP), Really Simple Syndication (RSS), Streaming, and Log collectors. Among these, log collectors include Flume, Scribe, and Chukwa. Big data distribution processing and storage technologies include disk or in-memory distributed processing techniques such as Hadoop Distributed File System (HDFS), Spark, and Storm. The Big Data database technology includes Relational Database Building (RDB) technologies such as MySQL, PostgreSQL and NoSQL database technologies such as Hbase, MongoDB, and Cassandra. Big data analysis techniques include various open sources such as Mahout, Zeppelin, and R. Finally, big data visualization techniques include Prefuse, D3.js, Node.js, and Matplotlib [3, 5–10] (Table 1).

Table 1. Development trend of H/W and S/W for building big data platform

		Major development trends
H/W	Processor core	Emerging processor core, High performance and low power for processor cores
	Processor-memory integrated computing	Increased the need for processor-memory integrated computing for big data processing
	Computer design using next generation memory	Development of high performance and low power next generation memory Commercialization of next-generation memory and market growth
	Computing solution based on artificial intelligence	Perceptual computing, Unstructured database management system, High performance computing Commercialization of artificial intelligence computer

(*continued*)

Table 1. (*continued*)

		Major development trends
M/W	Big data collection	Data collection technologies based on various types of data such as crawling, Open API, FTP, RSS, Streaming, Log collectors, etc.
	Big data storage and processing	Distributed processing technology based on disk or in-memory such as HDFS, Spark, Storm, etc.
	Database construction using big data	RDB (MySQL, PostgreSQL, etc.) NoSQL DB (Hbase, MongoDB, Cassandra, etc.)
	Big data analysis	Open sources for big data analysis (Mahout, Zeppelin, R, etc.)
	Big data visualization	Technologies for visualizing big data information (Prefuse, D3.js, Node.js, Matplotlib, etc.)

As mentioned above, the Big Data Platform H/W and S/W currently have sufficient technology to process large volumes of data. Especially, S/W is open source type and should be selected according to the type and structure of collected data and then further development. In addition, according to the service information provided to the end user, it is necessary to develop platform H/W specification and S/W for data collection and storage, processing and analysis, and information visualization.

2.2 Service Technology Using Big Data Platform

As shown in Table 2, the trend of service technology using big fata platform was examined in three aspects such as data collection, data analysis and prediction, and visualization. First, in terms of data collection, most of them are using their own data collection system that collects real-time data and external data. On the other hand, On the other hand, weather data collection is collected from external systems, not from their own systems. There is no use of vehicle sensing data collected from individual vehicle sensors. In terms of data analysis and prediction, information analysis and prediction are mainly performed based on historical data. However, the use of statistical analysis tools and fusion analysis of traffic data and weather data are insignificant. Finally, in terms of visualization, most information is visualized using GIS, tables, and graphs [3].

Therefore, this study develop a service platform to provide real-time and predicted driving environment information through analysis of fusion with various big data, including vehicle sensing data that are not utilized by existing systems. In addition, a statistical analysis tool that utilizes a variety of collection data and an information visualization system based on Web GIS for displaying analysis results are also needed.

Table 2. Technology trend of big data platform service

		Domestic system				Oversea system			
		NTIC	ROADPLUS	UTIC	TOPIS	RITIS	NPMRDS	VICS	TCC
Data collection	Real-time data collection	O	O	O	O	O	O	O	O
	Internal collection system	O	O	O	O	×	O	O	O
	Use of external data	O	O	O	O	O	O	×	O
	Weather data collection	×	×	O	O	O	×	×	O
	Vehicle sensing data	×	×	×	×	×	×	×	×
Data analysis and prediction	Historical data inquiry	O	O	O	O	O	×	×	×
	Statistical analysis tool	×	×	×	×	O	O	×	×
	Prediction/Forecasting	×	O	O	O	O	×	×	O
	Convergence Analysis with Weather Data	×	×	O	×	O	×	×	O
Information visualization	GIS visualization	O	O	O	O	O	O	O	O
	Table and graph visualization	O	O	O	O	O	O	×	O

※ NTIC (National Transport Information Center), ROADPLUS (Expressway Traffic Information Center), UTIC (Urban Traffic Information Center), TOPIS (Transport Operation and Information Service), RITIS (Regional Integrated Transportation System), NPMRDS (National Performance Management Research Data Set), VICS (Vehicle Information and Communication System), TCC (Traffic Control Center)

3   Development of Driving Environment Analysis Platform

3.1   Platform Development Concepts

This study is to develop a big data platform for the analysis of driving environment using various big data such as vehicle sensing data and public data. Vehicle sensing data is data collected from individual vehicle sensors (GPS, temperature and humidity sensor, radar, camera, etc.) and refers to coordinates, temperature and humidity, radar data, images, and the like. Public data is open data provided by public institutions and refers to traffic situation information, weather information, traffic accident and road construction information, and SNS information, etc. The driving environment information generated using vehicle sensing data is road surface temperature (°C), rainfall/snowfall (mm/h), and traffic density (veh/km). The development platform processes and analyzes the generated travel environment information to provide real-time driving environment service information such as road surface freezing, road rainfall/snowfall, incident situation, traffic congestion, etc.

The development platform consists of H/W and S/W. First, platform H/W was developed as a small platform server with parallel processing structure for big data distribution processing. Next, platform S/W was developed by each individual program for collecting, storing, processing, analyzing, and visualizing various types of big data. The overall structure of the development environment for this study is shown in Fig. 1.

3.2   Development of Platform H/W

Big Data Platform H/W is more efficient to scale-out multiple physical resources than scale-up one physical resource. This means that the scale-out method is easier to handle big data distribution than the scale-up method. When large servers are used for big data distribution processing, there are space problems of servers and high cost problems. If you use a regular PC, there is a processing speed problem for collecting and storing large amounts of data.

In this study, as shown in Fig. 2, 10 physical server nodes were installed and the small platform server with a parallel processing structure capable of running these server nodes was developed. Physical server nodes consisted of one master node and nine slave nodes. Physical server nodes can expand nodes according to the specifications that the user wants. If one or more of the ten nodes fail, a loss of node data occurs. In order to prevent such data loss, H/W is designed based on cluster. In other words, according to the hadoop cluster guideline, the name node is designed considering the disk backup system physically through redundancy like the secondary name node. Since data node can replicate three times in Hadoop by itself, we did not configure physical disk backup separately.

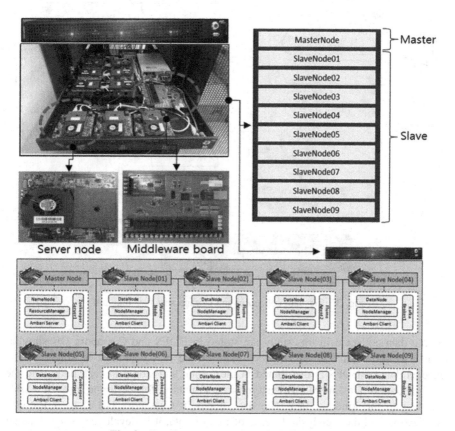

Fig. 2. Development structure of platform H/W

3.3   Development of Platform S/W

Big Data Collection and Storage

The data collected in this study can be classified into vehicle sensing data and public data as shown in Fig. 3. Vehicle sensing data is transmitted in the form of stream data through a wireless communication network. In order to efficiently collect such stream data, we have developed a collection interface using open source Kafka and Flume, which have the advantage of unstructured stream data collection. Public data is divided into real-time data and historical data. Real-time data is collected by Flume using Open API and historical data is collected by Sqoop.

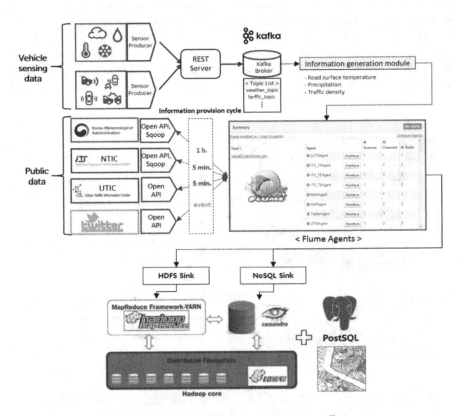

Fig. 3. Big data collection and storage structure diagram

The big data collected in this study are stored in HDFS and Cassandra DB simultaneously as shown in Fig. 3, and the usage of the two databases is different. The collected data of HDFS is distributed in file format and stored in each node, and the collected data of Cassandra DB is stored in a structured format in the form of a table according to the kind of data. The data stored in Cassandra DB is used for web visualization and data analysis. In case of loss of data, it is developed as a redundant structure to utilize the file type data stored in HDFS. In addition, PostSQL is applied to link with GIS spatial information, and S/W which can visualize to users the spatial information stored in PostSQL and the result analyzed by HDFS is developed.

Big Data Processing Based on Grid Index

Big data processing S/W can be divided into spatial unit matching and time interval interpolation and aggregation as shown in Fig. 4. Spatial unit matching refers to a method of matching coordinates (vehicle sensing data, etc.) and surface (administrative area, etc.) data to GIS link data. In order to match the vehicle sensing data of the coordinate unit collected in real time from the individual vehicle sensor to the GIS link unit, a fast spatial calculation is required. In this study, spatial unit matching is performed by applying spatial information processing technique based on grid index [11]. It is not a method to store and manage spatial big data in an internal system, but it is developed to manage various kinds of spatial big data efficiently by managing only an index for linking with other spatial information. Public data with different information providing cycles (1 h or 3 h) were applied linear interpolation method to interpolate the data at the aggregation time interval of this platform (every 5 min). The median value was used as a representative value for counting the previously matched link data in 5 min time intervals. The minimum number of samples is 13.4 by applying the central limit theorem at the 5-min aggregation interval, assuming that a sufficient number of samples are collected without assuming a specific distribution [12]. Therefore, the minimum number of samples for estimating the representative value at each 5-min time interval was 20 or more.

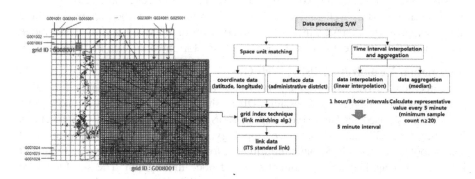

Fig. 4. Big data processing structure based on grid index

Driving Environment Analysis Tool Based on Platform DB

We developed an analytical tool based on a zeppelin notebook using a platform DB as shown in Fig. 5. This analysis tool is used for evaluation of development algorithm and development of future driving environment prediction model. Zeppelin provides the convenience of the development environment by providing Web-based spark envi-

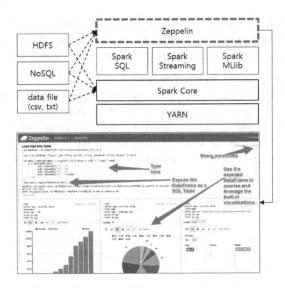

Fig. 5. Development concept of driving environment analysis tool

ronment developed in the existing shell command environment. It also provides the ability to visualize and share the results analyzed by spark code directly in Web GIS. Zeppelin can use the spark engine through the function called interpreter, and it is possible to link various frameworks such as Hive, Pig, MapReduce, and HDFS, HBase, and Cassandra DB as well as Spark. The spark framework can be deployed in three modes: Apache yarn, Apache mesos, and Standalone. In this study, a yarn-cluster mode was constructed for integration and operation with Hortonworks Data Platform (HDP).

Information Visualization Using Web GIS

We constructed a Web GIS engine to provide various driving environment information to users by using HDFS and Cassandra DB as shown in Fig. 6. The Web GIS engine framework is based on GeoServer, and PostGIS supports a variety of GIS functions including OpenGIS support, advanced topology, user interface tools, and Web-based access tools. The basic GIS spatial information for visualizing the driving environment information is applied to the national standard node link system. The shp files were converted into geometry information using PostgreSQL. The converted information is created in the national standard link table in PostgreSQL. A separate statistical table was constructed to effectively connect various big data and RDBMS, and a national standard node link table and dynamic view were created and visualized using Geo-Server. Finally, the web user interface is divided into three types such as real-time driving environment information, historical data analysis, and driving environment analysis tool.

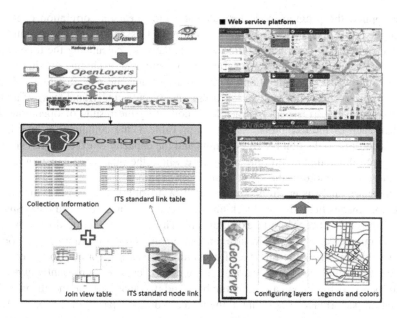

Fig. 6. Development concept for Web GIS-based information visualization

4 Performance Evaluation of Development Platform

Currently, this platform is not fully developed, and performance evaluation of Platform H/W and Big Data collection, storage, processing, analysis and visualization S/W will be performed in the future. We will compare and evaluate the advantages and disadvantages of open source applied to each development software with other open sources. In this section, we present the evaluation results of the performance of the operations performed so far as follows.

4.1 Computational Performance Test Based on Server Memory and Core in a Spark Environment

In order to examine the computational performance according to the memory capacity and the number of cores for the development server, we constructed a spark analysis environment using 10 physical nodes in the development server. The basic information for performance evaluation is as follows.

- Spark version: 1.6.1
- Spark operating mode: YARN-Cluster
- Cluster memory: 216 GB
- Cluster core: 27 EA
- Number of test data sets: 874,871 EA
- Test dataset size: 70 MB
- Test calculation model: Pearson correlation coefficient

We applied the program code for calculating the Pearson correlation coefficient between the traffic speed and the rainfall stored in the Cassandra DB for computational performance test. The computation time was calculated by calculating the time to start and end the calculation of the Pearson correlation coefficient, excluding the time to read the data. Spark performs operations with the driver and the executor of the logical execution unit. An executor is a work process that executes individual tasks of a given spark task, and the drive performs a role of collecting the results of operations performed in each of the executors. As shown in Table 3, the computation time is calculated according to the number of executors and the capacity of executor RAM. Also, we compared the computation time in the case where the number of executor cores is single and dual or more in the spark analysis environment.

In the spark analysis environment for big data distribution processing, the number of executors means the individual task units that can distribute collected data. Spark is executed in parallel according to executor unit. This executor can be driven by specifying the number of cores, RAM size, and so on. As shown in Table 3, the computation time decreases as the number of executors increases according to the executor RAM capacity, and the computation time does not change significantly when the executor becomes more than a certain number. And it is analyzed that the change of the computation performance according to the increase of the capacity of the exciter RAM is insignificant according to the number of each executor. As a result of comparing the computation time in the single core environment and dual core environment according to the number of the executors, the computation time decreases as the number of the executors increases by the number of cores, and it can be seen that the computation time increases if the number is more than a certain number. The computation time decreases when the number of cores is two than that of one executor, but the difference in computation time is small when the number of cores is two or three. Therefore, it is necessary to optimize the number of executors in the spark analysis environment in order to efficiently distribute big data when expanding or modifying collected data in the future.

Table 3. Computation time by capacity of executor RAM and number of executor core (sec)

		Capacity of the executor RAM (GB)				Number of executor cores		
		1	2	3	20	1 core	2 cores	3 cores
Number of executors (EA)	2	66.681	67.051	67.098	68.152	66.681	43.363	39.164
	4	42.224	41.083	41.615	43.240	42.224	31.461	27.580
	6	33.268	33.037	33.291	33.540	33.268	27.434	24.462
	8	29.109	29.271	29.662	29.204	29.109	25.044	25.884
	10	28.191	28.241	29.125	30.331	28.191	27.606	25.588
	12	28.907	27.965	28.622	–	28.907	27.720	27.287
	14	28.951	28.760	28.909	–	28.951	29.323	34.822
	16	29.038	29.507	29.810	–	29.038	30.667	29.237
	18	30.031	30.122	30.428	–	30.031	30.326	29.479
	20	32.591	30.935	32.187	–	32.591	33.019	35.840

4.2   Comparison of Computational Performance with Other Organization' Cloud Computing

In order to evaluate the performance of small platform server in this study, 10 virtual servers were created in cloud computing environment of other organization and cluster analysis environment was constructed as same as development server. The 10 instances (servers) were also created in the virtual server, and the big data cluster and software of this development server were installed in the same way. Table 4 shows the basic construction information of the development server and the other organization server for comparison of the computation performance.

Table 4. Basic construction information of two servers for compute performance comparison

	Development server	Comparison server
Operating system	Ubuntu 14.04	Ubuntu 14.04
CPU (node)	i5-4 core	Xeon-10 core
RAM (node)	32 GB	64 GB
DISK (node)	500 GB (SSD)	500 GB (SAS)
Number of nodes	10개	10개
HDP version	2.5	2.5
Spark version	1.6.2	1.6.2
Network bandwidth	1G	1G

We applied the kNN algorithm for performance evaluation of development server. This algorithm is based on the fact that the computation time becomes longer as the value of each data item increases. Figure 7 shows the flow diagram of the kNN algorithm for comparing the performance of the two servers [13]. The data size is 5 GB, and the performance of the two servers are compared according to the number of executor cores and executor RAM capacity, which are various environment variables of the spark analysis environment. Table 5 shows the evaluation results of the two servers. In the spark framework of the two servers, the executor was fixed to 10, and the kNN algorithm was performed while varying the various environment variables (RAM capacity, number of cores) of the executor and measuring each computation time. As a result, it is analyzed that the performance of the development server is better than the comparison server in all the analysis environments. When the same size of data and algorithm are executed in the comparison server of the other institutions constructed in the same environment as the development server, it is analyzed that the development server reduces the computation time more than the comparison server.

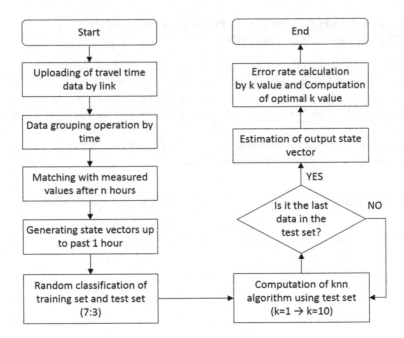

Fig. 7. Flowchart of the kNN algorithm

Table 5. Performance evaluation results of both servers

Number of core		1 core			2 core		
Capacity of RAM (GB)		1	8	16	1	8	16
Computation time (sec)	Development server(a)	125.29	123.28	132.86	78.62	74.48	71.87
	Comparison server(b)	214.96	213.18	207.70	102.24	89.81	90.86
	Gap((a)–(b))	–89.67	–89.89	–74.84	–23.61	–15.33	–18.98

5   Conclusion and Future Research

In this study, we developed a big data platform for forecasting and analyzing real-time driving environment information and future driving environment information using vehicle sensing data and public data. First, we developed a small platform server with a parallel processing structure for various types of large data distribution processing and applied it as a platform H/W. Second, we developed each platform S/W using open source to collect, store, process, analyze and visualize vehicle sensing data and public data. Currently, the development of the platform is not completed, and various studies on the platform H/W and S/W advancement are underway. As a result of performance evaluation of the platforms performed so far, it is evaluated that development platform is superior to other similar platforms in terms of operation speed.

For future research, performance evaluation of Platform H/W and Big Data collection, storage, processing, analysis and visualization S/W will be performed in the future. We will compare and evaluate the advantages and disadvantages of open source applied to each development software with other open sources. It is necessary to select the test bed area first and verify the information infrastructure technology of this study to construct the big data platform by using specific vehicles such as probe cars and business vehicles in the area. In addition, it is necessary to construct a pilot platform for collecting and storing large amounts of data considering the case where the individual vehicle sensing data is expanded nationwide, and it is necessary to support a standard protocol considering the IOT standard framework for data collection using the wireless communication network.

Through the future development platform, we want to quickly and accurately communicate information about the driving environment events (road surface freezing, road rain-fall/snowfall, incident situation, traffic congestion, etc.) that occur locally at every moment. The road managers are expected to provide basic data and analysis tools for real-time driving environment monitoring services and road operation evaluation.

Acknowledgements. This research was supported by a grant from a fusion research project (Development of Driving Environment Observation, Prediction and Safety Technology based on Automotive Sensors) funded by the Korea Institute of Civil Engineering and Building Technology.

References

1. Gantz, J., Reinsel, D.: The digital universe in 2020: big data, bigger digital shadows, and biggest growth in the far east, pp. 1–3. International Data Corporation Analyze the Future (2012)
2. National Information Society Agency, Open Data Portal. https://www.data.go.kr/
3. Jung, I., Chong, K.: Development of platform for driving environment analysis using vehicle sensing and public big data. Transp. Technol. Policy 14(4), 10–19 (2017)
4. Son, M.: Analysis of the domestic/foreign technology development and market outlook of promising industry related to artificial intelligence/big data, pp. 241–271. Knowledge Industry Information Institute (2016)
5. Lee, J.: Big data technology trend. Hallym ICT Policy J. 2, 14–19 (2015)
6. Kim, J.: Big data utilization and related technique and technology analysis. Korea Contents Assoc. J. 10(1), 34–40 (2012)
7. Shvachko, K., Kuang, H., Radia, S., Chansler, R.: The Hadoop distributed file system. In: Proceedings of the 2010 IEEE 26th Symposium on Mass Storage Systems and Technologies (MSST), pp. 1–10 (2010). https://doi.org/10.1109/MSST.2010.5496972
8. Lakshman, A., Malik, P.: Cassandra: a decentralized structured storage system. ACM SIGOPS Oper. Syst. Rev. 44(2), 35–40 (2010). https://doi.org/10.1145/1773912.1773922
9. Padhy, R., Patra, M., Satapathy, S.: RDBMS to NoSQL: reviewing some next-generation non-relational database's. Int. J. Adv. Eng. Sci. Technol. 11(1), 15–30 (2011)
10. Seo, Y., Kim, W.: Information visualization process for spatial big data. J. Korea Spat. Inf. Soc. 23(6), 109–116 (2015). https://doi.org/10.12672/ksis.2015.23.6.109

11. Singh, H., Bawa, S.: A survey of traditional and MapReduce based spatial query processing approaches. ACM SIGMOD Rec. **46**(2), 18–29 (2017). https://doi.org/10.1145/3137586. 3137590
12. Shim, S., Choi, K., Lee, S., Namkoong, S.: An expressway path travel time estimation using hi-pass DSRC off-line travel data. J. Korean Soc. Transp. **31**(3), 45–54 (2013). https://doi. org/10.7470/jkst.2013.31.3.045
13. Jung, I.: AADT estimation of unobserved road segments using GPS vehicle trip data, Seoul National University Ph.D. thesis (2016)

Application Track: BigData Practices

Application Track 1.1 Java Practice

Big Data Framework for Finding Patterns in Multi-market Trading Data

Daya Ram Budhathoki, Dipankar Dasgupta$^{(\boxtimes)}$, and Pankaj Jain

University of Memphis, Memphis, TN, USA
{dbdhthki,ddasgupt,pjain}@memphis.edu

Abstract. In the United States, multimarket trading is becoming very popular for investors, professionals and high-frequency traders. This research focuses on 13 exchanges and applies data mining algorithm, an unsupervised machine learning technique for discovering the relationships between stock exchanges. In this work, we used an association rule (FP-growth) algorithm for finding trading pattern in exchanges. Thirty days NYSE Trade and Quote (TAQ) data were used for these experiments. We implemented a big data framework of Spark clusters on the top of Hadoop to conduct the experiment. The rules and co-relations found in this work seems promising and can be used by the investors and traders to make a decision.

Keywords: Multimarket · Exchanges · Association rules
FP-Growth · Hadoop · Spark · TAQ · Clusters

1 Introduction

In Multimarket, securities listed in one exchange can also be listed in another exchange, and it can be traded on more than one exchanges [10]. Small liquidity traders generally trade in one trade but large traders split their trades across markets. In this work, we apply a data mining technique based on an FP-Growth algorithm to find out an interesting pattern in multimarket trading.

1.1 Security Information Processor (SIP)

US equities market is highly competitive and fragmented consisting of 13 exchanges and about 40–50 Alternative Trading System (ATS)/dark pools. Security Information Processor (SIP) was created to have a National Market System where investors and professionals can have access to the real time information related to quote (bid and offer) and trade (Fig. 1).

The SIP is operated by NASDAQ and New York Stock exchanges which creates a real-time consolidated record of every exchanges. The SIP is a central consolidated and live stream aggregator, where every stock exchange and ATS sends data stream of the best quotes (bid and offer) and updates public price

© Springer International Publishing AG, part of Springer Nature 2018
F. Y. L. Chin et al. (Eds.): BIGDATA 2018, LNCS 10968, pp. 237–250, 2018.
https://doi.org/10.1007/978-3-319-94301-5_18

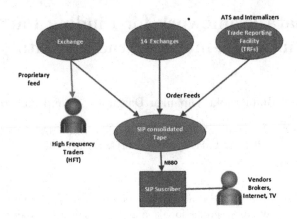

Fig. 1. Architecture of Security Information Processor (SIP)

quotes called the "National Best Bid and Offer" (NBBO) continually. In order to create NBBO, SIP compiles all of the bids and offers for all U.S. stock in one place. All of the exchanges piped the bids and offers into the SIP, and the SIP ultimately calculates the NBBO. SIP is a very easy way to get the current status of the market and acts as a benchmark to determine the NBBO.

Table 1. US stock exchanges

Code	Description	Code	Description
A	NYSE MKT LLC	P	NYSE Arca, Inc.
B	NASDAQ OMX BX, Inc.	S	ConsolidatedTape System
C	National Stock Exchange Inc. (NSX)	T	NASDAQ Stock Exchange, LLC (in Tape A, B securities)
D	Financial Industry Regulatory Authority, Inc. (FINRA ADF)	Q	NASDAQ Stock Exchange, LLC (in Tape C securities)
I	International Securities Exchange, LLC (ISE)	V	The Investors' Exchange, LLC (IEX)
J	Bats EDGA Exchange, INC	W	Chicago Broad Options Exchange, Inc. (CBOE)
K	Bats EDGX Exchange, Inc.	X	NASDAQ OMX PSX, Inc. LLC
M	Chicago Stock Exchange, Inc. (CHX)	Y	Bats BYX Exchange, Inc.
N	New York Stock Exchange LLC	Z	Bats BZX Exchange, Inc.

Table 1 shows the list of exchanges and their corresponding symbols. Exchanges can be classified in terms of Maker-Taker [1].

Table 2. Pricing models

Exchange	Pricing model
NYSE	Maker-taker
ARCA	Maker-taker
Nasdaq PSX	Maker-taker
Direct Edge X	Maker-taker
BATS	Maker-taker
NASDAQ	Maker-taker
Boston	Inverse-Maker-taker
BATS-Y	Inverse-Maker-taker
Direct Edge A	Maker-taker

2 Related Work

Several works have been done in the financial stock market using Association rule mining. Frequent patterns play important roles in association rule mining, finding correlations, and other interesting relationships among data stream [6, 12]. Asadifar [9] presents the application of association rules to predict the stock market. Other works [20, 22] used the Deep Learning approach for sentiment analysis and Data Mining of the financial Big Data. In addition, [22] presents Data Mining with Big Data. The focus of our work to build a Big Data framework to process the financial trading data efficiently and apply the association rules to seek the dominance patterns among exchanges (Table 2).

3 Background

3.1 Big Data Analytic

In earlier days, financial data were relatively small as most exchanges only reported Open, High, Low and Close (OHLC) at the end of each day. Now with the advent of high frequency trading in the financial market, the importance of Big Data in finance is increasing day by dayand such data [11] can be characterized by the 5V's of Big data as shown in Fig. 2).

1. **Variety:** It refers to the limitless variety of Big Data. Financial data can be either structured or unstructured. Structured data refers to the information which has fixed structure and length and can be easily represented in a tables in the form of rows and columns. The unstructured data can not be organized into a table (with rows and columns) and does not fall in a pre-determined model. Examples include gathered data from the social media posts, logs and even audio and videos.

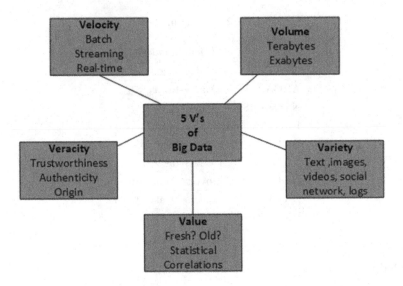

Fig. 2. 5V's of stock market big data [2]

2. **Veracity:** It refers to the truthfulness or accuracy of the data. It can be defined as the bias or abnormality in the data. Generally, 40 to 50% of the time is spent on data preparation cleansing.
3. **Volume:** It refers to the vast amount of data generated every seconds. It could be in Terabytes, Zettabytes or even higher. For example, single day NYSE quote file for August 24, 2017 is 203 GB. The high-frequency data of the financial stock market at each day consists of the information equivalent to 30 years of daily data [8].
4. **Velocity**: Velocity refers to the speed at which data is being generated or produced. As an example, we can think of social media messages that goes viral in few seconds. In the context of financial market, it can be thought of as a high-frequency trading data generated in microseconds. With the Big Data technologies, we can analyze these data as they are being generated very efficiently.
5. **Value:** It refers to the ability to turn the Big Data into business value. For businesses, it's really important to make use of cases before jumping to collect and store the big data. If we were not able to turn the Big Data into usable business value, it's useless.

3.2 Big Data Platform for Stock Market Analysis

In order to process huge amounts of both structured and unstructured financial data, Big Data technologies such as Apache Hadoop and Spark are essential since the conventional relational database and data warehousing system can not handle these efficiently. Big Data platform has been extensively used in the stock market to find the patterns or trends and ultimately predicts the outcome of the certain behavior in the financial market. Stock market data can be both structured and unstructured and have properties of 5V's of big data as in Fig. 2. The financial institution can extract information and process and analyze them to help investors in trading decisions. For example, by analyzing the changes in Google query volumes for big company names such as Apple, IBM, Microsoft, etc., we can find an interesting pattern that can be interpreted as a warning sign for stock market movements [18].

3.3 Apache Hadoop

Apache Hadoop is an open source implementation of Google File System and Map Reduce, which is a software platform used for distributed data storage and processing. Hadoop has mainly two components, MapReduce and Hadoop Distributed File System (HDFS). It was designed to store very large data sets efficiently and reliably in a cluster and to stream data with higher speed [19].

3.4 Apache Spark

Spark is a fast data processing engine for large-scale data which can run programs up to 100x faster than that of traditional MapReduce in Memory [3].

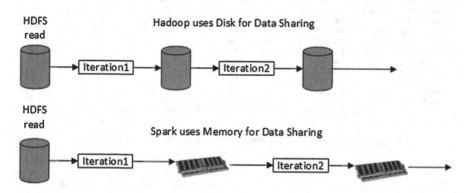

Fig. 3. Comparison of Spark and Hadoop

Since Apache Spark doesn't provide distributed storage like HDFS, we integrated it with HDFS into our system design. It can run on the top of HDFS and process the financial data. Spark also supports real-time data processing and fast queries. One major drawback of the Spark processing is, it needs more RAM as almost all of the data processing job is done in memory (Fig. 3).

3.5 Financial Data Sets

For our work, we used a popular database, the New York Stock Exchange (NYSE)'s Daily Trade and Quote (DTAQ), an academic market micro-structure research in U.S. equities [4]. DTAQ consists of a set of files containing all trades and quotes listed and traded in US regulated exchanges in a single day. The data sets are generally in a binary format derived from the output of CTA and UTP SIPs, tapes A, B, and C.

- **National Best Bid and Offer (NBBO) Appendix File:** NBBO file contains records when two different exchanges hold the best bid and best offer price in the securities information processors (SIPs).
- **Quote File:** It consists of the quoted price for all U.S equities and flags when both of the exchanges represent the NBBO.
- **Trade File:** It includes the exact trade happens in U.S. equities.

All quotes and trades files are time-stamped to the microseconds; however recent high-frequency trading is time-stamped in nanoseconds as well. According to Holden and Jacoben's work [15] on liquidity measurement problems in fast, competitive markets and expensive and cheap solutions, the NBBO file is incomplete and in order to get the complete NBBO it needs to be merged records flagged as NBBO of the quote file (Fig. 4).

```
+------------+------+------------+-----------+-----------+-----------+---------------+---------------+
|            | Time|Ticker|BestBidPrice|bestbidsize|bestaskprice|BestAskSize|BestBidExchange|BestAskExchange|
+------------+------+------------+-----------+-----------+-----------+---------------+---------------+
|32400.168105|  SPY|     190.31|         4|     190.32|         7|             Z|             P|
|32400.566107|  SPY|     190.31|         8|     190.32|         3|             Z|             P|
|32400.566185|  SPY|     190.31|         8|     190.32|         1|             Z|             P|
|32400.566328|  SPY|     190.32|         1|     190.32|         1|             T|             P|
| 32400.56634|  SPY|     190.32|         2|     190.32|         1|             T|             P|
|32400.566643|  SPY|     190.32|         2|     190.32|         3|             T|             P|
| 32400.56714|  SPY|     190.32|         2|     190.32|         1|             T|             P|
|32400.604909|  SPY|     190.32|         1|     190.32|         1|             T|             P|
|32400.628667|  SPY|     190.31|         4|     190.32|         1|             Z|             P|
|32400.629165|  SPY|     190.31|         2|     190.32|         1|             P|             P|
+------------+------+------------+-----------+-----------+-----------+---------------+---------------+
```

Fig. 4. NBBO in DataFrame processed by the Spark Clusters

For multimarket data analysis using association rules, we used the TAQ Trade, Quote and NBBO files of a particular month (August 2015). We first calculated the trading frequency and then used the Association rules based on FP-Growth to find the interesting pattern in the exchanges. The pattern is also analyzed in the perspective of price maker and price taker in the financial market.

3.6 Association Rules Mining

In this section, we describe the unsupervised data mining technique called Association rule mining. Association rules are a very popular rule in machine learning to find out interesting relations among variables in a large data sets. Association rules are used to find the pattern (sub-sequence or substructure of a set of frequent items that occur together). The pattern represents intrinsic and important properties of large data sets and is very useful in business for making a decision. Formally, an association rule is an implication of the form $X \rightarrow Y$ [5]. Where $X = \{x_1, x_2, \ldots, x_n\}$ and $Y = \{y_1, y_2, \ldots, y_n\}$ are the set of distinct items in a transaction T. The X is commonly known as the antecedent and Y as the consequent. The association rules simply says that whenever there is a possibility of happening event X there is also likely to happen event Y as well. One of the earliest applications of the association rules is the market basket analysis done on a large set of costumer transactions. In order to formulate the rules for association the notion of support and confidence are used. Support and confidence are used to measure the strength of the rule. The quality measure of the association rules are represented by three terms: **Support, confidence and lift.**

Support is a fraction of a transaction that that contains both item sets X and Y. It determines how frequent a rule is applicable to a given data set. Formally support for association rule $X \rightarrow Y$ is $support(X \rightarrow Y) = \frac{\sigma(X \cup Y)}{N}$ Where N is the number of instances and $\sigma(X \cup Y)$ is the number of instances covering both X and Y [21]. The idea is to take the number of items that cover both X and Y and divide that by the total number of instances in the database under consideration.

Confidence measures how often items in Y appear in a transaction that contains X. Formally confidence can be defined as $confidence(X \rightarrow Y) = \frac{\sigma(X \cup Y)}{\sigma(X)}$ [21].

The third quality metric used in the association rule is lift. The association rule $X \rightarrow Y$ is interesting with high support and confidence but sometimes it tends to be misleading and give false positives. In such cases, the correlation between X and Y is considered and is called lift. Three types of correlations are considered from the lift value.

- Positively correlated when $lift(X \rightarrow Y) > 1$
- Negatively correlated when $lift(X \rightarrow Y) < 1$
- X and Y are independent if lift is nearly equal to 1.

Support, confidence, and lift can be expressed in terms of probability.

$$support(X \rightarrow Y) = P(X \cup Y)$$

$$confidence(X \rightarrow Y) = P(Y|X)$$

$$lift(X \rightarrow Y) = \frac{P(X \cup Y)}{P(X)P(Y)}$$

Support, s is the probability that a transaction contains $\sigma(X \cup Y)$ and Confidence, c is the conditional probability that a transaction containing X also contains Y [13]. We count the number of instances which covers both X and Y, and divide that by the number of instances covering just X. Both support and confidence have their own significance. Confidence indicates the strength of the rules however support has a statistical significance. The higher the confidence, the more strength the rule. Another motivation for support is to suppress the rules that are the minimum threshold for business reasons. In order to formulate the meaningful rules and pattern, the optimum value of minimum support (min-sup) and minimum confidence (minconf) are chosen. Both minsup and minconf varies within the range of 0 to 1 [6].

In association rule mining we will consider all the rules, $X \rightarrow Y$ with minimum assumed support and having maximum confidence.

3.7 Scalable Mining Methods

There are three major approaches to mining Level-wise, Join-based approach: Apriori [7], Vertical data format approach: Eclat [24] and Frequent pattern projection and growth: FP-Growth [14]. In this research, we adopted the FP-Growth technique because it is efficient, has less memory usage, takes shorter execution time, and scalable and can be used in the distributed system [17].

The example of Association rule in the context of the financial stock market can be:

- When exchange X is dominant, at 70% of probability exchange, Y is also dominant on the same day.
- If price change occurs in exchange X, then exchange Y simply copies that exchange by 85% probability.
- If the price of X goes up, then, 50% of the time, the price of Y goes down.
- If exchange X is dominant, Y immediately follows it most of the time.

It is clear that investors and traders are more interested in the above rules.

The main purpose of the study is to apply the association rule to find out the conditional dominance in the US stock market. This paper utilizes the concept of inter-transaction rule mining [16]. In order to find the association rules and apply the FP-Growth algorithm we have to convert the data into the transactional format. The example of a transactional format is as shown in Table 3. Here, individual stock exchanges are listed with their corresponding name. The term rise, fall, low and high are used to represent their status in dominance analysis.

4   Implementation of Big Data Framework

In this section, we describe how we used big data framework such as Apache Hadoop and Spark to process and analyze the high-frequency stock trading data. The TAQ data is generally in zipped format. They are unzipped and are kept in the HDFS in the clusters.

Our system consists of the clusters of 4 Nodes; one acts as a master or namenode and other 3 acts as a slave or datanode. The cluster's configuration consists of the following environment:

Hardware: Intel(R) Xeon(R) CPU E5320 @ 1.86 GHz Operating System: Ubuntu 16.04, kernel: 4.4.0-87-generic Hadoop Version: Hadoop-2.7.3, Spark Version: Spark 2.0.0, Memory: 16 GB per node, Hard Drive: 2 TB in Master Node and 1 TB in each data node.

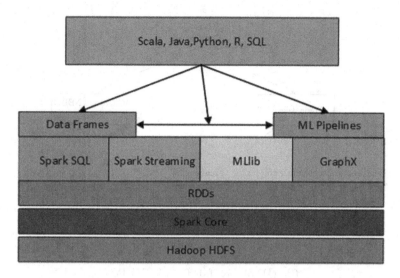

Fig. 5. System architecture showing major big data components

Figure 5 shows the overall system architecture. The lowest block represents the Hadoop File System and the next block represents the Spark Core. The third is Spark resilient distributed data sets (RDD), which is essentially a read-only collection of objects distributed across the set of machines or clusters [23]. The upper block represents the higher level API used to access the spark cores and RDDs. The Spark core takes these data and coverts them to RDD and then ultimately to data frames by applying filtering and certain rule sets. Once a data frame, it is processed by a high-level visualization engine such as R and Python. In addition to visualization, Python also generates transactional rules set for applying the association rules algorithm called FP-Growth.

5   Experimental Results

In this section we describe several experiments that have been done to process the financial data using Big data framework such as Spark and FP-Growth algorithm.

By using association rules on Trade file, Quote and NBBO interesting relations can be formulated. Figure 6 shows different ways to find the inter relations between U.S. exchanges.

5.1   Dominance Pattern Using Trade File

In order to find the dominance pattern, we used the trade file provided by the NYSE. Each trade frequency that happened in the exchange is calculated during time t_i and compared with time t_{i-1}.

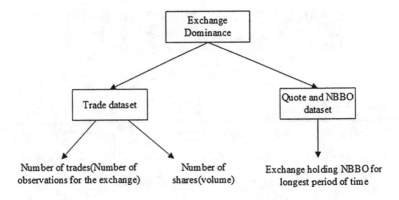

Fig. 6. Finding exchange dominance

Table 3. Transactional data format

fB	rD	rJ	rK	fM	fP	rT	fX	fY
rB	fD	rJ	fK	fM	rP	fT	rX	fZ
fB	rD	fJ	fK	fP	fT	rX	rY	fZ
rB	fD	rJ	fK	rP	fT	fX	rY	rZ

The transactional data set is created using the simple mapping techniques. If the trade frequency in exchange X is rising in time t_i, then it is indicated with rX; similarly, if trade frequency in exchange Y is falling in time t_i, then it is marked with ry. Additionally, if exchange X is the dominant trade with a heavy traded frequency in particular time t_i then it is marked with mX.

Once the transactional data set is ready, we apply it the Association rules based on FP-Growth in Apache Spark clusters.

Table 4 lists the sample association rules produced by applying the FP-Growth algorithm in the trade file of the corresponding day of August 2015. The rules are a conditional probability in the form $X \Rightarrow Y$ meaning, if X is true, Y is also true with a probability of c, the confidence.

Here in this rule, as shown in the Table 3, r implies rising, and f implies falling. The next letter in upper case denotes the name of the Exchange. Here, the rules with higher confidence greater than 0.85 are more significant and carry a stronger meaning in terms of financial interpretations. For example, the rule $[rT, rP] \Rightarrow [rZ]$ with confidence of 0.88 says that whenever NASDAQ Stock Exchange, LLC (**T**) and NYSE Arca, Inc. (**P**) are rising there is a possibility of a rise in Bats BZX Exchange, Inc. (**Z**).

5.2   Dominance Pattern

In this experiment, we took one month of TAQ trade data from August 2015 and conducted the experiment to find out the the dominance pattern. The result is

Table 4. Sample Association Rules with minimum support 0.3

August 21	Confidence	August 24	Confidence	August 25	Confidence
$[fT, fZ] \Rightarrow [fP]$	0.84	$[rT, rP] \Rightarrow [rZ]$	0.88	$[fT, fZ] \Rightarrow [fP]$	0.84
$[fT, fZ] \Rightarrow [fD]$	0.81	$[fZ] \Rightarrow [fT]$	0.81	$[fY, fP] \Rightarrow [fZ]$	0.90
$[fY, fP] \Rightarrow [fZ]$	0.90	$[rP, rZ] => [rT]$	0.85	$[rK, rT] \Rightarrow [rZ]$	0.86
$[rK, rT] \Rightarrow [rZ]$	0.86	$[fZ, fT] \Rightarrow [fP]$	0.81	$[fP, fD] \Rightarrow [fZ]$	0.83
$[fP, fD] \Rightarrow [fZ]$	0.83	$[fZ, fP] \Rightarrow [fT]$	0.86	$[rP, rZ] \Rightarrow [rT]$	0.84
$[rP, rZ] \Rightarrow [rT]$	0.84	$[rT] \Rightarrow [rZ]$	0.81	$[fK, fP] \Rightarrow [fZ]$	0.86
$[fT, fP] \Rightarrow [fZ]$	0.87	$[fP, fT] \Rightarrow [fZ]$	0.86	$[rZ, rK] \Rightarrow [rT]$	0.83
$[fZ, fP] \Rightarrow [fT]$	0.81	$[fB, fT] \Rightarrow [fZ]$	0.86	$[fZ, fD] \Rightarrow [fP]$	0.83
$[rP, rT] \Rightarrow [rZ]$	0.84	$[fZ, fB] \Rightarrow [fT]$	0.88	$[rZ, rT] \Rightarrow [rP]$	0.80
$[fY, fZ] \Rightarrow [fP]$	0.87	$[rT, rZ] \Rightarrow [rP]$	0.81	$[fZ, fP] \Rightarrow [fT]$	0.81

Table 5. Dominance Pattern on Trade, Aug 2015

Day	Dominance Patterns	Day	Dominance Patters
03	D P T Z K Y J B X M	18	D P Z T K Y J B X M
04	P Z D T K Y B J X M	19	D P Z T K Y J B X M
05	D P T Z K J B Y X M	20	D P T Z K J Y B X M
06	D P T Z K Y J B X M	21	D Z T P K J B Y X M
07	D P T Z K Y J B X M	24	D P Z T K J Y B X M
10	D P T Z K Y J B X M	25	D Z T P K J B Y X M
11	D P Z T Y K J B X M	26	P D T Z K J Y B X M
12	D T N P Z Y K J B X A M	27	T Z P D K J B Y X M
13	D P T Z K B J Y X M	28	D P T Z K Y J B X M
14	D P Z T K J Y B X M	31	D P Z T K J B Y X M
17	D P T Z Y K J B X M		

presented in Table 5. The results show an interesting pattern where Dark pools trading Finra (**D**) is immediately followed by the NYSE Arca (**P**).

5.3 Price Analysis

The trade file reflects the exact trade that happens during the day. Here we present the price analysis of the SPY ticker during the flash crash day (August 24, 2015) and normal day (just a day before the flash crash day August 21, 2015). Figure 7 shows a comparison of the price between normal day (Aug 21) and flash crash day (Aug 24) of 2015 for SPY ticker. It was found that there was a sharp fall in SPY ticker price of about 10% in flash crash day than that of a normal day.

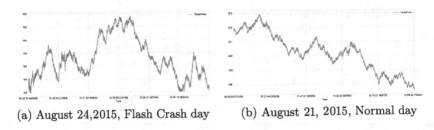

(a) August 24,2015, Flash Crash day   (b) August 21, 2015, Normal day

Fig. 7. Price Analysis of SPY Ticker

It can be clearly seen that the SPY declined to more than 5% below its closing price on the previous day (Friday, August 21, 2015).

6   Performance Test of the Big Data Platform

To test the performance of the spark clusters, we used the TAQ quote data of Aug 24, 2015 with uncompressed size of 203 GB. In the test, we measured the execution time of the clusters for given data vs. the number of nodes. The Table 6 shows the total time taken by the clusters to process the quote data of 203 GB with a decreasing number of clusters. It is clear that more data nodes have a shorter execution time for processing large volume of data.

Table 6. Execution time in Apache Spark Cluster

Number of data nodes	Execution time
3	17 min 44 s
2	21 min 17 s
1	33 min 56 s

7   Conclusion

In this work, we developed a Big Data framework based on Apache Spark to process the financial Daily Quote and Trade file to find out the hidden relations among the stock exchanges using data technique called FP-Growth algorithm. In addition, we also performed the dominance analysis based on the NYSE's DTAQ NBBO and quote data sets. The results appear promising and can be used by investors and other companies to find in which exchange they can trade. Moreover, the framework can be reused to find the future stock prices of certain companies and predict flash crash.

References

1. Exchange pricing model (2011). http://www.nomura.com/europe/resources/pdf/ExchangePricingModels_20110614.pdf
2. 5vs big-data (2015). http://www.ibmbigdatahub.com/blog/why-only-one-5-vs-big-data-really-matters
3. Apache spark (2018). https://spark.apache.org/
4. Nyse daily taq (trade and quote) (2018). http://www.nyxdata.com/Data-Products/Daily-TAQ
5. Agarwal, R.C., Aggarwal, C.C., Prasad, V.: A tree projection algorithm for generation of frequent item sets. J. Parall. Distrib. Comput. **61**(3), 350–371 (2001)
6. Agrawal, R., Imieliński, T., Swami, A.: Mining association rules between sets of items in large databases. In: ACM SIGMOD Record, vol. 22, pp. 207–216. ACM (1993)
7. Agrawal, R., Srikant, R., et al.: Fast algorithms for mining association rules. In: Proceedings 20th International Conference on Very Large Data Bases, VLDB, vol. 1215, pp. 487–499 (1994)
8. Aldridge, I.: High-Frequency Trading: A Practical Guide to Algorithmic Strategies and Trading Systems, vol. 459. Wiley, Hoboken (2009)
9. Asadifar, S., Kahani, M.: Semantic association rule mining: a new approach for stock market prediction. In: 2017 2nd Conference on Swarm Intelligence and Evolutionary Computation (CSIEC), pp. 106–111. IEEE (2017)
10. Chowdhry, B., Nanda, V.: Multimarket trading and market liquidity. Rev. Financ. Stud. **4**(3), 483–511 (1991)
11. Fang, B., Zhang, P.: Big data in finance. In: Yu, S., Guo, S. (eds.) Big Data Concepts, Theories, and Applications, pp. 391–412. Springer, Cham (2016). https://doi.org/10.1007/978-3-319-27763-9_11
12. Han, J., Cheng, H., Xin, D., Yan, X.: Frequent pattern mining: current status and future directions. Data Mining Knowl. Discov. **15**(1), 55–86 (2007)
13. Han, J., Kamber, M., Pei, J.: Data Mining: Concepts and Techniques. The Morgan Kaufmann Series in Data Management Systems. Morgan Kaufmann, Burlington (2000)
14. Han, J., Pei, J., Yin, Y.: Mining frequent patterns without candidate generation. In: ACM SIGMOD Record, vol. 29, pp. 1–12. ACM (2000)
15. Holden, C.W., Jacobsen, S.: Liquidity measurement problems in fast, competitive markets: expensive and cheap solutions. J. Financ. **69**(4), 1747–1785 (2014)
16. Luhr, S., Venkatesh, S., West, G.: Emergent intertransaction association rules for abnormality detection in intelligent environments. In: Proceedings of the 2005 International Conference on Intelligent Sensors, Sensor Networks and Information Processing Conference, pp. 343–347. IEEE (2005)
17. Mythili, M., Shanavas, A.M.: Performance evaluation of apriori and fp-growth algorithms. Int. J. Comput. Appl. **79**(10), 279–293 (2013)
18. Preis, T., Moat, H.S., Stanley, H.E.: Quantifying trading behavior in financial markets using google trends. Sci. Rep. **3**, 1684 (2013). https://doi.org/10.1038/srep01684
19. Shvachko, K., Kuang, H., Radia, S., Chansler, R.: The hadoop distributed file system. In: Proceedings of the 2010 IEEE 26th Symposium on Mass Storage Systems and Technologies (MSST), MSST 2010, pp. 1–10. IEEE Computer Society, Washington, DC (2010). https://doi.org/10.1109/MSST.2010.5496972

20. Sohangir, S., Wang, D., Pomeranets, A., Khoshgoftaar, T.M.: Big data: deep learning for financial sentiment analysis. J. Big Data 5(1), 3 (2018)
21. Tan, P.N., et al.: Introduction to Data Mining. Pearson Education, India (2006)
22. Wu, X., Zhu, X., Wu, G.Q., Ding, W.: Data mining with big data. IEEE Trans. Knowl. Data Eng. 26(1), 97–107 (2014)
23. Zaharia, M., Chowdhury, M., Franklin, M.J., Shenker, S., Stoica, I.: Spark: cluster computing with working sets. HotCloud 10(10–10), 95 (2010)
24. Zaki, M.J., Parthasarathy, S., Ogihara, M., Li, W., et al.: New algorithms for fast discovery of association rules. In: KDD, vol. 97, pp. 283–286 (1997)

Who's Next: Evaluating Attrition with Machine Learning Algorithms and Survival Analysis

Jessica Frierson[✉] and Dong Si[✉]

University of Washington, Bothell, WA 98011, USA
{jess2018, dongsi}@UW.edu

Abstract. Every business deals with employees who voluntarily resign, retire, or are let go. In other words, they have employee turnover. Employee turnover, also known as attrition can be detrimental if highly valued employees decide to leave at an unexpected time. This paper aims to find the employee(s) that are most at risk of attrition by first identifying them as someone who will leave. Second, identify if their department increases the probability of them leaving. And third, identify the individual probability of the employee leaving at a given time. This paper found Logistic regression to consistently perform well in attrition classification compared to other Machine Learning models. Kaplan-Meier survival function is applied to identify the department with the highest risk. An attempt is also made to identify the individual risk of an employee leaving using Cox proportional hazard. Using these methods, we were able to achieve two of the three goals identified.

Keywords: Machine learning · Attrition · Logistic regression
Survival analysis

1 Introduction

All companies experience a degree of attrition and high attrition rates are concerning. However, what is more concerning is having experienced, knowledgeable, high performing employees leave. This leads to a company having to hire and train someone new in the domain of the previous employee, at possibly a higher market rate. This could lead to a period of low productivity while the new employee ramps up with company policies and trainings [1]. Therefore, it is vital that any business, big or small know what contributes to their employees leaving. In addition, being able to identify those red flags can lead to identifying the potentially at-risk employees. Remedial steps can then be taken to balance out the contributing factors leading to attrition and lessen the probability of the employee leaving. This project will aim to identify those contributing factors based on sample data provided by IBM which was found on Kaggle, and to identify employees at risk of attrition. In addition, through survival analysis, attempt to answer: Who's next?

This paper aims to use the best model based on experimental runs to identify employees that are at risk of attrition. Principal component analysis (PCA) will be performed to choose the top N feature components that contribute the most to attrition. In addition, survival analysis is used on each department to identify which one has the highest probability of its members leaving the company at a given time. Lastly, we will use a different survival analysis algorithm to find the employee most likely to leave from the department with the highest attrition rates.

1.1   Related Work

Previous literature, primarily in management, social, and other fields, has explored the issue of attrition. Limited research has been done with identifying employees that have higher risks of attrition with machine learning. For instance, literature that identifies a better subset of attributes to give a better predictor of attrition has been explored by Chang [2]. He found that it can be difficult to choose and limit the features to use specifically when pulling data straight from HR databases since there is a variety of information available. Identifying and selecting a good set of attributes from databases is a start.

A problem with HR datasets is that it can become difficult to not overfit models [1]. One method to reduce overfitting is to forgo the accuracy and make the model more generic to accommodate a more varied dataset. Extreme Gradient Boosting is found to be more robust than other models due to its implementation of regularization [3]. A previous study from Singh et al. [1] went one step further than just identifying attrition risks. They not only identified employees at a higher risk of attrition, but also explored if proactive steps, such as a salary increase, lowered the attrition rates.

In addition, there are papers that explored the aspect of employee retainability. In the case of Ramamurthy et al. [4], interest was in creating models to help employers reduce attrition by identifying employees that are good candidates to be retrained in a new skill based on their current skill set. In another case, Saradhi and Palshikar [5], drew parallelisms between customer turnover and employee turnover. They evaluated customer models and applied them to employees. They found that some of those models can be used for employee attrition prediction.

There is also a lot of literature available for survival analysis, but those literatures do not cross over to the machine learning fields too frequently. Goli and Soltanian sought to compare the performance of support vector regression (SVR) with Cox model on both a simulated and real survival problem. They found that with certain parameters, the SVR outperformed the Cox model and with others– the performance increase was diminished [6].

2   Methods

Previous papers that use human resource data to explore attrition have mainly concentrated on comparing the performance of various machine learning models and not so much on applying the model. In other words, we have not found papers that apply their models to predict if there are specific departments, or employees that are at a higher risk of turnovers. This is what sets this work apart from other work. In addition, other work does not combine different statistical models to the issue of attrition in combination with machine learning models. Palshikar [5] came close by recommending as part of future work to use survival analysis models in combination with their learning models but did not implement it. As of this time, we have not come across another paper who has.

The approach for this paper, is to use various learning models to find the best performant model for classification. Figure 1 shows the process pipeline. The process begins with pre-processing. After preprocessing, the data is split 70/30 for training and testing sets. With this split, we use and compare six different classification models. These models are, Decision tree, Logistic regression, Support vector machines, Naïve Bayes, K Nearest Neighbors, and Neural Networks. All models will be evaluated based on their accuracy, precision, and recall. Once the models have been implemented, we apply two different survival analysis algorithms.

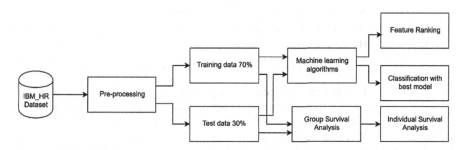

Fig. 1. This shows the ML pipeline for this term project. Preprocessing is needed to remove attributes that provide no value. The dataset is split into a training set and a test set. The training data is used for data familiarity, machine learning, and survival analysis. Test data is only used for machine learning and survival analysis.

Survival analysis is defined as: "A set of methods for analyzing data where the outcome variable is the time until the occurrence of an event of interest" [7]. In our case the event of interest is attrition. Two survival analysis models are explored. The first model, Kaplan-Meier estimator, is to evaluate each department group. The Kaplan-Meier estimate involves computing the probabilities of an event occurring at a specific point in time [8]. The second algorithm is the Cox proportional hazards. This model will help identify the probability that an employee will leave, ultimately answering "Who's next."

2.1 Data Preprocessing

The dataset used is the IBM_HR simulated dataset found on Kaggle. This dataset contains 1470 data rows and 34 features. Data preprocessing is done by removing features from the dataset as seen in Table 1. These attributes contained the same value for all the data rows or contained non-relevant identifying information such as Employee number. In addition, there are some attribute fields that where re-mapped from string text to enumeration values. Such fields can be seen in Table 2. This was done in consideration of the machine learning algorithms that will be applied since it's easier to process numbers than text.

Table 1. Attributes removed from the dataset.

Attribute	Reasons
Employee count	Employee count
All records had the same value of (1), no valuable information could	All records had the same value of (1), no valuable information could
Be gained from this attribute.	Be gained from this attribute.
Employee number	Employee number

Table 2. Mapping of attributes that need to be enumerated.

Attribute	Source values	Target values
BusinessTravel	Non_Travel, Travel_Rarely, Travel_Frequently	0, 1, 2
Department	Human resources, research & development, sales	1, 2, 3
EducationField	Human resources, life sciences, marketing, medical, other, technical degree	1, 2, 3, 4, 5, 6
Gender	Female, male	1, 2
JobRole	Healthcare representative, human resources, laboratory technician, manager, manufacturing director, research director, research science, sales executive, sales director	1, 2, 3, 4, 5, 6, 7, 8, 9
Marital Status	Divorced, married, single	1, 2, 3
Overtime	No, yes	0, 1

2.2 Factors Leading to Attrition

One of the research goals was to identify top contributing attributes for the simulated data. To get those top features, we used an ensemble tree classifier to extract the feature importance rankings. The algorithm for the ensemble is based on a perturb-and-combine technique, which is designed for trees [9]. The rankings can be seen in Table 3. The feature that holds the highest ranking is Overtime with a ranking of .79.

Table 3. Top ten features ranked based on the results from the tree classifier. The name of the attribute is followed by the ranking.

No.	Feature	Ranking
1	Overtime	0.791443
2	Age	0.512488
3	YearsWithCurrentManager	0.427158
4	MonthlyIncome	0.416188
5	TotalWorkingYears	0.414962
6	DistanceFormHome	0.403594
7	YearsAtCompany	0.368960
8	WorkLifeBalance	0.360206
9	DailyRate	0.347088
10	JobRole	0.346335

The following feature only holds a ranking of .51. If we consider features ranked .5 or higher as significant, we would only have those two features. All other features ranked below that arbitrary threshold value. However, these results are not consistent. The attribute Overtime is always the highest ranking and the last few features remain in their ranked position with each subsequent run. However, the features in between vary by one or two positions with multiple runs. After exploring the data and finding the rankings, we then focused on selecting a model for the classification objective.

2.3   Machine Learning Algorithm Comparison

The models we compare in this section are: Decision Tree, Logistic regression, Support vector machine, Gaussian Naïve Bayes, K-Nearest Neighbors, and Neural Networks. These models are selected because they are appropriate for supervised binary classification. The label to identify is Attrition, which contains 0 for data rows that remain in the company, and 1 for data rows that have left the company.

Python's scikit learn library contains implementations of the models we target to explore. We consume the library models by using the standard pattern by first fitting the training data and the training label. Then the fitted model is used to predict using the test data. From the predicted model, metric scores can be calculated. Table 4 contains each model's respective metrics. The model that best performed is Logistic regression. Logistic regression contains the highest accuracy with (.8752), precision (.8603), and recall scores (.8752). The least performant model is Neural Networks. It is assumed that this is because our data is not ideal for neural networks, which works best with sequential data such as image and audio files.

Table 4. Model comparison table.

Model	Accuracy	Precision	Recall
Decision tree	.7823	.7950	.7823
Log regression	.8752	.8603	.8752
SVM	.8458	.7153	.8458
G. Naïve Bayes	.8140	.8318	.8141
KNN	.8344	.7867	.8344
NN	.6938	.7504	.6939

The receiver operating characteristic curve (ROC) of all models can be seen in Fig. 2 along with the area under the curve (AUC) value. From this plot, Gaussian Naïve Bayes algorithm performs almost as well as Logistic regression. After conducting these runs, the best classifier model for the IBM_HR dataset is consistently Logistic regression.

After identifying the Logistic regression as the best model, we wanted to see if it can still be improved by including feature dimension reduction since the data contained the full feature set after the initial pre-processing of the data. We combined Principal Component Analysis with Logistic regression using sklearn pipeline make_pipeline library. The results are seen in a combined plot with standalone Logistic regression in

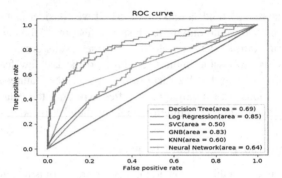

Fig. 2. Receiver operating characteristic curve of all six ML models with their area under the curve score.

Fig. 3. There is a slight almost negligible improvement using PCA with Logistic regression. Figure 3 shows the output from the best performing run of the piped model. When modifying the number of components for PCA, the model performance decreased if the number of components is reduced from the full feature set.

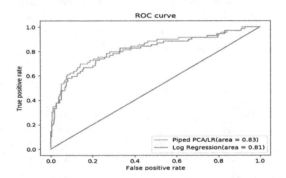

Fig. 3. Receiver operating characteristic curve of piped and not piped Logistic regression with their area under the curve score.

2.4   Survival Analysis

This survival analysis section aims to identify the department with the highest probability that its personnel will leave the company at a given time. To do this we used two different Python survival packages: 1. Sksurv and 2. Lifelines. Each package requires the data to be in a specific format. Therefore, we had to do more pre-processing to get the data in the right formats. Learning models typically contain a minimum of two parameters that need to be provided, 1. the data matrix/array and 2. the label array. With the survival packages, it is still expected to provide a data matrix/array, but instead of a label file, a censored event and time needs to be provided. We therefore created a censored event and time file using the Attrition label as the event (either 0 or 1) and added the YearsAtCompany attribute as the time parameter. With this modification we were able to get the overall probability of survival for the dataset as well as the probabilities of survival for each department using Kaplan-Meier model. From the plots provided in Fig. 4, we can estimate the probability or survival (no attrition) by choosing a time. For example, the probability of someone who just joins (t = 0) has a survival rate of or near 100. As time goes by the probability of survival decreases. Figure 4(b) contains the survival step functions for each of the three departments present in the data (refer to Table 2 to identify the departments). It's a bit tough to clearly see if department 1 or department 3 has more people leaving. However, it looks like department 3 has steeper and more frequent steps than department 1. Each step indicates that the event (attrition) was observed.

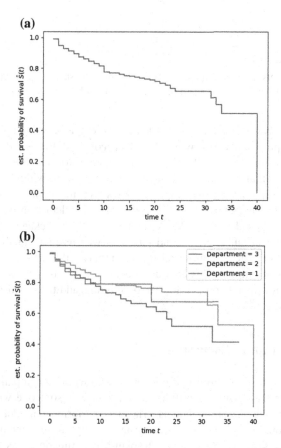

Fig. 4. Kaplan-Meier plotted as step functions. (a) Is the function for the overall dataset, whereas (b) is the plot for each department. The probability can be deduced by evaluating a specific time (t).

Figure 5 helps to distinguish between department 1 and department 3 as the department with the highest probability of attrition. In combination with the above Kaplan-Meier plots and the bar graph, it can be determined that Department 3 is the department that has the highest probability of employees leaving.

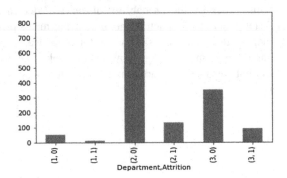

Fig. 5. Bar graph sets representing the number of people who stay in the company vs those that leave. Labels (1, 0) represent the department (1, 2, or 3) and Attrition (1, 0).

The last objective of this paper is to identify the employees from department 3 that have the highest probability of leaving next. To accomplish this, we decided to use the Cox proportional hazard model. The documentation for using the scikit-survival package mentioned that this function can be used to get the individual probability based on the "patient" or in our case, the employee. However, the algorithm kept failing with our dataset. The failing point was on an issue where a calculation received a NaN or infinite value. We decided to look for another survival package and came across Lifelines. After we installed the second package, we were still unable to get results using the Cox model. The Lifelines package also reported the same error of an expected value being NaN or infinite. This lead us to believe the issue was with our data. The data was either not in the proper format or missing additional information for the functions to yield results.

3   Conclusion and Discussion

This paper explored the possibility of predicting who within a company can be next to leave. This was done by first identifying an employee as someone who will leave or stay using binary classification models. The best performant model for this data is Logistic regression. Logistic regression consistently outperformed the other models and retained an AUC score of 85%. Next, Kaplan-Meier function was used to get the survival function for the overall company and each department. Through the Kaplan-Meier step function plot and bar graph, department 3 was shown to contain the least survival rate. Lastly, the Cox's proportional hazard model was attempted with two different libraries ultimately both failing to complete and reported the same issue. Since the same issue was reported, this lead us to conclude that the data was not in the right state for these functions.

The primary challenge we faced in this work, was the lack of in depth knowledge around survival functions. With a stronger background knowledge of the functions used, we may not have reached a blocking stop. Another challenge was getting the feature rankings to produce a consistent ranking. Each run to produce a ranking gave slightly different results.

Future work will require processing the data into the right expected format to complete the last objective of this paper. An alternative to using the library functions could be to implement our own survival functions. In addition, these methods can be applied to different datasets and measure the fragility of the approach taken in predicting who will be next to leave the company.

References

1. Singh, M. et al.: An analytics approach for proactively combating voluntary attrition of employees, In: IEEE ICDM, pp. 317–323 (2012)
2. Chang, H.: Employee turnover: a novel prediction solution with effective feature selection. In: Information Science and Applications, pp. 417–426 (2009)
3. Ajit, P., Punnoose, E.: Prediction of employee turnover in organizations using machine learning algorithms. Int. J. Adv. Res. Artif. Intell. 5(9) (2016). http://dx.doi.org/10.14569/IJARAI.201.050904
4. Ramamurthy, K., et al.: Identifying employees for re-skilling using an analytics-based approach. In: IEEE ICDM, pp. 345–354 (2015)
5. Saradhi, V., Palshikar, G.K.: Employee churn prediction. Expert Syst. Appl. 38, 1999–2006 (2011)
6. Goli, S., et al.: Performance evaluation of support vector regression models for survival analysis: a simulation study. Int. J. Adv. Comput. Sci. Appl. (IJACSA), 7(6), 381–389 (2016)
7. Cornell University. Cornell Statistical Consulting Unit
8. Khanna, P., Kishore, J., Goel, M.K.: Understanding survival analysis: Kaplan- Meier estimate. Int. J. Ayurveda Res. 1(4), 274–278 (2010)
9. Breiman, L.: Arcing Classifiers. Annals Stat. 26(3), 801–849 (1998)

Volkswagen's Diesel Emission Scandal: Analysis of Facebook Engagement and Financial Outcomes

Qi An[1], Morten Grimmig Christensen[1], Annith Ramachandran[1], Raghava Rao Mukkamala[1,2(✉)], and Ravi Vatrapu[1,2]

[1] Centre for Business Data Analytics, Copenhagen Business School, Frederiksberg, Denmark
{rrm.digi,vatrapu}@cbs.dk
[2] Westerdals Oslo School of Arts, Communication and Technology, Oslo, Norway

Abstract. This paper investigates Volkswagen's diesel scandal with a focus on the relationship between their Facebook engagement and financial performance during the period of 2012–16. We employ the big social data analytics approaches of visual and text analytics on Volkswagen's Facebook data and financial reporting data. We specifically analyze the potential effects on the company in the diesel emission scandal years of 2014–2016. We find that the diesel emission scandal had the most impact in the short-term period immediately after its occurrence resulting in Facebook users reacting negatively against Volkswagen but also some users defending the company. In the long-term, it seems that the scandal has not impacted the company based on the analysis of both their financial data and their social media data.

1 Introduction

Volkswagen (VW) was ranked as the biggest car company measured on cars produced in 2016 and reports from 2017 shows that they will keep this position surpassing competitors such as Toyota and Nissan-Renault[1]. The brand portfolio of the Volkswagen group includes brands such as Audi, Porsche, Seat, Skoda, Bentley, Lamborghini and VW[2]. Volkswagen kept their place as the biggest automaker in the world even though they were subject to a global emissions scandal, which broke out in September 2015. The crisis entailed a total of 11 million diesel cars, as admitted by VW, fitted with a special software that made it possible them to show lower emissions during tests than they would if they where used regularly on the road[3]. The software was detected on multiple of VW Group's car brands this meant that VW, Audi Skoda and Seat had installed the software on their diesel cars. The cars used the same types of motors and motor technology as a way to obtain synergy affects across the multiple car brands[3]. A scandal at this level also

---

[1] Volkswagen Is World's Largest Automaker.

[2] Volkswagen annual report.

[3] Volkswagen: The scandal explained.

© Springer International Publishing AG, part of Springer Nature 2018
F. Y. L. Chin et al. (Eds.): BIGDATA 2018, LNCS 10968, pp. 260–276, 2018.
https://doi.org/10.1007/978-3-319-94301-5_20

affected the stock price, which plummeted in the period after the announcement of the software. It fell from a level of 160 EUR to a level of 110 EUR in November[4]. The fall can be attributed investors concerns of how VW's brand impairment will affect future sales, the possible litigation cost, and the cost of the 11 million recalls. VW has prepared to spend up to $25 billion in the US to take care of the emission problem this includes legal fees, buyback of cars, repairs etc. The diesel scandal had broader ramifications than just the negative impact on the stock price. Therefore it becomes relevant to analyze sentiments on social media, revenue, profits and sales in the aftermath of the diesel scandal, in order to measure whether the diesel scandal have had any affect on these metrics. This paper explores whether the emission scandal was just a media news story or whether the scandal affected:

1. The revenue and profits of VW in the period of 2012–2016.
2. The production of cars or delivering of cars from VW in the period 2012–2016
3. The engagement on social media. This consists of sentiment, consumer decision, personality traits and keyword analysis to analyze the development sentiments and sentiments expressed before and after the crisis.

By analyzing the above we aim to obtain a holistic view of how the diesel scandal has impacted the VW Group.

1.1   Problem Statement

We want to investigate the VW crisis in 3 ways. First, to explain the revenue and profits fluctuations during the diesel scandal. Second, to establish whether the production of cars and cars delivered dropped after the diesel scandal broke out. Third and last, to establish what sentiments and keywords were expressed in social media in the aftermath of the diesel scandal. The overall research question is stated below:

> *What were the primary volumetric and linguistic characteristics of social media reactions towards Volkswagen during the diesel scandal and to what extent, if any, did the diesel scandal in itself and as manifest on Facebook impact VW financially?*

This paper will analyze data from social media and annual reports in order to answer the research question. The analysis will focus on answering the 6 propositions stated below. The 6 propositions will be analyzed with the use of big social data in order to either confirm or reject them. Proposition 1 is about the consumers reaction to the scandal. Proposition 2 is about the effectiveness of the corporate apology to the scandal. Proposition 3 is about consumers' reactions to the scandal solutions. Proposition 4 is about the Volkswagens social media community. Proposition 5 is on the long lasting effect, if any, to consumer purchase decisions.

---

[4] Timeline of Volkswagen's tanking stock price.

Proposition 1: We expect that the negative sentiment outweigh positive sentiment after the diesel scandal breaks out on September 18th, 2015.

Proposition 2: The negative brand sentiment outweigh positive brand sentiment after the diesel scandal breaks out on September 18th, 2015.

Proposition 3: The positive sentiment outweigh negative sentiment after Volkswagen apologize for the scandal on September 25th, 2015.

Proposition 4: Consumers' positive sentiment will outweigh the negative after VW announces the recall and Goodwill package, because Volkswagen is actively trying to fix the problem.

Proposition 5: The product recall would cause an emotional reaction from consumers and therefore we expect that the personality trait of neuroticism is the dominant personality trait.

Proposition 6: The scandal would have negative influence to consumers' purchase behavior.

2   Conceptual Framework and Related Work

This section is divided into four parts. The first part describes the conceptual underpinnings of social media and the second part focus on the current literature on corporate crisis and negative word-of-mouth In the third part we discuss brand equity and product recall and in the last part we provide background of VW Crisis.

2.1   Social Media

Social media can provide individuals with increased power of voice. Instead of vocalizing ones opinion to a few close friends, social media has the possibility for individuals to share their opinions to thousands of even millions of people. 2,46 billion is currently users of social media webpage[5]. Social media can be a place where firms are able to manage their brand equity. Brand equity defined as "Brand equity is the added value endowed on products and services" [1, p. 492]. This company page enables a firm to communicate with followers but does also include posting news about products, posting announcements. Thereby social media can be used as a way to strengthen the brand image of a firm, and manage relationship with stakeholders [2]. Socio-technical interactions take place on social media. This includes that people interact with technologies and individuals. Social-technical interaction leaves a digital trace; this could be liking someone's photo, writing a comment to a profile picture etc. This digital trace can be defined as social data [3]. These unstructured volumes of data can be challenging for many firms as it can be difficult to extract meaning out of these

---

[5] Social media - Statistics & Facts.

unstructured volumes [4]. This implies that the data cleaning process is important when working with big data as it contributes to make data meaningful. In social-technical interaction there is a distinction between social graph analytics and social text analytics [3]. Social graph is a communication network of people who interact. Social graph includes Actors, actions, activities and artifacts [3, 4] Social text focuses on the content of what is being communicated [4]. Social text includes: topics, keywords, Pronouns, Sentiments. Social text is can therefore be a useful tool when analyzing social media activity as it then becomes possible for this project to analyze the sentiment expressed in the comment post on Facebook.

2.2   Crisis and Word-of-mouth

Crisis management becomes important when enduring a crisis like the diesel scandal. The typical traits of a crisis is as follows: (1) severe consequence(s), (2) threats to the fundamental value of an organization, (3) limitations in response time, and (4) unexpectedness of the event [5, p.372]. Once a company crisis emerges the news of it can spread vastly and quickly through the use of word-of- mouth on social media but also how it affect the brand equity of firms. Social media can provide power to people whenever they express their opinion or thoughts about a given brand or product. The power relies in the fact that people are able to connect and communicate with vast amounts of other users. Corporate firms have only limited possibilities in altering the discourse of the conversation that people are having and thereby negative word-of-mouth has the possibility to spread very quickly. Social media can provide negative branding effect for a firm, if consumers react negatively to a product or service, it is then possible for the users to spread the news quickly among them [2,6]. Possible reasons why people share their thoughts and opinions and initiate negative word-of-mouth can be to obtain emotional support from other members in a social network another reason could be that that consumers is left with a feeling of firm injustice making business with a firm [6,7]. Negative word-of-mouth do have consequences for a business as it can hurt the purchase intention and brand image of the company [6]. The further implication of such is volatile stock prices and uncertainty about long-term outcomes [6]. It can be difficult for firms to stop an outburst of negative word-of-mouth once it is initiated through social media. Therefore, it becomes important to respond with a well articulated crisis strategy in order to respond to the attacks from dissatisfied customers. These attacks can be difficult for firms to stop because customers will often listen more to each other in search of advice on products and services than listen to firms [8]. Therefore consumers come to rely on each other when they review products or services. In Pace et. Al. [9] they argue that when a consumers is exposed to a brand crises on social media it causes a more negative reaction towards the brand than if it was through traditional media. Personal experiences with the products of a brand and the experiences of others determinate of what people think of a particular brand [10]. This proves that negative word-of-mouth can change people perceptions of brands. Therefore it becomes important that a firm is able to control the outburst on social media. Especially stakeholders become central in a potential crisis. Some stakeholders can be very engaged in the particular issue and react strongly. Furthermore, they also becomes important for the firms because they can inhibit a position as brand creators [9] and they describes "Stakeholders can amplify and extend the crisis, providing further meanings and resonance to the critical event" [9, p.136]. Thereby stakeholders can play a critical part in the event of a crisis and it is important to execute a precise communication strategy with these stakeholders. It becomes important to deal with negative word-of-mouth quickly and effectively because by delaying the problem it can become even worse [6]. If the firm as failed to live up to their promise in terms of e.g. product quality, it becomes important that they apologies quickly and acknowledges the problem, and in that way it can soften the potential backlash from the stakeholders [6,11]. Often a crisis result in a increasing brand awareness and it is important that firms use this heightened attention to invest in communication, which can restore the brand credibility for stakeholders [11].

2.3   Background of VW Crises

It all started in 2006 when VW had a low market share in the US and the cars had difficulty passing the American diesel test. A short time after, VW executives decided in 2006 that illegal software should be installed on cars in order to obtain lower emission level during testing[6]. In 2008 VW pushes out a new marketing campaign for their low emission vehicles. In 2013 a small team from West Viginia University receives a grant and the purpose of the grant was to test whether diesel cars had a higher emission during normal driving than under tests. The test revealed that VW had a far higher emission level than during regular road use than under tests (See footnote 6). In 2014, after the publication of the report from the team in Virginia, VW receives a memo, which states that VW cars may be subjected to further investigation due to the report and an inquiry is opened towards VW. VW responds to the regulators by providing with inaccurate data, which they apologize for later on. In early 2015, VW starts recalling diesel cars, VW is postulating that a software update will fix the problem, even after adjusting on the software the problem is not fixed (See footnote 6). In the summer of 2015, VW starts destroying documents that could be incriminating to them in a potential legal battle. In the fall of 2015, VW admits that their diesel cars are equipped with illegal software. The 23rd of September, Martin Winterkorn, then CEO resigns. In the summer of 2016, VW settles the lawsuits against them at $14,7 billion. In January 2017, VW pleads guilty and enter a settlement with the US justice department for $4,3 billion. Furthermore makes a $1,3 billion settlement to owners of the 3.0 liters diesel motors that where placed in Audi, Volkswagen and Porsche[8]. VW decided to recall upwards of 8,5 mio of the cars in Europe alone, which were affected by the illegal software[7]. People experienced a loss of power and acceleration as well

---

[6] Engineering a Deception: What Led to Volkswagen's Diesel Scandal.

[7] Volkswagen recalls 8.5 million diesel vehicles in Europe.

as worse fuel consumptions after recalls meaning that cars where also affected even after the removal of the software[8]. This has provided further challenges to VW who in the start said that a software update could fix the problem but as it turns out people have experiencing problems with the cars even after the recall.

VW has responded to the crises through social media and they have also opened an Internet site vwdieselinfo.com where the consumers can seek guidance if you are included in the range of cars, which has the illegal software, installed. However VW did experience some problems with their communication. A few instances did occur where they where denying any wrongdoing and fed regulators with wrongful information and then a few weeks later admitted the misdemeanor. Their social media tactics has been orientated towards being reactive rather than proactive. This is evident in the in the time between that the Viginia team publishes the report in May 2014 to the point when VW actually admits wrongdoing in September 2015. They did not admit nor acknowledge the problem right away even after different reports where published which proved that the cars did have lower emission during test than during normal use. Later on the have corrected this behavior. Despite their social media strategy it becomes interesting to analyze what reaction the crisis caused on social media.

3   Research Methodology

In the methodology section we will be focusing on the data and the analysis techniques we have used. First, we collected four different datasets related to financial measures e.g. stock prices, sale numbers etc. Facebook data was collected using the Social Data Analytics Tool SODATO [12]. The purpose of Facebook data was to measure the social media engagement and the social media conversations about Volkswagen. This data was collected for the years 2015 and 2016 where Volkswagen admitted to placing fraudulent software in 11 million diesel vehicles. The data was further transformed into a file that contains only the posts and comments without any likes, specifically made to process through Text Analysis Tools. We used this file but also split it into two different CSV files which were called: *Before scandal* and *After scandal*. The distinction between these two were made by the dates 17th of September 2015 and 18th of September 2015. Every comment that was posted before the 17th of September where classified as *before the scandal* and every comment from the 18th of September and onwards was classified as *after the scandal*. For the social media data we had to clean for empty text comments and other gibberish comments meaning the data had dirty noise that needed to be removed before we could continue using the data.

3.1   Data Transformation

In order to analyse the comments, we have chosen four domain specific models (Information Type, Sentiment, Consumer Decision and Personality Traits)

---

[8] Up in smoke: the VW emissions fix has left our car undriveable.

based on their relevance and importance to our dataset and field of inquiry. The description of the models are shown in Table 1. We wanted to identify the comments which are actually about the scandal and therefore we have used supervised machine learning algorithm to build a classifier that can label the comments according to our models. Therefore, we started manually labelling different comments according to the models described in Table 1, to prepare a training-set that can be used to train the classifier.

As part of *Information Type* model, carRace was labelled based on the comments that were made specifically for the Red Bull Global Rallycross race that was during the year 2015. Moreover, 'Information' was labelled based on most of the general comments. This included a whole variety of comments which did not have anything to do with the scandal or the Red Bull race. The label 'Scandal' includes all the comments that revolve around the diesel scandal and these comments are highly relevant to our paper since we want to know how much social media attention the scandal received. These comments contained both positive and negative sentiments in which some people react negatively and "attack" Volkswagen and others that decide to defend Volkswagen. This often leads to discussions between the two groups. Being able to analyze how many of the comments that were aimed towards the scandal would be very helpful in understanding if and how much attention the scandal received outside of the news. Lastly, the label 'Irrelevant' entailed the comments where people tag each other, so the comment is only a name. In the end, 1500 comments were labelled to prepare training set for the classifier.

The second classification model *Sentiment* contains the labels positive, neutral and negative for the attitudes expressed in the social media data. The comments that had a positive attitude towards Volkswagen, the scandal or just in general were marked as positive. Negative attitudes were based on when people were angry or sad about the scandal or against Volkswagen in general. Neutral attitudes happened when the data was just informative, were objective or just asking questions without having a specific attitude towards Volkswagen. The third classification model (Table 1) is based on the Hierarchy of Effects model which contains 6 stages of a consumer's decision making. These stages are: 1. Awareness, 2. Knowledge, 3. Liking, 4. Preference, 5. Conviction, 6. Purchase. The first two steps are also called the cognitive stage which is where the consumer processes as much information as possible about his/her need and the product. The next two steps are also called the affective stage which is where the consumer figures out his/her attitude towards the brand. These steps are can also be more emotional than the cognitive stage where you only look at information. The affective stage is where you look at how you actually feel about the brand and if you would prefer to buy that brand over another brand. The last stage contains the last two steps in which you combine the first two stages and make your decision based on them. This is called the Conative stage [13].

The fourth model (Table 1) is the five factor model which is based on personality traits, which are Extroversion, Neuroticism, Openness, Agreeableness, and conscientiousness. Extroversion is characterized as being sociable and having

Table 1. Domain specific text classification models

Label	Definition
Text Classification Model: **Information Type**	
CarRace	Comments talking about Volkswagen car race team
Information	Comments discussing about cars but not related to diesel scandals, including the functions of their cars and new car models etc
Scandal	Comments directly talking about diesel scandals, including positive, negative and neutral attitude
Irrelevant	Comments like names or information can't be classified into any above
Text Classification Model: **Sentiment**	
Positive	Comments shows supportive, joyful and/or encouraging attitude
Negative	Comments shows unsupportive, sad and/or bad attitude
Neutral	Comments shows attitude that is neither positive or negative
Text Classification Model: **Consumer Decision**	
Awareness	Consumer is aware of the presence of your brand in a particular product segment
Knowledge	Consumer have certain knowledges about the product and will evaluated product against other brands
Liking	Consumers like the product and started to consider emotional benefits
Preference	Consumers maybe convinced to try out the products but may like other brands too
Conviction	Consumers doubt about buying the product might be converted into action. Consumers at this stage would decide whether stick to the brand
Purchase	Consumer decided to buy the product
Text Classification Model: **Personality traits**	
Openness	Describe a general openness to new ideasm, experiences and is related to curiosity, adventure and imagination
Conscientiousness	Describe an individual who aims for achievements and expresses a propensity to be thoughtful, thorough
Extraversion	Extraversion is often opposed to introversion and the extravert is often the center of attention, out-going, socially comfortable, energetic and likes to talk
Agreeableness	Describes an individual is focused on establishing consensus to achieve social harmony. Such individuals often conform to social norms and are usually generous, trustworthy, optimistic, caring and emotionally supportive
Neuroticism	This trait is linked to emotional instability, anxiety and depression. Individuals labeled with neuroticism will be vulnerable and emotionally reactive

interest in others and being confident in new environments. Neuroticism based on stability, anxiety and volatility. The word has a negative connotation to it but can mean both low and high stability, anxiety and volatility. In our paper we will be looking at Neuroticism as a negative trait towards VW. Openness is about how welcoming you are towards new ideas and situations. Agreeableness is how you get along with others. Conscientiousness is based on the high amount of consideration you have towards others before making decisions [14]. Furthermore, Keyword analysis was used on both the before and after scandal comments. With this we were able to see if there were any big differences which words were used after the scandal hit. For the social media data the same techniques were used with more focus on specific time periods regarding different events of the scandal e.g. the news about the scandal, the recalls and goodwill packages etc. the focus was then put on the different sentiment analysis that we had to see how the sentiments changed because of these events.

3.2   Financial and Sales Overview

During the diesel scandal Volkswagen went through a rough media storm and with a decreasing stock price. Overall there has been as slow increase of cars delivered in the period. In the period from 2012–2017 the trend of the revenue has been increasing as shown in Fig. 1, which means that there has not been a major setback in revenue during this time period. The profit development as seen in Fig. 1 in the same period has been volatile and with a decreasing trend. This can be attributed to the severe litigation cost that VW has been subject to, which lies in the range of $25 billion[9]. The two periods with negative profit where incurred in Q3 (−$1,6 billion) and Q4 (−$1,3 billion) of 2015. Different factors can be attributed to this fact, such as litigation cost, recall, cost of fixing the cars and these provisions caused a deficit. VW Group has not incurred any deficit in 2016 and 2017, meaning that they once again started earning profits. The performance of Volkswagen has been positive during the scandal, both when

Fig. 1. Distribution of profits for VW brand

---

[9] Volkswagen falls to biggest annual loss in its history.

looking at the revenue as well as the number of delivered cars. The only area where the financial performance of the VW group was compromised was in terms of the profits, which were severely affected by the litigation cost and the Recall cost as shown in Fig. 1. Even though the number of delivered cars increased for the sub-brands Audi, Skoda and Seat, the number of delivered cars did decrease for the main brand VW.

4   Results

In this section, we will present results of our data analysis on VW dataset.

4.1   Proposition 1

Figure 2 shows the topic distribution of VW after the scandal. As we could see in Fig. 8, more than 26,000 comments are specifically talking about the scandal. This is twice as much as the second most frequent topic which is 'information'. 'Irrelevant' and 'CarRace' were not mentioned very frequently.

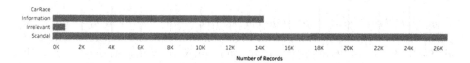

Fig. 2. Distribution of comments for information type

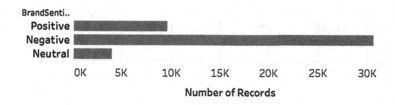

Fig. 3. Distribution of comments for sentiment

We have classified the dataset according to text classification model *Sentiment* to analyze people's attitude toward the scandal. As we could see in Fig. 3, comments that show positive sentiments and negative sentiments generally have the same frequency, which is different from what we expected. When we look into the details of different sentiments, the main points varied differently. The data was manually examined in order to clarify what topics were dominant in the positive, negative, and neutral.

For the positive attitudes, people were defending Volkswagen from different angles. Many of the defenders are diesel car owners or loyal customers.

They defend Volkswagen because of the high-quality cars and also the accumulated brand reputation of Volkswagen. However, among the main points in the scandal discussion, what surprised the most is that many diesel car owners did not seem to care about the emission. These owners showed intense dislike to environmental friendly cars like Prius. Furthermore, some people compared VW to other car manufacturers who had recalled cars due to safety issues in 2015, many Volkswagen owners expressed that these safety issues were worse than that of VW because this was not a safety issue but emissions issue. Their point was that increased emission does not risk your life when driving but safety issues do. People with negative attitudes were generally people concerned about the environment and many of them were also loyal Volkswagen customers who felt that the company had cheated them. Neutral opinions tend to be more rational and concerned more about side effects by the scandal. These comments were concerned more about how Volkswagen would fix the problem and why it happened in such a high reputation company. If Volkswagen were doing this, then there might be more car manufacturers doing the exact same thing, tricking people into believing that they were buying low-emission vehicles.

4.2   Proposition 2

The negative brand sentiment outweigh positive brand sentiment after the diesel scandal breaks out on Sep 18th, 2015. Fig. 4 shows the brand sentiments for 2015 and 2016. The overall tendency is that the negative brand sentiments outweighs the positive ones. However, during the outbreak of the diesel scandal in September 2015 there were far more negative brand sentiments than before and after. After the scandal the negative brand sentiments dropped. In the period after the diesel scandal the negative brand sentiments still outweighed the positive. This is perhaps due to the fact that people were not reacting to the news about the goodwill package positively it could also be that people did not react to the news of the recalls positively.

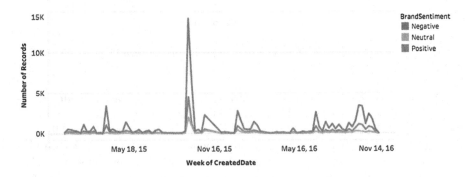

Fig. 4. Distribution of sentiment after the scandal

4.3   Proposition 3

The positive sentiment outweighs negative sentiment after Volkswagen apologize for the scandal on Sep 25th, 2015 Fig. 5a shows the reaction from the 18th when the news came out and to the 24th, which was the day before the apology. The figure shows that Facebook users are generally reacting negatively to the news about the illegal software, in the time before VW apologizes. There are almost twice as many negative sentiments as positive.

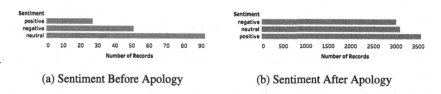

(a) Sentiment Before Apology   (b) Sentiment After Apology

Fig. 5. Sentiment before and after apology

In the time after the scandal the discourse in the comments of VW turned to focus on the scandal. Therefore, it becomes relevant to analyze the timeline from the 25/09/2015 to 01/10/2015 this entails the first week after VW apologized. The data from the Facebook sentiments in the week after VW apologized is depicted in Fig. 5b.

The stock price adjusted itself after the news where the lowest stock price was reached 28th of September just around the time where VW apologizes. This is accordance to the theory of Chen et. al. (2009) that when uncertainty and the issue of a product recall arises, evidence show that the general impact is a decreasing of stock price. A further reason which could explain that the stock price stopped decreasing shortly after the news broke, can be attributed to the fact that stakeholder was informed of the crisis and was provided with an apology. This is in accordance to the theory of Balaji et. al (2016) and Dawar (1998) that it is important for a firm to quickly acknowledge and admit to problems when things go wrong.

4.4   Proposition 4

Consumers' positive sentiment will outweigh the negative after VW announces the recall and goodwill package, because Volkswagen is actively trying to fix the problem. During the process of fixing the problem, Volkswagen provided a goodwill package, which was a $500 prepaid visa card, to all diesel car owners (motortrend, 2017). It aimed to compensate for the inconvenience of the diesel problem for the VW customers. The people generally did not respond well to the goodwill package. The negative sentiments far exceeded both the positive and the neutral sentiments. Some of the worries that users expressed were related to the resale value of their cars.

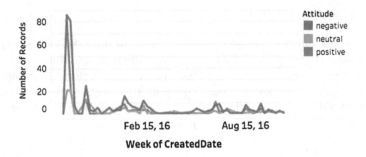

Fig. 6. Sentiment after recall

The next area to investigate is how people reacted to the recall of cars. Figure 6 shows that a lot of negative sentiments were expressed after recall. This indicates that the reactions to the recall confirms what the theory assumes in the sense that people in general show a negative attitude towards recalls.

4.5   Proposition 5

Product recall would cause an emotional reaction from consumers and therefore we expect that the personality trait of neuroticism is the most dominant personality trait. The distribution of personality trait model shown in Fig. 7, most people show sign of neuroticism and extraversion, which indicate that they might be 'vulnerable' and 'emotionally reactive', or 'out-going' and 'like to talk'. This can explain why people are commenting frequently when the news broke about the diesel scandal and again when VW apologized.

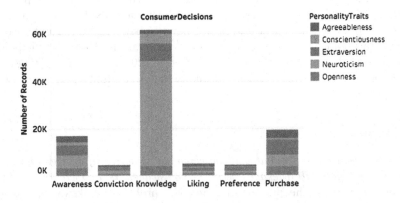

Fig. 7. Consumer decision versus personality traits

4.6   Proposition 6

The scandal would have negative influence to consumers' purchase behavior. The Fig. 8 shows that there are more 'conviction' than there is regarding 'liking' and 'preference'. The 'purchase' data also increased during the scandal. It might be caused by the Volkswagen owners who defended the company. However, just like the scandal related comments and brand sentiment, it soon returned to the normal level which indicate that the scandal did not have long lasting influence to consumer purchase behaviors. Furthermore, when looking at the Sales Overview, it is noticed that the sub-brands of Audi, SEAT and Skoda were all increasing the number of cars which they delivered and VW was the only brand that experienced a decreasing number of cars delivered. This could indicate that the consumer purchasing behavior did not change the sales much in the aftermath of the scandal. Therefore the data does not show that the scandal had negative influence on the people purchasing behavior.

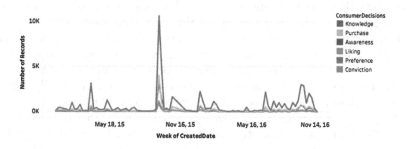

Fig. 8. Distribution of consumer decision process labels

5   Discussion and Conclusion

Facebook users expressed mainly negative brand sentiment during and after the crisis and they did also express negative sentiments toward the recall and goodwill package. Furthermore the research also proved that scandal was the dominant topic. Therefore at lot of negative sentiments was expressed regarding sentiment, recall and goodwill package but during the last 2 quarters of 2015 the negative sentiments decreased severely again, and the brand sentiments reached approximately the same levels as before the crisis. People also were positive towards the apology of VW. This all points in the direction of why the VW revenue was not affected much by the scandal as it showed an increasing trend from 2012–2016. The number of delivered cars by the VW Group decreased in the third and fourth quarter of 2015 when comparing to the same numbers from 2014. In the same period negative brand sentiments increases greatly during the scandal and decreased to normal level suggesting that VW recovered relatively quickly from the scandal. The profit was however affected greatly by the scandal due to litigation cost and the recall cost (Table 2).

Proposition 1 was not supported because the findings did not suggest that the negative sentiment outweighed the positive sentiment after the crisis broke out as shown in Fig. 3, where both positive and negative sentiment was almost equal. Evidence suggests that proposition 2 is supported because the negative brand sentiment outweighs the number of positive brand sentiment in period after the

Table 2. Propositions overview

Proposition	P1	P2	P3	P4	P5	P6
Empirical findings	Not supported	Supported	Supported	Not supported	Supported	Not supported

news of the diesel scandal breaks out. It can be a sign of high brand equity that the negative outburst decreases so quickly after crisis. This could indicate that VW has a very strong brand equity as sales fairly stable throughout the diesel scandal, where the total number of cars for the VW Group experienced a small decrease in 2014 compared to 2015 but recovered quickly in 2016 with sales numbers which were even higher than in 2014. This is evidence of strong brand equity that VW were able to recover those sales numbers so quickly after the diesel scandal. This was proven by Hsu and Lawrence [11, 15] that high brand equity firms were able to withstand negative word-of-mouth better than low equity firms, where firms can use the high equity as a buffer in relation to crisis. Proposition 3 is supported because the positive sentiments outweighs the negative sentiments after VW apologizes. This could indicate that VW managed to minimize the information asymmetry between VW and their stakeholder, by providing are clear statement where they acknowledges the that the installed the illegal software and in the same time apologized for their actions. The shareholders did also seem to receive the news well as the stock price reached its lowest level on 28th of December before stabilizing and going upwards. This is also in accordance to theory which predicts when a firm acknowledges a problem the potential backlash from stakeholders can be decreased [11].

The negative brand sentiments, which were observed in relation to the recalls, is also in accordance with what the theory predicts. This is evident that when a product recall is materialized the financial cost increases, the uncertainty increases and the stock price decreased as consequence, this was also evident in the VW case [16]. The findings from this paper suggest that the fourth proposition is not supported because the positive sentiments did not outweigh the negative sentiments after VW announces the recall and goodwill package. The findings show that the as the neuroticism is the dominant personality trait and therefore the fifth proposition is supported. The findings also show that Proposition 6 is not supported because the scandal did not have a continually negative influence on consumers' purchasing behavior. This is also evident according to two different metrics, the revenue and the number of cars delivered. Both these metrics a higher in 2016 than in 2015.

5.1   Implications for Research and Practice

Our findings are in line with the extant literature on big social data analytics of corporate crisis [17–19] which show an volumetric increases in engagement during the crisis with a proportionate increase in negative or positive sentiments depending on the crisis type and a regression to the mean user engagement. For researchers, this empirically demonstrated lack of persistent negative social media engagement, implies that slacktivisim needs to accounted for in the analysis of big social data for corporate crisis and other social movements. For companies, this implies that reputational risks from corporate crises due to social media crises can be managed with a suitable crisis communication and management strategy and in the final analysis might not be a determining factor. Further research is needed to further understand and better estimate these effects.

References

1. Kotler, P., Keller, K.L., Brady, M., Goodman, M., Hansen, T.: Marketing Management, 2nd edn. Pearson Education, London (2012)
2. McGriff, J.A.: A conceptual topic in marketing management: the emerging need for protecting and managing brand equity: the case of online consumer brand boycotts. Int. Manag. Rev. 8(1), 49 (2012)
3. Mukkamala, R.R., Hussain, A., Vatrapu, R.: Towards a set theoretical approach to big data analytics. In: proceedings of 3rd International Congress on Big Data (IEEE BigData 2014), June 2014
4. Vatrapu, R.: Understanding social business. Emerging Dimensions of Technology Management, pp. 147–158. Springer, India (2013)
5. Xu, K., Li, W.: An ethical stakeholder approach to crisis communication: a case study of foxconn's 2010 employee suicide crisis. J. Bus. Ethics 117(2), 371–386 (2013)
6. Balaji, M., Khong, K.W., Chong, A.Y.L.: Determinants of negative word-of-mouth communication using social networking sites. Inf. Manag. 53(4), 528–540 (2016)
7. Chung, J.Y., Buhalis, D.: Information needs in online social networks. Inf. Technol. Tour. 10(4), 267–281 (2008)
8. Cheung, C.M., Lee, M.K.: What drives consumers to spread electronic word of mouth in online consumer-opinion platforms. Decis. Support Syst. 53(1), 218–225 (2012)
9. Pace, S., Balboni, B., Gistri, G.: The effects of social media on brand attitude and wom during a brand crisis: evidences from the barilla case. J. Mark. Commun. 23(2), 135–148 (2014)
10. Keller, K.L., Lehmann, D.R.: Brands and branding: research findings and future priorities. Mark. Sci. 25(6), 740–759 (2006)
11. Dawar, N., Pillutla, M.M.: Impact of product-harm crises on brand equity: the moderating role of consumer expectations. J. Mark. Res. 37(2), 215–226 (2000)
12. Hussain, A., Vatrapu, R.: Social data analytics tool: Design, development, and demonstrative case studies. In: IEEE 18th International Enterprise Distributed Object Computing Conference Workshops and Demonstrations (EDOCW), pp. 414–417, September 2014
13. Proctor, T.: Creative problem solving for managers. Developing Skills for Decision Making and Innovation. Routledge, London (2010)
14. Digman, J.M.: Personality structure: emergence of the five-factor model. Ann. Review psychol. 41(1), 417–440 (1990)
15. Hsu, L., Lawrence, B.: The role of social media and brand equity during a product recall crisis: a shareholder value perspective. Int. J. Res. Mark. 33(1), 59–77 (2016)
16. Chen, Y., Ganesan, S., Liu, Y.: Does a firm's product-recall strategy affect its financial value? an examination of strategic alternatives during product-harm crises. J. Mark. 73(6), 214–226 (2009)

17. Flesch, B., Vatrapu, R., Mukkamala, R.R., Hussain, A.: Social set visualizer: a set theoretical approach to big social data analytics of real-world events. In: Proceedings of Mining Social Data Workshop at the 2015 IEEE International Conference on Big Data (IEEE Big Data)., IEEE Xplore (2015). in press
18. Mukkamala, R.R., Sørensen, J.I., Hussain, A., Vatrapu, R.: Detecting corporate social media crises on Facebook using social set analysis. In: IEEE International Congress on Big Data (BigData Congress), pp. 745–748. IEEE (2015)
19. Mukkamala, R.R., Sørensen, J.I., Hussain, A., Vatrapu, R.: Social set analysis of corporate social media crises on Facebook. In: IEEE 19th International Enterprise Distributed Object Computing Conference (EDOC), pp. 112–121. IEEE (2015)

LeadsRobot: A Sales Leads Generation Robot Based on Big Data Analytics

Jing Zeng[1,2,3]([✉]), Jin Che[2], Chunxiao Xing[1], and Liang-Jie Zhang[2,3]

[1] Research Institute of Web Information, Tsinghua University, Beijing, China
jerryzengjing@gmail.com
[2] Kingdee International Software Incorporation, Shenzhen, China
[3] National Engineering Research Center for Supporting Software
of Enterprise Internet Services, Shenzhen, China

Abstract. Sale leads are the essential concern of salesman and marketing staffs, who may seek them in blindly searching by using search engine in a substantial of online information. Unfortunately, it is tricky to extract useful and valuable leads from such huge online data. To address this issue, in this paper, we present a leads generation robot-LeadsRobot, which is a software enabled robot. It can intelligently understand the requirements of leads for salesman and then automatically mine the leads from web big data to recommend them to salesman. A robot architecture is devised with service based technologies, it can accomplish the automatic understanding, crawling, analysis and recommendation. To achieve the task, we use automatic web crawling to gain the raw data from web data. Natural language processing is employed for extract leads from them, then intelligence recommendation is proceeded for salesman via word2vec based text analysis. Finally we demonstrate our proposed robot in a real application case and evaluate performance of system to show its efficiency and effectiveness.

Keywords: Sale leads · NLP · Big data · Robot

1 Introduction

Salesman plays a critical role in a company due that any products need finding their target customers to achieve the performance. Therefore, salesman or marketing staffs pay great attentions in discovering the sale leads. Using search engine to find the leads is a common channel, for instance, salesman may search a keyword with specific product, company, industry or location in search engine to find potential customers; however, it is a non-trivial task for the salesman in a varied market environment. First, it will cost a large amount of time and energy for a salesman to extract information from massive web data, which greatly hinders the progress of salesman. Second, it is relative inefficiency and insufficient to mine useful and available leads from data. Third, many leads may be unconsciously missed by the salesman due to the limitation of human attentions.

© Springer International Publishing AG, part of Springer Nature 2018
F. Y. L. Chin et al. (Eds.): BIGDATA 2018, LNCS 10968, pp. 277–290, 2018.
https://doi.org/10.1007/978-3-319-94301-5_21

Traditional business intelligence (BI) [3] focuses on offering the decision support for enterprises via inner data created in production procedure. For example, leveraging analyzing the sale data of enterprises, the user can find the law and predict future sale trend for improving product sale strategies. However, little works mention how to automatically discover sales leads by intelligent technologies from external data environment (web data).

To address these challenges, we present a software robot named LeadsRobot, which can discover and mine the leads from web big data according to the requirements of the salesman. It can enable the real-time leads generation via big data analytics technologies and natural language process (NLP) [6], and it can provide the customized leads recommendation based on the salesman's personas and the history feedback leads. Specifically, we attain the keywords of leads by semantic understanding technologies for the salesman's natural language input, and then employ the keywords to automatically the crawling the raw leads from public web data leveraging TextRank and Word2Vec technologies. Subsequently, the crawled data is automatically analyzed to mine potential leads by Name Entity Recognition (NER) for Chinese company name.

The main contributions of this paper can be summarized as follows:

- We propose LeadsRobot, which is a framework of leads discovering robot. It can achieve the automatic mining sales leads from the public web data.
- Multiple NLP technologies including TextRank, Word2Vec and NER are combined to enable the raw web data processing and the leads data extraction.
- The robot can accomplish the intelligent recommendations according to user' personas and feedback of leads data from the salesman.

The reminder of this paper is organized as follows: Sect. 2 introduces the related work about sales intelligence and robot. In Sect. 3, we present a framework of leads robot and the details. Section 4 gives the application of LeadsRobot. For Sect. 5, the performance and evaluation for LeadsRobot is described. Finally the conclusion is summarized in Sect. 6.

2   Related Work

Recent years robot has attracted many intensive attention in academia and industry with the development of artificial intelligence. Software robot is an emerging direction for service based applications. In the work of [20], the authors presented a human pose tracking for service robot applications by a SDF representation model. The work in [19] proposed an approach to modeling social common sense of an interactive robot, which provides customers in a queue. The work of [12] proposed an indoor navigation service robot system according to vibration tactile feedback. All these service robots are physical robot.

Another type of robot is software robot, which seeks to provide a virtual service for the customers by embedding into a software, such as chat robot [17], business agent robot [8], shopping robot [5], crawling robot [18], etc. All these works in software robot are very attractive, however little works in sale intelligence which focuses on using big data analytics and artificial intelligence technologies to provide services for the salesman to improving the performance.

Different from traditional BI, sales intelligence advocates that using multiple analytics to find prospects from external data sources. Existing BI seeks to provide decision support for company managers based on existing data of companies. However, little attentions are paid on the salesman.

Many products of sales intelligence have been developed to bridge the gap. RainKing [14] offers IT sales intelligence solutions, which can transform data into actionable, forward-looking intelligence and help salesman speedily to discover prospects. It can generate sales leads via a substantial of web data. Besides, it can identify the needs of potential customers such as pain point, new funding, management change, project scoop, contract scoop, etc.

In the real scenarios, there are many sale intelligence products that are similar with RainKing, for instance, LinkedIn sales navigator [10] help users to discover the prospects and to understand the intention of buyers. Insideview [7] allows to collect social and business sales from various sources than competitors, and to appropriately deliver to your sales rep's computers and smart cellphone - online or inside your CRM. Other sales intelligence products, such as Discoverorg [4] SaleLoft [16], Zoominfo [21], etc, are developed to provide the promising function to address the pain point of the salesman.

Even though various sale intelligence products have been developed in market, currently it lacks effective technical framework and down-to-earth applications for enabling sale intelligence in sales leads discovering domain. To address this issue, we use big data analytics and NLP to create a leads robot that allows gaining the insight about sales leads from public web data.

3   The Framework of LeadsRobot

To address the issue of leads automation generation, in this section, we present a framework for designing leads robot. The robot can understand the requirements of salesman and then automatically crawl the public web data to analyze and to extract the sales leads for recommendation to the salesman. The salesman is able to give the feedback to the robot, which improves the accuracy of leads recommendation.

The framework is demonstrated in Fig. 1, firstly, the salesman can propose their requirements for leads by natural language, for example, the salesman can say he wants to find some medical industry companies and sell ERP products to them. By extracting the keywords from the natural language, the basic requirements words can be attained. They also called leads keywords which will be used for further mining from public web data. The requirements descriptions usually are simple and ambiguous that is insufficient to mine much more information from public web data.

Therefore, the second step is to discover much more support words in order to automatically crawling. The support words are the similar words with the leads keywords, which can be enabled by support words discovery model.

Subsequently, the model generates knowledge base of semantic support words. All the words in knowledge base are the features for the salesman.

Leveraging these knowledges, we can learn the demands of the salesman, and realize the customized crawling for him meanwhile recommend proper leads via leads analysis engine. The automatic web crawling is able to gain the raw leads data from public web data by putting the words of the knowledge base into an automatic crawling queue, such as news website, recruitment website, funding website, project scoop website, etc. The crawled data are analyzed to create raw leads in real-time and then they are proceeded for word segment and indexed for a quick query. Lead analysis engine is responsible for recommending leads to salesman according to the knowledge base of semantic support words, user personas, history leads data from raw leads. Meanwhile it can achieve the automatic contact crawling for each leads for the salesman to reach the customers.

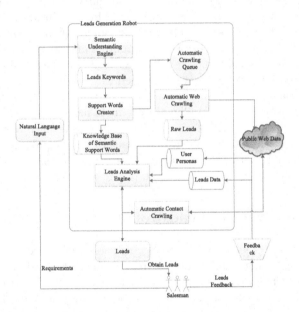

Fig. 1. The framework of LeadsRobot

3.1 Semantic Understanding Engine

Semantic understanding engine aims to understand the questions proposed by the salesman and further to extract the keywords included in the text. The result of the semantic understanding is leads keywords which are extracted via NLP approaches.

To conduct the keywords extraction, firstly segmenting word for a sentence or a passage is the basis in Chinese keyword recognition due that Chinese segmentation is quite different from English word segmentation, which uses blank to segment word. In the real application of Chinese word segmentation, it heavily relies on corpus resources. We use Ansj[1] as a tool to proceed the Chinese word segmentation, which supports efficient word segmentation by optimized Trie Tree, however its memory usage is relatively lower. By using it, a sentence can be conversed into a set of words.

The second step is to recognize keywords according to the result of word segmentation. TF-IDF approach [15] is employed for keywords recognition. It combines the term frequency and inverse document frequency to discover the keywords in a document. The basic idea is to filter the common use words and to remain the vital word for a document. Regarding to a term frequency, for a given corpus D, word i, document j, it can be defined as: $TF_{ij} = \frac{n_{ij}}{\sum_k n_{kj}}$, where TF_{ij} denotes the numbers of word i occurs in document j, n_{ij} denotes the number that i occurs in document j, $\sum_k n_{kj}$ denotes the sum that all words occur in the document j. For inverse document frequency, $IDF_{ij} = log \frac{N}{|\{j \in D: i \in j\}|+1}$, where $|N|$ is the number of document in corpus D, $|\{j \in D : i \in j\}|$ is the document number where the word i appears in corpus D. Therefore, the $TFIDF_{ij} = TF_{ij} \times IDF_{ij}$, where $TFIDF_{ij}$ represents the importance for each word in a document.

We use TF-IDF to extract the keywords from the words given by word segmentation from natural language input. These keywords are closely related to the sales leads, which refer to the industry, products name, targeted sales distract, etc. The leads keywords are the basis to conduct leads mining from public web data. They to some extent express the expected leads intentions for the salesman.

3.2   Support Words Creator

Based on the semantic understanding engine, various leads keywords can be extracted from the salesman expressions. However, it is tricky and insufficient to use these limited keywords to leads from public web data. Therefore, support words creator is devised for discovering much more words about the leads keywords, these generated words are called support words which allow us to find much more similar and close words for the leads keywords.

The framework of support words creator is demonstrated in Fig. 2, a substantial of public web data is firstly gained from the Internet to be considered as a training dataset, these data is unstructured data in the initial period of gaining the data, and from multiple sources with diverse data structure, therefore the body of data is extracted and then data sharing is conducted by word segmentation processing.

Then TF-IDF and TextRank algorithm [9] are jointly used to model the data after word segmentation due to the poor performance of TF-IDF algorithm in the extraction of semantic support words regarding to short text. Meanwhile it does not consider the semantic context. TextRank algorithm has much higher complexity in the specific application context. It is suitable for combine TF-IDF

---

[1] https://github.com/NLPchina/ansj_seg.

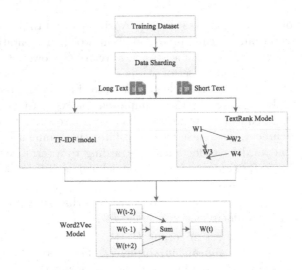

Fig. 2. The framework of support words creator

and TextRank algorithm to extract semantic support words. There are two types of text, which is short text and long text. The former is more suitable for using TextRank algorithm, while the later is appropriate for using TF-IDF. By using the combination of the two approaches, the framework can guarantee the quality of extraction meanwhile ensuring the extraction speed.

TextRank is a keyword extraction algorithm for short text, which is employed to identify the keywords within the news that is closely related to leads context. We use TextRank algorithm [9] to perform the keyword extraction for short text. Similar to PageRank which is a well-known search engine ranking algorithm successfully used in Google search, TextRank elaborately borrows the same idea in keyword extraction under long text context. The algorithm brings in the concept of the weight value of edge, and determines relationships among words via a sliding window. A good advantage is that the performance of algorithm only relates to input corpus without extra training cost.

The model of TextRank can be represented as a directed weighted graph, in the context of sentence; the edge represents the relationship between the words appearing in different window. Specifically, we use Algorithm in Fig. 3 to extract keywords of external data.

For the algorithm in Fig. 3, first, we use the HMM model by training to segment word for input data and get words. Second, we use Pos model [11] to filter the words, and merely remain the word list of assigned Pos to assign words Then, each word in words is considered as a node, and the window is set (i.e, 3), built the graph model where the weight of edge is the sum of distance between two words. Furthermore, we use TextRank Algorithm (getWeight) to gain the importance of each word. Finally, we choose the first n words which have the highest weight as keywords. In the getWeight approach, for each input word and graph model word_graph, we will output its importance. First, we

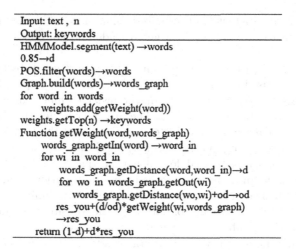

```
Input: text , n
Output: keywords
HMMModel.segment(text) →words
0.85→d
POS.filter(words)→words
Graph.build(words)→words_graph
for word in words
      weights.add(getWeight(word))
weights.getTop(n) →keywords
Function getWeight(word,words_graph)
      words_graph.getIn(word) →word_in
      for wi in word_in
           words_graph.getDistance(word,word_in)→d
           for wo in words_graph.getOut(wi)
                words_graph.getDistance(wo,wi)+od→od
           res_you+(d/od)*getWeight(wi,words_graph)
                →res_you
      return (1-d)+d*res_you
```

Fig. 3. Algorithm of keywords extraction by TextRank

gain all the word set word_in which haves a path to word, and get the weight d from word to word_in. Then, for each node wi in word_in, we iteratively find the reachable node for wi, and calculate the distance sum of them to assign value for od. Subsequently, we calculate the sum of (d/od)*getWeight(wi, words_graph), and assign value to res_you. Lastly, we gain the (1−d)+d*res_you, which is the importance of input data.

Due that the data for training and modelling is quite limited, the discovered support words is insufficient by using above approaches, therefore, word2vec model is used to generate some words which are similar with the discovered semantic support words. The word2vect model uses a substantial of public dataset for training, it consists of a number of Chinese words and phrases, which are mapped into fifty-dimensional float vector. The close words in semantic are also close in vector distance. The similarity between words can be gained via calculating the cosine value between vectors. Meanwhile, to accelerate the speed of iteration, all the vectors are portioned and take the close words into the same partition to speed the discovery of semantic support words. By using word2vec model to generate semantic support words can discover a large number of semantic support words for benefiting to mine external enterprise data mining from existing extracted semantic support words.

3.3 Data Crawler

Data crawler aims to gain the raw leads data from public web data, which achieve the automatic web crawling and automatic contact crawling. The former is to gain the raw leads data from a lot of open websites (business news, recruitment, funding, project scoop, etc), while the later is to discover the contact for the corresponding leads, such as phone number, email, leads address, etc. It benefits for the salesman to directly reach targeted leads and converses them into sale performance.

Therefore, based on the generated semantic support words by support words creator, it can be used as the features for describing the required leads of the salesman. By using them, we employ data crawler to mine much more similar data from potential targeted websites. The architecture for the data crawler is shown in Fig. 4, many existing websites have provided the potentially possible interface for data searching, but they usually set the strict restriction for visitors. Even if the result set is limited for each time query, we can build different search conditions to mine much more valuable information, semantic support words and website search interface can be used for data mining. Specifically, by examining the updated frequency of data of public websites, the timed scheduler can accomplish the reliable and efficient crawling of data by using IP resource pool. The public web data is parsed into raw leads database by data parser that extracts the raw leads data from the web page format files. We use the Beautiful Soup[2] to achieve the process. It is an open source html parser for web pages process.

3.4   Leads Analysis Engine

Leads analysis engine aims to analyse the leads from raw leads data from data crawler. The core part for leads is to recognize Chinese company name from a substantial of text. It is a tricky task for company name recognition, which is a typical NER task.

As the Fig. 2 shown, the input for leads analysis engine is raw leads, user personas, knowledge base of semantic support words and leads data from

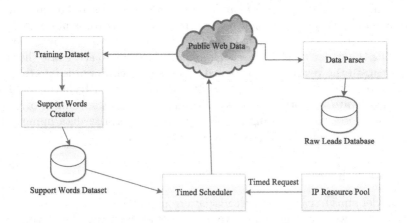

Fig. 4. The architecture for data crawler

history record of the salesman. Leads analysis engine proceed two types of tasks. First, it can achieve the company name recognition from the raw leads data. NER approaches are employed for addressing the Chinese company name recognition. Second, it allows recommending the leads for salesman according to his customized needs.

Regarding to the company name recognition, a NER service is developed to deal with it.

To proceed the NER, various libraries are needed to assist the recognition of company name entity. In our framework, it includes three libraries, which are company library, corpus library and rule library. Company library plays a critical role in recognizing the company name entity, all the recognized potential company names are compared with company library to verify the results. The corpus library can help to improve the accuracy of NER via company products and projects information due that most news mentioned are oriented to specific projects or products of companies. By using them, we can perform the search of a company to find potential companies. The other important issue is rule library. The task of NLP is quite tricky; usually NLP algorithms are difficult to tailor all the conditions. To address the incorrect recognition results, we need to establish a rule library to help filter the noise in the actual recognition. Assuming take the Chinese enterprises as an example, our system builds the following rules: (1) we choose several high-quality companies referred to specific location or geography. In these positions, more sales leads have relatively existed than other districts; (2) we set a series of blacklists to filter the words that seriously affect the accuracy of recognition. Especially for the noise in the company library, we shield them in recognition process; (3) we directly filter some unexpected characters occurring in the company name, such as digital, too short or too long company name, etc; (4) We also deleted the foreign names in appearing in the company name and some cartoon names in the company name. All these steps are to reduce the noise as far as possible.

To achieve the recognition of company entity, we use CRF based name entity algorithm via corpus library. The corpus In our framework is used to assist the word segmentation by using two-dimensional hidden Markov model (HMM) [2,13] in a substantial of labelled Internet data. Based on this, word segmentation dictionary is built to help the algorithms for NER. Then filtering by rule library, we can calculate the certainty factor and association factor.

As Algorithm in Fig. 5 demonstrates, the input is the text ready for recognition and output is the possible company name recognizing from the text. First, cleared bi-dimensional word transition matrix and generated probabilistic matrix are used to train for attaining HMM model, and the labelled training set by labelling name entity is used for training to gain the CRF model. Subsequently, HMM model is employed to segment the word of input data to get the words list. Furthermore, CRF model is utilized to mark the sequences of words for getting possible company name entity. Then, all possible name entities are used to match the company list built by using inverted index, and all matched company name entities are remained for gaining company_entities. Finally, we use all the

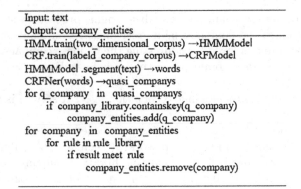

```
Input: text
Output: company_entities
HMM.train(two_dimensional_corpus) →HMMModel
CRF.train(labeld_company_corpus) →CRFModel
HMMModel .segment(text) →words
CRFNer(words) →quasi_companys
for q_company in quasi_companys
        if company_library.containskey(q_company)
                company_entities.add(q_company)
for company in company_entities
        for rule in rule_library
                if result meet rule
                        company_entities.remove(company)
```

Fig. 5. Algorithm of Chinese company name recognition

name entities in company_entities, to match the rules in rule_library in turn, and remove the name entities by rule and get the final results.

For the convenient of searching, we extract the keywords for indexing the leads (company name) by using the TextRank approach given in Sect. 3.2.

However, there are some inaccuracy recognized company names due to the context of text. Therefore, to improve the precision of company name recognition, we use the chi-square verification model [1] to improve the precision of final results. The basic idea is to calculate the divergence degree between theoretic value and observation value. By using the NER service and keyword extraction service, we can attain the company name and corresponding keywords. Based on them, we use manual to mark the company name and keywords partly, and take them as observation values, the generated results are considered as theoretic value, the chi-square model is used to match them and output the most

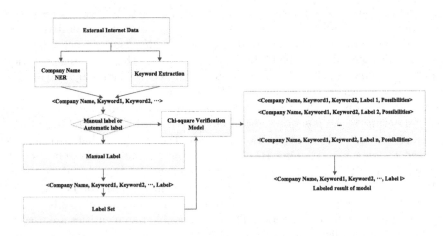

Fig. 6. The chi-square cerification model

significant company name and corresponding keywords list. Concretely, for each generated result, it can be represented as a two-tuple <company, keyword>, we mark it as a class, and then we calculate the chi-square value for each keyword. Furthermore, for each input two-tuple, we calculate its chi-square value to ensure its class (Fig. 6).

Based on the leads and keywords given by the above approaches, the leads recommendation for salesman can be conducted according to the following steps, firstly the list of semantic support words should be gained from the knowledge base of semantic words. Then we get the user personas of a salesman from user personas database and get the N users which have the similar action with the salesman. Furthermore, we attain the lead data about these users from the conversed leads data, and proceed statistic model to gain the statistic characteristics. Finally, we find the most closest M companies with these characteristics from leads database to as the recommendation leads for the salesman.

4   Case Study

To demonstrate the feasibility and effectiveness of our proposed robot framework, we employ a case about ERP sales in a real scenario to verify it. For the case, we collaborate with Kingdee International Software Group Company, which is a leading ERP solutions supplier in China. There are multiple product lines in ERP for different industry solutions. ERP sell is a non-trivial problem due that it is closely related to the information requirements for enterprises. The sales leads area critical for improving the ERP sale performance, however, it is inefficient and insufficient for sales to searching valuable leads from a substantial of web pages.

In our case, we recommend the high-quality leads to Kingdee product lines according to their diverse requirements. Taking Kingdee Cloud as an example, it is the flagship product of Kingdee ERP and provides the ERP service based on Cloud mode. The ERP is sold in various districts of China. According to their requirements, we recommend leads to the leads management staff, then he distributes them to the salesman in Kingdee Cloud. The leads are the prospects in buying ERP, which include all the information about the prospects (full company

Table 1. Critical N values

Recommendation date	District	Phone call number	Valid leads	Valid lead rate
2017/9/18	Shen Zhen	118	34	28.81%
2017/9/18	Shang Hai	96	24	25.00%
2017/10/12	Guang Zhou	141	9	6.38%
2017/10/12	Shang Hai	182	28	15.38%
2017/10/12	Shen Zhen	264	39	14.77%

name, industry, staff, register capital, address, contact email, phone number, etc.). All the leads are distributed to the salesman for following up (Table 1).

As shown, we use feedback results of three cities (Shen Zhen, Shang Hai, Guang Zhou) from sales in China. The sales sell the ERP by dialling the phone number by our provided leads. From the results, it can be found that Shen Zhen and Shang Hai have much higher valid leads. The average valid leads rate has exceeded 20%, while Guang Zhou is relatively lower than other two cities. Usually, according to the salesman feedback, the valid leads collected from web page are about 1%–2% by them, which is totally insufficient to satisfy their demands of need. The leads provided by our service can at least increase triple times in valid leads rate than manual searching by the web. It can significantly improve the conversion rate in ERP product sales.

According to the feedback from their CRM, the sale performance for Cloud ERP by using LeadsRobot is greatly improved, the number of signed contract from the recommendation leads is 135, the contract number is 256 and contract amount is 5.682171 million RMB in 2017.

5   Performance and Evaluation

To test the performance of LeadsRobot, we carry out the evaluation from the speed of leads generation and throughput of recommendation leads. The configuration for the cluster of leads collection and generation are indicated in Table 2.

Table 2. The configuration for the cluster

Computer ID	CPU	Memory	Bandwidth
1	Eight cores Intel (R) Xeon(R) CPU E5-26xxv3	32 GB	10 Mb
2	Eight cores Intel (R) Xeon(R) CPU E5-26xxv3	32 GB	10 Mb

The configuration of computer for throughput test is given in Table 3. From Oct. 10, 2017 to Feb. 28, 2018, we have collected the 150,7681 leads, and generated 13706 leads every day in average. In other words, the speed is that each lead is generated an average of 6.3 s.

Table 3. The configuration for throughput test

CPU	Memory
Four cores Intel(R) Core(TM) i5-6500 CPU @3.20GHz	32 GB

All the leads are generated relying on the crawling due to the automatic raw web data crawling and contact crawling. To test the throughput of the recommendation by LeadsRobot, we randomly selected 1000 from the leads database to test it, as the Fig. 7 is demonstrated, the average throughput of LeadsRobot is 689 per hour. According to the salesman, they can collect 100 leads every day at most; the LeadsRobot generates the leads for one hour that can be equal to the leads collected by two sales, which can greatly improve the efficiency of the salesman.

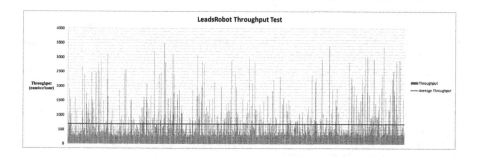

Fig. 7. LeadsRobot throughput test

6   Conclusions

The sales leads are critical for the salesman to carry out his business. Currently, it is inefficient for salesman to blindly search leads from a substantial web pages. Besides, the leads analysis by manual is for the salesman which greatly occupies his time.

To address the issue of lead gaining, in this paper, we propose a LeadsRobot which is a software robot and seeking to address the automatic leads discovery via analyzing the web big data on the Internet. Multiple text processing approaches are used for mining the web big data, web mining is proceeded by using TextRank and TF-IDF model, and a NER service is presented and developed for Chinese company name recognition. According to the feedback from the ERP sell scenario, the recommendation by LeadsRobot can greatly help the company to improve its performance of sales.

Acknowledgement. This work is partially supported by the technical projects No. c1533411500138 and No. 2017YFB0802700. This work is also supported by NSFC (91646202) and the National Hig-tech RD Program of China (SS2015AA020102).

References

1. Bryant, F.B., Satorra, A.: Principles and practice of scaled difference chi-square testing. Struct. Eqn. Model. A Multidiscip. J. **19**(3), 372–398 (2012)
2. Btoush, M.H., Alarabeyyat, A., Olab, I.: Rule based approach for arabic part of speech tagging and name entity recognition. Int. J. Adv. Comput. Sci. Appl. (IJACSA) **7**(6), 331–335 (2016)

3. Chaudhuri, S., Dayal, U., Narasayya, V.: An overview of business intelligence technology. Commun. ACM **54**(8), 88–98 (2011)
4. DiscoverOrg (2017). https://discoverorg.com/ . Accessed 8 Dec 2017
5. Gross, H.M., Boehme, H.J., Schröter, C., Müller, S., König, A., Martin, C., Merten, M., Bley, A.: Shopbot: progress in developing an interactive mobile shopping assistant for everyday use. In: IEEE International Conference on Systems, Man and Cybernetics, 2008. SMC 2008, pp. 3471–3478. IEEE (2008)
6. Hirschberg, J., Manning, C.D.: Advances in natural language processing. Science **349**(6245), 261–266 (2015)
7. Insideview (2017). https://www.insideview.com/. Accessed 10 Dec 2017
8. Masuch, N., Küster, T., Fähndrich, J., Lützenberger, M., Albayrak, S.: A multi-agent platform for augmented reality based product-service systems. In: Proceedings of the 16th Conference on Autonomous Agents and MultiAgent Systems, pp. 1796–1798. International Foundation for Autonomous Agents and Multiagent Systems (2017)
9. Mihalcea, R., Tarau, P.: Textrank: bringing order into text. In: Proceedings of the 2004 Conference on Empirical Methods in Natural Language Processing (2004)
10. Navigator, L.S. (2017). https://business.linkedin.com/sales-solutions/sales-navigator. Accessed 8 Dec 2017
11. Niehues, J., Kolss, M.: A pos-based model for long-range reorderings in SMT. In: Proceedings of the Fourth Workshop on Statistical Machine Translation, pp. 206–214. Association for Computational Linguistics (2009)
12. Peng, H., Song, G., You, J., Zhang, Y., Lian, J.: An indoor navigation service robot system based on vibration tactile feedback. Int. J. Soc. Robot. **9**, 1–11 (2017)
13. Rabiner, L., Juang, B.: An introduction to hidden markov models. IEEE ASSP Mag. **3**(1), 4–16 (1986)
14. RainKing (2017). www.rainkingonline.com/. Accessed 19 Dec 2017
15. Ramos, J., et al.: Using tf-idf to determine word relevance in document queries. In: Proceedings of the First Instructional Conference on Machine Learning, vol. 242, pp. 133–142 (2003)
16. Salesloft (2017). https://salesloft.com/. Accessed 8 Dec 2017
17. Sato-Shimokawara, E., Nomura, S., Shinoda, Y., Lee, H., Takatani, T., Wada, K., Yamaguchi, T.: A cloud based chat robot using dialogue histories for elderly people. In: 2015 24th IEEE International Symposium on Robot and Human Interactive Communication (RO-MAN), pp. 206–210. IEEE (2015)
18. Thangaraj, M., Sivagaminathan, P.: A web robot for extracting personal name aliases. Int. J. Appl. Eng. Res. **10**(14), 34954–34961 (2015)
19. Uotani, K., Kanbara, M., Nishimura, S., Kanbara, T., Satake, S., Hagita, N.: Social common sense modeling of a spread-out queue in public space for a service robot. In: Proceedings of the Companion of the 2017 ACM/IEEE International Conference on Human-Robot Interaction, pp. 311–312. ACM (2017)
20. Vasileiadis, M., Malassiotis, S., Giakoumis, D., Bouganis, C.S., Tzovaras, D.: Robust human pose tracking for realistic service robot applications. In: Proceedings of the IEEE Conference on Computer Vision and Pattern Recognition, pp. 1363–1372 (2017)
21. Zoominfo (2017). https://www.zoominfo.com/. Accessed 8 Dec 2017

Research on the High and New Technology Development Potential of China City Clusters Based on China's New OTC Market

Liping Deng[1,2(✉)], Huan Chen[1,2], Liang-Jie Zhang[1,2],
and Xinnan Li[1,2]

[1] National Engineering Research Center for Supporting Software of Enterprise
Internet Services, Shenzhen, China
[2] Kingdee Research, Kingdee International Software Group Company Limited,
Shenzhen, China
liping_deng@kingdee.com

Abstract. This article builds the index of enterprise competitiveness based on the annual reports of the New Third Board of China, and then proposes the index of city competitiveness, which reflects the competitiveness of cities in high and new technology field. We select the critical values to get the city high and new technology development level, combined with the heat map analysis of the status quo of China urban high and new technology development and forecast the future development trend of Beijing-Tianjin-Hebei, Yangtze River Delta, Pearl River Delta and Chengdu-Chongqing science and technology ecosystem. Finally, the development potential of science and technology in china urban clusters is compared and analyzed.

Keywords: Enterprise competitiveness index
Urban high technology development level
Science and technology ecosphere · China city clusters

1 Introduction

This article builds the enterprise competitiveness index [1–3] based on the talent reserve and financial status of new three board enterprise of China, reflecting the overall competitiveness of high and new technology enterprises in sustainable profitability. And according to the city where the enterprise is registered, the enterprise competitiveness index [4–6] is summed up to obtain the Chinese city competitiveness index, which reflects the city's sustainable competitiveness in high and new technology field. Combined with the rank of high and new technology urban development in China, the status quo of high and new technology city development in China during 2013–2016 is analyzed. The development potential of high and new technology in major science and technology ecosphere [7–9] is predicted, which will provide reference for the city's high and new technology industry layout.

2 Research Background

China's new OTC (Over the Counter) Market [10] originated in the "National Equities Exchange and Quotations" in 2001, the earliest to undertake two companies and delisting companies, commonly known as the "old three boards". In 2006, Zhongguancun Technology Park unlisted shares into the agency transfer system to transfer shares quoted, known as the "new three boards". With the gradual improvement of the new three board market, China will gradually form a multi-level capital market system, including motherboard, gem, over-the-counter counter trade network and property market. In 2012, with the approval of the State Council, it decided to expand the pilot share transfer of unlisted joint-stock companies. The first batch of pilot projects was to expand Shanghai's Zhangjiang High Technology Industrial Development Zone, Wuhan East Lake New Technology Industrial Development Zone and Tianjin Binhai High Technology Zone. Since December 31, 2013, the share transfer system has been receiving application for listing enterprises in the country. As of December 31, 2016, the number of new three board listed companies was 10,163, with a total share capital of 585.15 billion shares, and the total market capitalization of listed companies reached 4.055811 trillion yuan.

The competitiveness of enterprises [11] refers to the comprehensive ability of enterprises to realize their own value through creating their own resources and capabilities, acquiring externally-addressable resources and using them comprehensively under the competitive market conditions. The core competitiveness of China's new OTC business is the line of endowment of talent; the number of employees in this paper reflects the number of employees in the enterprise reserve situation. The accounting identity reveals the operating results of a company in a specific period. This article chooses the total operating costs, total operating income and total operating profit as indicators to reflect the long-term sustainable profitability of the enterprise.

High and new technology [12] refers to the advanced technology group that has a far-reaching impact on the politics and economy of a country or a region, and can form an industry. National high and new technology zones and industrial bases have become important growth poles for promoting regional economic development and provide strong support for the cultivation and development of strategic emerging industries. The purpose of this paper is to study the status quo, trends and laws of high and new technology development in cities in China through China's new OTC enterprise competitiveness index and to forecast the potential of high and new technology in major science and technology ecosphere.

3 China's New OTC Enterprise Competitiveness Index

This paper through the crawler technology, accesses China's new OTC corporate annual report from the online authoritative data source. After finishing the collection of a total of China's new OTC annual report number as follows:

From the enterprise annual report, select the number of employees (a), total operating income (million), total operating costs (million) and total operating profit (million) four indicators of numerical values (Table 1). In this article, the 100-point system, the indicators of various years the value of data cleaning, extreme value processing and normalization, and then in accordance with the various dimensions of "equal rights" proposes enterprise competitiveness index [13].

Table 1. National enterprise annual report 2013–2016.

Year	2013	2014	2015	2016	Total
Annual report number (copies)	356	1572	5129	10163	17220

3.1 Data Cleaning

Uniform data types and handle missing values. Data cleaning follows the principle of "total operating revenue − total operating costs = total operating profit". After cleaning, the number of indicators is 17,100.

3.2 Extreme Value Processing

For a given indicator, if the indicator value is higher than the mean plus twice the standard deviations, we define it as positive outliers; if the indicator value is smaller than the mean minus twice the standard deviation, we define it as the negative outliers. These outliers are replaced by the respective maximum and minimum values observed over all the years and all enterprise.

3.3 Normalized

Given a target value of x, it is linearly transformed $(x - min)/(max - min)$, where min is the minimum and max is the maximum. After normalization, the minimum is equal to 0 and the maximum is equal to 1.

3.4 Enterprise Competitiveness Index

Adopting the 100-point system and the "Index Equal Power Law", the result of normalizing four indicators is multiplied by 100 and averaged to obtain the index of competitiveness of China's new OTC enterprise.

In accordance with the stock code order, given some China's new OTC enterprise competitiveness index as follows:

The index of enterprise competitiveness is between 0 and 100, and the greater the value, the stronger the overall competitiveness of enterprises for sustainable profitability (Table 2).

Table 2. National enterprise annual report 2013–2016.

Number	Stock code	Enterprise abbreviation	City	2013	2014	2015	2016
1	430002. OC	Sinosoft Co., Ltd.	Beijing	100.0	100.0	100.0	100.0
2	430003. OC	Beijing Time Technologies Co., Ltd.	Beijing	40.37	43.40	45.83	27.49
3	430004. OC	Beijing Greentec Equipment Holdings Co., Ltd.	Beijing	15.51	13.97	12.08	13.45
4	430005. OC	HTA Co., Ltd.	Beijing	76.99	79.80	82.55	85.85
5	430009. OC	Beijing Huahuan Electronics Co., Ltd.	Beijing	30.71	38.22	34.19	34.80
6	430010. OC	Modern Agricultural Equipment Co., Ltd.	Beijing	75.00	75.00	75.00	75.00
7	430011. OC	Beijing Compass Technology Development Co., Ltd.	Beijing	10.44	28.27	80.11	80.84
8	430014. OC	Beijing Hengye Century Technology Co., Ltd.	Beijing	27.76	26.97	36.01	34.59
9	430015. OC	Beijing Gateguard Information & Technology Co., Ltd.	Beijing	10.24	8.71	10.82	13.42
10	430016. OC	Shenglong Science & Technology Co., Ltd.	Beijing	10.70	11.55	12.00	11.17

The research shows that the national average enterprise competitiveness index shows a steady upward trend, and the profitability, research and development ability and innovation ability of enterprises have been gradually increasing (see Fig. 1).

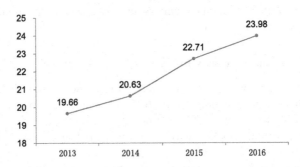

Fig. 1. Average enterprise competitiveness index 2013–2016.

4 Urban High Technology Development Level

China's new OTC Market Services Innovative, entrepreneurial, growth of small and medium-sized micro-enterprise coverage has been expanding. As of the end of 2017, among China's new OTC listed companies, the proportion of small, medium and micro-enterprises reached 94%. China's new OTC enterprise is an important constituent of the urban center of the city's high and new technology development. China's new OTC enterprise competitiveness indicator is used to construct the metropolis competitiveness index and measure the sustainable competitiveness of cities in high and new technology field.

In this paper, according to the city where China's new OTC enterprises registered, the enterprise competitiveness index is summed up and the city competitiveness index score is obtained. Using the function $f(x) = 10 \cdot 2^x$ to select the critical value, the city competitiveness index is divided into corresponding intervals, each corresponding to an urban high and new technology development level [14] (Table 3). The corresponding relationship is shown in the following table:

Table 3. Comparison table of urban high and new technology development level

Number	Critical value	Interval	Level
1	0	-	0
2	20	(0, 20]	1
3	40	(20, 40]	2
4	80	(40, 80]	3
5	160	(80, 160]	4
6	320	(160, 320]	5
7	640	(320, 640]	6
8	1280	(640, 1280]	7
9	2560	(1280, 2560]	8
10	5120	(2560, 5120]	9
11	10240	(5120, 10240]	10
12	20480	(10240, 20480]	11
13	40960	(20480, 40960]	12
...	...	...	...

According to the levels from high to low in 2016, the table below shows the top 10 cities in high and new technology development indicators:

The stage of urban high and new technology development reflects the development potential of cities in high and new technology industries (Table 4). The higher the stage, the greater the potential for development. In 2016, the high and new technology development level of Beijing and Shanghai was the highest, reaching 12 level. Shenzhen, as the next, reached 11. Suzhou, Guangzhou, Hangzhou and Nanjing reached level 10.

Table 4. Indicators of high technology development in major cities

Number	City	City competitiveness index score				Urban high and new technology development level			
		2013	2014	2015	2016	2013	2014	2015	2016
1	Beijing	4869.95	7460.20	17499.32	33581.24	9	10	11	12
2	Shanghai	971.62	3486.40	10706.50	20915.37	7	9	11	12
3	Shenzhen	12.71	1694.58	8560.85	19052.30	1	8	10	11
4	Suzhou	0.00	1603.81	5429.19	9959.43	0	8	10	10
5	Guangzhou	0.00	956.68	3939.44	9319.67	0	7	9	10
6	Hangzhou	0.00	579.18	4151.59	8832.68	0	6	9	10
7	Nanjing	37.32	871.40	2902.23	5208.63	2	7	9	10
8	Wuxi	0.00	1003.45	2523.83	5059.89	0	7	8	9
9	Chengdu	0.00	529.83	2479.08	4986.40	0	6	8	9
10	Wuhan	621.64	1437.97	2938.40	4918.82	6	8	9	9

5   China Urban High Technology Development Status Quo

Since 2013, the stock transfer system has been receiving application for listing in the country and the new OTC market of China has entered a golden stage of rapid development. This section mainly studies and analyzes the status quo of China city high and new technology development in 2013–2016.

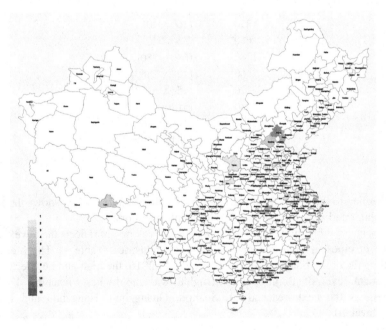

Fig. 2. Thermodynamic diagram of high and new technology urban development in China in 2013

In 2013, Beijing ranked the nation in terms of the level of high and new technology development in cities with a rank of 9. Shanghai, Wuhan and Tianjin followed. Shenzhen and Guangzhou have just started (Fig. 2).

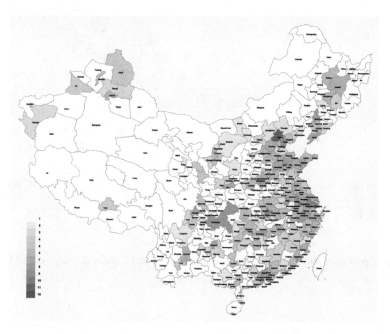

Fig. 3. Thermodynamic diagram of high and new technology urban development in China in 2014

In 2014, Beijing reached level 10. Shanghai second, to 9. Shenzhen and Guangzhou to achieve a qualitative leap, respectively, reached 8 and 7 (Fig. 3).

In 2015, the three companies covering 266 cities across the country. Beijing and Shanghai reached 11, followed by Shenzhen and Suzhou, level 10 (Fig. 4).

In 2016, China's new OTC enterprises cover 299 cities nationwide. Beijing and Shanghai reached 12, Shenzhen level 11, Suzhou, Guangzhou, Hangzhou and Nanjing reached 10 (Fig. 5).

In the past four years, the new OTC market of China has undergone a start-up, development and maturity stage. China's new OTC companies gradually expanded from Beijing and Shanghai to cover most of the country's small and medium-sized metropolis. In 2013–2016, China high and new technology city achieved leapfrog development. From the coverage of the breadth, from the pilot Zhongguancun Technology Park quickly covered the entire country. From the depth of development, high and new technology development levels in first-tier cities such as Beijing, Shanghai and Shenzhen have been increasing year by year. Beijing has been gradually raised from the 9th level to 12 levels. Shenzhen has become even more remarkable. It has grown from level 1 to level 11. Other major cities, such as Suzhou and Guangzhou, also enjoyed explosive and rapid urban high and new technology growth.

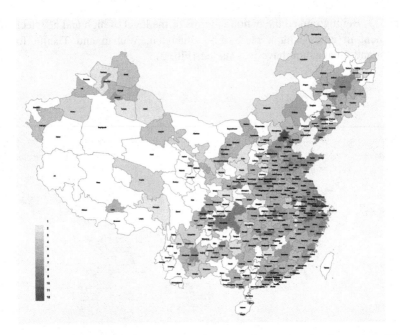

Fig. 4. Thermodynamic diagram of high and new technology urban development in China in 2015

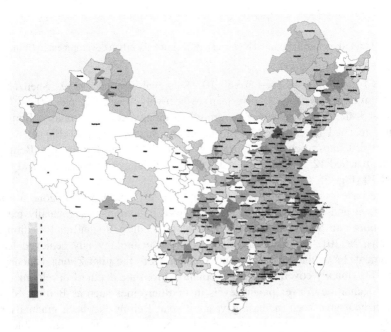

Fig. 5. Thermodynamic diagram of high and new technology urban development in China in 2016

6 Science and Technology Ecosystem Development Potential Forecast

This section examines the urban competitiveness index of 2017 in China and studies the potential and trends of high and new technology development in major cities in China from the perspective of science and technology ecosystem. Assuming that the average enterprise competitiveness index of cities in 2017 is consistent with that of the previous year, a regression model will be used to predict the number of China's new OTC enterprises [15] in each city in 2017 and the product of the number of enterprises and the average enterprise competitiveness indicant to obtain the 2017 China urban competitiveness index number. Science and technology ecosystem is based on high and new technology, geographical space adjacent to urban agglomeration. The following is a combination of heat maps to predict the potential of high and new technology urban development in Beijing-Tianjin-Hebei, Yangtze River Delta, Pearl River Delta and Chengdu-Chongqing science and technology ecosystem in 2017.

6.1 Beijing-Tianjin-Hebei Science and Technology Ecosystem

With Beijing as the core, China will lead the development of science and technology in the entire country. The two core cities in the ecosphere co-exist and the neighboring cities have limited capacity to accept core science and technology radiation (Figs. 6 and 7).

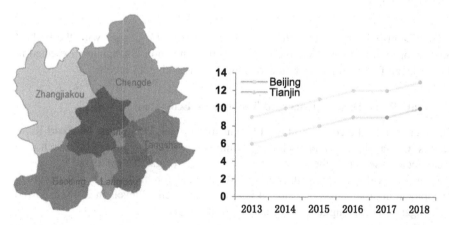

Fig. 6. Thermodynamic diagram of the Beijing-Tianjin-Hebei science and technology ecosystem in 2017

Fig. 7. Beijing-Tianjin-Hebei science and technology ecosystem development level forecast chart

Beijing ranks the highest in the development of high and new technology cities in the country and will reach 13 in 2018. Tianjin and Beijing echo each other, reaching 10 in 2018.

6.2 Yangtze River Delta Science and Technology Ecosystem

With Shanghai as the center, the high and new technology industries in cities such as Suzhou, Hangzhou, Nanjing and Wuxi have been developing rapidly. The ecosystem shows a trend of multi-polarization (Figs. 8 and 9).

Fig. 8. Thermodynamic diagram of the Yangtze River Delta science and technology ecosystem in 2017

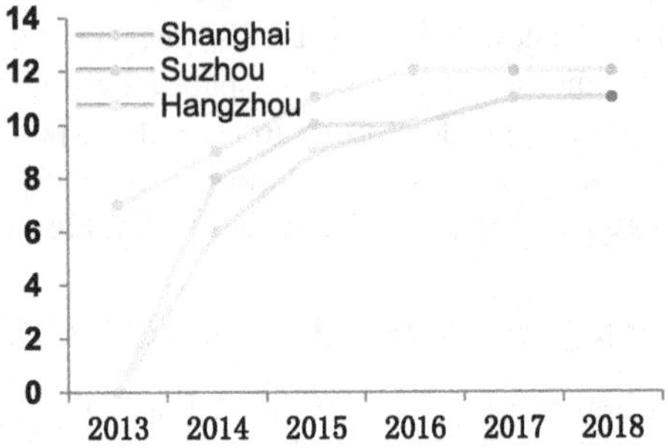

Fig. 9. Yangtze River Delta science and technology ecosystem development level forecast chart

The Yangtze River Delta technological ecosphere is the region with the highest potential for high and new technology development in China. Shanghai is expected in 2018 12 level, followed by Suzhou and Hangzhou, reaching 11 levels.

6.3 Pearl River Delta Science and Technology Ecosystem

With Shenzhen as the center, followed by Guangzhou and Dongguan, the Pearl River Delta is China science and technology innovation and technology research and development base and is the most dynamic eco-circle (Figs. 10 and 11).

Shenzhen is the region with the highest potential for high and new technology development in urban areas in China. It is forecast to reach 12 levels in 2018, followed by Guangzhou and Dongguan at 11 and 10 respectively.

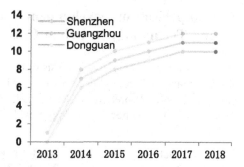

Fig. 10. Thermodynamic diagram of the Pearl River Delta science and technology ecosystem in 2017

Fig. 11. Pearl River Delta science and technology ecosystem development level forecast chart

6.4 Chengdu-Chongqing Science and Technology Ecosystem

Focusing on Chengdu and Chongqing, we will foster the leading role in leading the development of high and new technology in the western region and strengthening the strategy of developing the western region (Figs. 12 and 13).

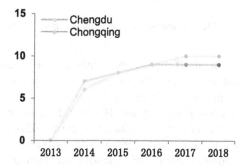

Fig. 12. Thermodynamic diagram of the Chengdu-Chongqing science and technology ecosystem in 2017

Fig. 13. Chengdu-Chongqing science and technology ecosystem development level forecast chart

Chengdu and Chongqing are the dual-core of the Chengdu-Chongqing science and technology ecosphere. In 2018, the city's high and new technology is expected to reach 10 and 9 levels, respectively.

7 Comparison of Development Potentiality of Science and Technology in Urban Cluster

This section is a comparative analysis of the growing potential of the main cities in the above science and technology ecosphere. Figure 14 shows the urban competitiveness index of 2016 and forecast in 2017. Finally, it illustrates the advantages and characteristics of high and new technology development in each urban cluster.

Fig. 14. Comparison of development potentialities of science and technology in urban cluster

Beijing-Tianjin-Hebei urban cluster: Zhongguancun radiation led role has continued to highlight the deepening of regional scientific and technological cooperation. Beijing is dominated by modern service industry. Tianjin is the dominant mode of processing industry and Hebei is the resource-based mode of development.

Yangtze River Delta urban cluster: Zhangjiang is the source and core of scientific and technological innovation in the Yangtze River Delta region, promoting the integration of technological innovation chain and industrial chain. Jiangsu efforts to build Sunan into an internationally influential industrial technology innovation center, Zhejiang around Hangzhou to build a national Internet platform, Anhui actively undertake the industrial transfer Jiangsu, Zhejiang and Shanghai developed regions.

Pearl River Delta urban cluster: Shenzhen is the first national independent innovation demonstration zone featuring cities as its basic unit. Guangzhou speeded up the construction of a national innovation center city and an international science and technology innovation hub. The other 7 cities in the PRD promoted innovation and development in light of their own advantages and industrial realities.

Chengdu-Chongqing urban cluster: Chengdu High Technology Zone is a state-level high and new technology industry development zone and an important engine for the rise of western China. High Technology Zone in Chongqing adhere to the principle of differentiation and sustainable development, is building a leading western "scientific development model window."

Acknowledgement. This work is partially supported by the technical projects No. 2017YFB 1400604, No. 2016YFB1000803, No. 2017YFB0802703, No. 2012FU125Q09, No. 2015B0 10131008 and No. JSGG20160331101809920.

References

1. Lin, L., Tang, J.: Medium and small-sized enterprises' financing status and performance analysis of the new three board market. J. Hebei Univ. Econ. Bus. 71–76 (2015)
2. Fang, X., Wu, Y.: A research on financing efficiency of SMEs in NEEQ market. Econ. Manag. J. 42–51(2015)
3. Shen, C.: A research on financing efficiency of SMEs in NEEQ market. J. Audit Econ. 78–86 (2017)
4. Ni, P.: An empirical test about contribution of infrastructure to urban competitiveness. China Ind. Econ. 62–69 (2002)
5. He, T.: Progress of urban competitiveness study at home and abroad. Inquiry Into Econ. Prob. 21–24 (2005)
6. Yu, T.: Progress of urban competitiveness study abroad. Urban Plann. Overseas 28–34 (2004)
7. Ma, Y.: Build a corporate open innovation ecosystem. China Ind. Rev. 22–26 (2016)
8. Wang, F., Lv, J.: The evaluation and spatial-temporal evolvement of the city competitiveness of Zhongyuan Urban Agglomeration. Geogr. Res. 49–60 (2011)
9. Xu, X., Cheng, Y.: Spatial-temporal changes of urban competitiveness in urban cluster of Pearl River Delta. Scientia Geographica Sinica 257–265 (2006)
10. Song, X.: The characteristics, motives and implications of the development of the new third board market. Secur. Mark. Herald 4–12 (2015)
11. Zhang, X., Yu, W., Hu, Y.: Quantitative evaluation method of enterprise competitiveness. Manag. Rev. 32–37 (2003)
12. Fu, X.: Present situation and developing tendency of high and new technology. Mod. Chem. Ind. 2–6 (2001)
13. Jiang, R., Wu, Y.: Research on the competitiveness of China's top 500 enterprises. Res. Fin. Econ. Issues 80–87 (2012)
14. Wang, X.: High-tech comprehensive assessment. Stud. Sci. Sci. 5 (1998)
15. Wang, M., Li, B.: Seamless linear regression and prediction model. Acta Geodaetica et Cartographica Sinica 1396–1405 (2016)

Short Paper Track: BigData Analysis

Analysis of Activity Population and Mobility Impacts of a New Shopping Mall Using Mobile Phone Bigdata

Kwang-Sub Lee[(⊠)] [iD], Jin Ki Eom, Jun Lee, and Dae Seop Moon

Korea Railroad Research Institute, Uiwang-si, Gyeonggi-do 16105, Korea
leeks33@krri.re.kr

Abstract. This paper is an explorative research, examining the impact of the new shopping mall, Hyundai, in Seongnam city of Korea. The focuses are on the activity population and mobility around the major shopping malls in the city, including three pre-existing major shopping malls. For this purpose, we analyzed mobile phone records in 2015 and 2016, before and after the new shopping mall. The data represent total mobile users in Korea. The number of activity population in the Hyundai mall increased by 103%. The new shopping mall negatively impacted the nearest shopping mall, AK, decreasing activity population by 3%. Internal and external mobility showed that visitors coming from all regions to the Hyundai mall were increased, while internal population visiting to the SK mall showed mixed results. Local government and urban planners will find this case study to be of interest with regards to mitigating the impacts of a new shopping mall development in similar urban situations.

Keywords: Impact analysis · Shopping mall · Activity population
Mobility

1 Introduction

Shopping malls are one or more buildings from a complex of shops representing merchandisers with interconnecting walkways that enable customers to walk from unit to unit (Wikipedia). Nowadays they are not just for shopping, but are places where consumers visit for doing diverse activities for social life, economic activities, entertainment, recreational, and cultural functions. Shopping malls have become an inseparable element of spatial functions in large cities. In particular, most urban cities in Korea have mixed used developments, and shopping malls offer consumers with an attractive and integrated community in which to live, work and shop.

Shopping malls attract buyers, sellers and customers. They choose a shopping mall based on store assortment, convenience, distance to malls, economic advantage and leisure facilities (Rajagopal 2009). Previous studies of the impacts of shopping malls have focused on the economic impacts (ICSC 2013; McGreevy 2016), the social impact on a city (Heffner and Twardzik 2014), the change of the functional structure of city (Dickinson and Rice 2010), and the choice of shopping malls (Borgers et al. 2013; Rajagopal 2009). It is inevitable that the opening of a new shopping mall has an impact

© Springer International Publishing AG, part of Springer Nature 2018
F. Y. L. Chin et al. (Eds.): BIGDATA 2018, LNCS 10968, pp. 307–311, 2018.
https://doi.org/10.1007/978-3-319-94301-5_23

on pre-existing shopping malls as well as on activity patterns in the surrounding area of shopping malls.

The purpose of this paper is to examine the impact of a mega shopping mall, Hyundai Pangyo department store in Seongnam city of Korea, on the surrounding business district, and the pre-existing and competitive shopping malls located in the city. The focuses are on the activity population around the major shopping malls in the city, and on the spatial impacts and mobility of the visitors to the area. This study is an explorative research, and attempts to address those issues by examining mobile phone bigdata before and after the opening of the new shopping mall. Although when using mobile phone data it is hard to precisely distinguish shoppers or visitors inside the mall from those outside the mall, an analysis of changes in the number of activity population and mobility can provide valuable information on the planning issues that emerge from such major shopping mall developments.

The paper reviews the study area and the dataset in the next section. It then presents the analysis of activity population and mobility surrounding four major shopping malls in the city. Conclusions and future researches are discussed in the final section.

2 Study Area and Mobile Phone Bigdata

2.1 Study Area

Seongnam is the second largest city in Korea's Gyeonggi Province and the 10th largest city in the country. Its area is 141.72 km^2 and population is about one million. Seongnam is a satellite city located immediately southeast of Seoul, and largely a residential city. Bundang, one of three districts in Seongnam, is a planned city and was developed in the 1990s to accelerate the dispersion of Seoul's population to its suburbs and relieve the congested Seoul metropolitan area. All four major shopping malls analyzed in this paper are located in the Bundang district of Seongnam.

The Hyundai shopping mall has opened in August, 2015. It is a big, complex shopping mall with about 900 shop units, 210 brands, 100 restaurants, and parking space for 2,700 cars. The building has 13 floors above ground and 6 underground. The total floor area is about 233,518 m^2. The lot area is about 22,905 m^2, which is equivalent to two soccer fields. Before the opening of the Hyundai shopping mall, there were big three shopping malls, including AK Plaza, Lotte, and NC, in Bundang-gu. Activity population and mobility of four shopping malls are compared in this paper.

2.2 Description of Mobile Phone Data

As of 2014, about 57 million people owned mobile phones in Korea, which is equal to 1.13 mobile phones per person on average. Thus, most of adults and teenagers have a mobile phone. For this analysis, we used the SKT's mobile phone data. The original mobile phone records were collected and preprocessed by SKT. First, every mobile phone signal is received by a nearby cellular tower, and the existing location and time information is stored in a server. Mobile traffic is identified by phone calls, SMS messages, notifications or updates from apps, or Internet connections. A location of mobile phone is initially identified by a cellular tower, albeit covering a large area of at least 2,500 m². Thus, SKT further segregated the identified location into a square cell area (50 m by 50 m), based on a spatial correlation of land use and the weights of the total floor areas of building types. SKT, one of three mobile phone telecommunication operators in Korea, consistently held about half of the country's total mobile phone memberships. Before providing the dataset, SKT further expanded their mobile phone records to include the total mobile phone users in Korea, using the country's market share rate. This is a huge advantage for the population and mobility analysis because the dataset represents the total population of mobile phone users.

The daily records consist of 16 columns (Table 1). The data include information on the number of users at each cell grouped by age and gender of users, the coordinates of the cell location, and the user's home location. It should be noted that it is a daily-based records, and double counts are not allowed in the same cell in the same day. The data do not store a person logs in order to protect privacy. For this study, we analyzed two weeks of data in two years (March 16–22, 2015 and March 14–20, 2016). The dataset include about 160 million records per day on average.

Table 1. Structure of daily mobile phone data.

Columns	Variable	Description
1	Date	Date of data
2–3	X, Y	Position of the center of a cell based on the UTM-K coordinate system
4–9	Number of men by age groups	Daily mobile phone users (men) by 6 age groups (10s, 20s, 30s, 40s, 50s, over 60s)
10–15	Number of women by age groups	Daily mobile phone users (women) by 6 age groups (10s, 20s, 30s, 40s, 50s, over 60s)
16	Home	Code of home location of users

3 Analysis

3.1 Changes of Activity Population

We analyzed one week of mobile phone data in each year of 2015 and 2016. Activity population is defined as the number of mobile phone users in this paper, who are located in a 50 m cell. To compare activity population before and after the opening of a new shopping mall, we need the number of population visiting to a shopping mall. However, when using mobile phone data, it is very difficult to identify whether people are inside a shopping mall with purposes of vising the mall, because the data does not have an activity or trip purpose and the exact destination. Thus, we extract the number of activity population located within a certain catchment area instead of in a cell. This study specifically interested in the comparison of changes of activity population, including shoppers in the mall as well as visitors around the mall. For this reason, we analyzed activity population in catchment areas ranged from 100 m to 300 m with an interval of 50 m, and we determined 300 m as the reasonable catchment area in this paper.

First, we examined activity patterns by day of the week, and found that they were very similar to all shopping malls. The number of activity population gradually increased from Monday to Friday, and dropped on weekends. The peak activity population was appeared on Friday. However, activity patterns in 2016 showed slightly different patterns in 2015. The number of activity population on Saturday was lower than that of Monday in 2015. However, the number of population on Saturday was very similar that of Monday in 2016. All shopping malls showed these patterns.

We compared average daily activity population in four malls in 2015 and 2016. We compared average daily activity population within 300 m before and after the new shopping mall. Activity population in the Hyundai shopping mall increased by 103% in 2016, while that of the AK shopping mall slightly decreased by 3%. The AK mall is the nearest shopping mall from the Hyundai mall, locating in 1.3 km away from the east-south of the Hyundai mall. Thus, the opening of the Hyundai mall negatively impacted the AK mall. On the other hand, activity population in both Lotte and NC malls increased by 8% and 78%, respectively in 2016. The distances are 1.6 km and 2.4 km, respectively, away from the Hyundai mall. So, it is interesting that the NC mall was not negatively impacted by the Hyundai mall.

We analyzed shows the population differences of three shopping malls in each catchment areas. In all catchment areas, activity population in 2016 in the AK mall were decreased compared to those in 2015, while those in the Hyundai mall were increased. The activity population in the Lotte mall decreased within 250 m, while they are increased within 300 m.

3.2   Changes of Mobility

Shoppers are willing to sacrifice longer distances and travel times in order to go to larger shopping malls with more options. According to the spatial interaction theory, the distance a consumer is willing to travel to a shopping center is proportional to the size of the shopping center even though the shopping center is far (Openshaw 1975; Kanoga et al. 2015). Therefore shoppers spend more on travel cost and time to get what they want from a larger mall as long as it is accessible.

We analyzed visitors to each shopping mall, and particularly focused on the place where they are coming from, using the home location information from mobile phone data. For this purpose, we divided activity population into two groups depending on the home location: (1) internal population, whose home is in Seongnam, and (2) external population whose home is outside of Seongnam, such as Seoul and Gyeonggi. Internal population in all administrative dongs of Seongnam visiting to the Hyundai mall were increased on both weekdays and weekends in 2016. External population were also increased in all outside regions, except Gangwon-do. Compared to the Hyundai mall, internal population visiting to the AK mall showed mixed results: that is, the number of internal population visiting from Seohyun-dong and Backhyun-dong was increased, while the number was decreased visiting from Jungja-dong. External population visiting to the AK mall from most of regions outside Seongnam were decreased.

4 Conclusion

This paper is a preliminary research attempting to assess the impact of the opening of the new shopping mall, using mobile phone data. It compared the new shopping mall and the pre-existing shopping malls, focusing on activity population and mobility surrounding the shopping malls.

First, activity patterns by day of the week were very similar to all shopping malls, gradually increased from Monday to Friday and dropped on weekends. However, the number of activity population on Saturday in 2015 was much lower than on Monday, while its on Saturday in 2016 was very similar on Monday in 2016. Second, we compared activity population within 300 m of each shopping mall, and found that the new shopping mall, Hyundai, negatively impacted to the nearest shopping mall, AK. The number of activity population of the Hyundai mall increased by 103%, while that of the AK shopping mall decreased by 3%. Third, we investigated visitor's mobility. Internal and external population visiting to the Hyundai mall were increased in 2016, compared to 2015, while the population visiting to the AK mall showed mixed results.

It is difficult to separate actual mall visitors from people not actually visiting to the mall when using mobile phone records. Nevertheless, local government and urban planners will find this case study to be of interest with regards to mitigating the impacts of a new shopping mall development in similar urban situations.

More works in the future are expected for the analysis of the impact of a new shopping mall such as identifying the factors affecting activity population around shopping malls and interconnecting the reason for vising shopping malls with land use information and network service.

References

Borgers, A., Swaaij, S., Janssen, I.: Assessing the impact of peripheral mega retail centers on traditional urban shopping centers. Belgeo (2013)

Dickinson, A., Rice, M.D.: Retail development and downtown change: shopping mall impacts on Port Huron, Michigan. Appl. Res. Econ. Dev. 7(1), 2–13 (2010)

Heffner, K., Twardzik, M.: Shopping malls and its social impact on the outer metropolitan zones. In: 5th Central European Conference in Regional Science, pp. 238–247 (2014)

ICSC: Economic Impact of Shopping Centers. International Council of Shopping Centers (2013)

Kanoga, S., Njugana, R., Bett, S.: The effect of place on performance of shopping malls in Nairobi County Kenya. J. Soc. Sci. Humanit. 1(4), 381–390 (2015)

McGreevy, M.: The economic and employment impacts of shopping mall developments on regional and peri-urban Australian towns. Australas. J. Reg. Stud. 22(3), 402 (2016)

Openshaw, S.: Some theoretical and applied aspects of spatial interaction shopping models. Issue 4 of concepts and techniques in modern geography. Geo Abstracts (1975)

Rajagopal: Growing shopping malls and behavior of urban shoppers. J. Retail Leis. Prop. 8(2), 99–118 (2009)

Wikipedia Homepage. https://en.wikipedia.org/wiki/Shopping_mall. Accessed 18 Jan 2018

Activity-Based Traveler Analyzer Using Mobile and Socioeconomic Bigdata: Case Study of Seoul in Korea

Jin Ki Eom$^{(\boxtimes)}$ (ID), Kwang-Sub Lee, Jun Lee, and Dae-Seop Moon

Korea Railroad Research Institute, Uiwang-si, Gyeonggi-do 16105, Korea
jkom00@krri.re.kr

Abstract. This paper introduces a pilot study of developing Activity-BAsed Traveler Analyzer (ABATA) using the Big Data. The mobile phone bigdata is used to estimate total activity population in a case study area, Gangnam, Seoul. The pilot system estimates the activity population and the derived travel demand from the activities taken into account for individual schedules and activity categories (home, work, shopping, and leisure) with respect to land use types based on the various data inventory. The transportation planners will find this case study to be of interest with regards to the simulation results on the socio-demographic factors and land use changes.

Keywords: Mobile phone data · Activity-based model · Activity population Mobility

1 Introduction

The transportation planning evaluates the current transportation situation and predicts the future travel demand, and the effect of the facility investment accordingly. The transportation policy diagnoses the current transportation situation, solves the problems, and efficiently uses the given resources to provide the public convenience. In this process, both microscopic travel pattern and precise demand prediction and accuracy of analysis are required.

Recent developments and changes in science and technology, such as the 4th Industrial Revolution, AI, IoT, advanced ICT, ITS, and Big Data, have given a lot of implications to transportation planners. With the development of data collection technology and the accumulation of various big data, transportation planning is still sticking to the traditional four-step transportation demand estimation method. This is a method of sequential application of four independent steps of trip generation, trip distribution, mode choice, and trip assignment, which are widely used due to easy understanding and logical analysis. However, it is premised on the aggregate spatial unit (city district or administrative district) called Traffic Analysis Zone and time unit (daily traffic volume). As a result, there are limitations in analysis of various effects for analyzing the population and social characteristics of individuals. It does not reflect the interrelationships between the steps, and travel behavior from transport policies. In addition, the temporal analysis unit is O/D demand aggregated in daily units, which neglects the direction of

© Springer International Publishing AG, part of Springer Nature 2018
F. Y. L. Chin et al. (Eds.): BIGDATA 2018, LNCS 10968, pp. 312–318, 2018.
https://doi.org/10.1007/978-3-319-94301-5_24

travel, and does not consider the trip chain [1]. It is difficult to analyze changes in travel patterns and changes in land use and population characteristics.

On the other hand, in the activity-based model, which has been actively studied recently [2–5], it is assumed that travel demand is not generated for the purpose of "trip" itself but that trip is derived from each individual's "activity" do. In the activity-based model, it is possible to use a microscopic analysis unit rather than an aggregated analysis unit of the existing 4-step demand forecasting technique [6]. It can utilize the socioeconomic indicators of the more detailed county (about 1/25 of the size of the village), and the 24 h activity schedule reflecting the individual characteristics and the trip chain for the travel impact analysis, micro-transportation planning, and traffic management services.

The activity-based traveler analyzer uses a variety of data such as household travel data, mobile phone bigdata, socioeconomic data, and land use data.

2 Traveler Analyzer

2.1 Concept of Traveler Analyzer

The Activity-BAsed Traveler Analyzer (ABATA), based on the Big Data, is a system that estimates the activity population and the derived travel demand from the activities taken into account for individual schedules and activity categories (home, work, shopping, and leisure) with respect to land use types (Fig. 1).

The ABATA system estimates the existing population by time and space (aggregate district or administrative unit) of the analysis area based on the statistics of the National Statistical Office (NSO), the survey data on the actual condition of households, and other microscopic spatial data, then establish the O/D trips. Especially, the purpose of this study is to develop a system to analyze changes in travel behavior due to changes in socioeconomic population or land use and travel related policies. For example, analysis of changes in travel behavior and travel pattern when population composition ratio of a specific city changes due to aging, analysis of transportation system change due to transportation policy implementation such as time lag, analysis of travel behavior according to change of school hours, construction of department store, and analysis of change of transportation system and influence according to land use change.

The ABATA system utilizes a variety of basic data. We utilize survey data on the actual condition of households to establish individual activity schedules. The household travel survey is conducted every five years on a nationwide basis. Among the survey data, household status, characteristics of household members, and individual travel characteristics are important data for establishing a 24-h activity schedule. However, because the individual trips are recorded in the survey on the households, it is necessary to convert the trips into activity schedules. Household travel survey data have individual characteristics and travel information, but they are limited by small sample data (about 3% to 5%). Therefore we use a mobile phone data which represent all population engaged into a certain activity at study area. The mobile phone data defined by the number of people in each age group by 50 m × 50 m grid cells. The mobile phone data do not include individual data for personal information protection, however, it is a valuable data because it can identify the existent population by time and space.

Socioeconomic data of the National Statistical Office (total population by household composition, number of workers by industrial classification, etc.) are reliable for providing information on various socioeconomic populations of micro spaces although there is a limitation that they do not have information by time frame. In ABATA system, we utilize building association area data and student number data of Nice National Service.

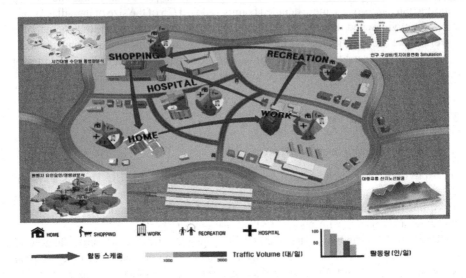

Fig. 1. Concept of activity-based traveler analyzer.

2.2   Description of Mobile Phone Data

In this study, we used the SKT's mobile phone data. The original mobile phone records were collected and preprocessed by SKT, one of three mobile phone telecommunication operators in Korea. First, every mobile phone signal is regularly received by a nearby cellular tower, and the existing location and time information is stored in a server. The SKT consistently held about half of the country's total mobile phone memberships. Before providing the dataset, SKT further expanded their mobile phone records to include the total mobile phone users in Korea, using the country's market share rate. This is a huge advantage for the population and mobility analysis because the dataset represents the total population of mobile phone users.

The daily records consist of 16 columns (Table 1). The data include information on the number of users at each cell grouped by age and gender of users, the coordinates of the cell location, and the user's home location. It should be noted that it is a daily-based records, and double counts are not allowed in the same cell in the same day. The data do not store a person logs in order to protect privacy. For this study, we analyzed two weeks of data in two years (March 16–22, 2015 and March 14–20, 2016). The dataset include about 160 million records per day on average.

Table 1. Structure of daily mobile phone data.

Columns	Variable	Description
1	Date	Date of data
2–3	X, Y	Position of the center of a cell based on the UTM-K coordinate system
4–9	Number of men by age groups	Daily mobile phone users (men) by 6 age groups (10s, 20s, 30s, 40s, 50s, over 60s)
10–15	Number of women by age groups	Daily mobile phone users (women) by 6 age groups (10s, 20s, 30s, 40s, 50s, over 60s)
16	Home	Code of home location of users

3 System Structure

3.1 Total Activity Population

The ABATA system is first to calculate total activity population based on the mobile phone data recorded on the study area. The mobile phone data provide the real number of people presenting in study area at each time. The existing population is a population that exists in a specific space without regard to the purpose of activity, and the active population means the population that is performing specific activity. Since the mobile phone data provides total existing population by time, if data can be secured, it can be used directly to calculate activity population. If it is not available, however, the statistics of the National Statistical Office (NSO) are used to estimate the total activity population in the ABATA system.

3.2 Construct Activity Schedule

The ABATA system construct individual activity schedule by using household travel survey data. The seven activity categories (home, work, shopping, leisure, school, education, and others) are defined to construct each activity profile. To do this, the all trips data of households are converted into individual activity schedules in 10-min increments, and an activity schedule and an activity profile for each hour are developed. The activity profile represents the composition ratio of the active population per activity purpose. Figure 2 is an example of a comparison of activity profiles when the elderly population ratios increased from 6% to 30%. The hourly total activity population is combined with activity profile, then the each activity population by hourly is calculated.

3.3 Develop Activity Attractiveness

The hourly activity population needs to combine with land use data. Since a certain activity is occurred at a certain place or area. For example, the shopping activity has to be occurred at market or department store. The land use data easily obtained from the statistics of the National Statistical Office (NSO). The land use type and job categories connect to activity type. Based on various data, the multiple regression models are

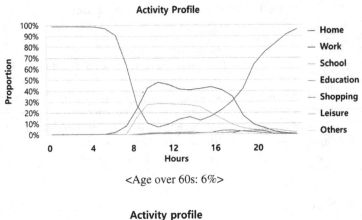

<Age over 60s: 6%>

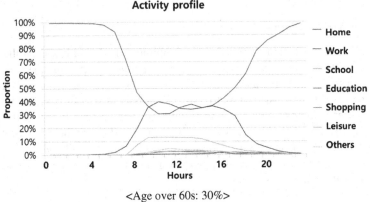

<Age over 60s: 30%>

Fig. 2. Comparison of activity profiles between the proportions of the older people.

constructed and estimated to activity attractiveness. From results, the activity attractiveness for each block is estimated based on total activity attractiveness in study area. Then the ratio of individual block against the whole study area by each activity represents the attractiveness for the activity type.

3.4   Estimate Activity Population

The ABATA system calculates the activity population and distribute it onto each space and hourly manner (Fig. 3). To develop O/D, first calculate the amount of travel and the destination. Some of the people who are doing a certain activity at the present time can make a trip for the next activity. In order to find out this, we extract the probabilities (conditional travel probabilities by activity type and group) from the household travel data by time and activity, and then apply the probability of trip occurrence to the time, activity, then calculate the amount of travel generated per hour. The destination choice for the generated trip is decided from the spatiotemporal activity attraction. Finally, the O/D is developed for 24 h, 7 activities, and counties (Fig. 3). Based on the estimated O/D, the choice of model considering the difference in the choice of travel mode for each

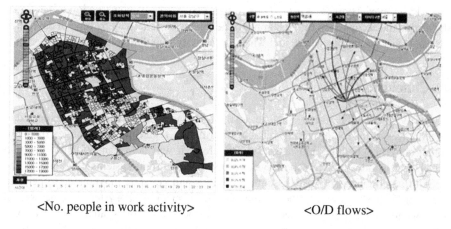

<No. people in work activity>   <O/D flows>

Fig. 3. Activity population and origin-destination flows.

demographic characteristic, the ABATA system examines the explanatory variables (trip purpose, travel time, population characteristics (sex, age, occupation) and runs the decision tree model.

3.5   Simulation Activity Population

ABATA system is currently developing for Gangnam in Seoul. The system has the ability to visualize the results of various scenarios. It is constructed to allow the user to select scenarios and compare the changed results, and it is configured to enable scenario-specific simulations (such as land use change, population composition ratio, policy change, etc.). Figure 4 shows the graphic maps of simulation results from the scenario of

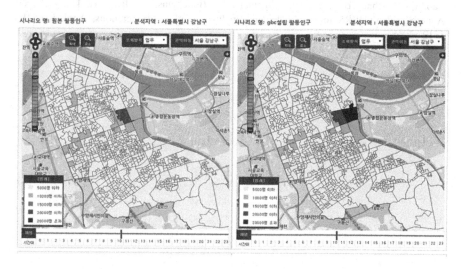

Fig. 4. Activity population changes from the Hyundai Global Business Center Construction.

new skyscraper, Hyundai GBC (Global Business Center), construction. The figure shows that the comparison of the number of people in work activity by time of day. Then we can easily figure out where and when the activity population changes occurred.

4 Conclusion

The Activity-BAsed Traveler Analyzer (ABATA), based on the Big Data is expected to be reliable system for providing reliable information regarding the changes of population, activity schedule, and land use. Therefore, it will help transportation decision makers out to introduce transport systems and make current systems efficient.

All human beings perform essential activities and generate movements to perform certain activities. Therefore, the developed passenger analysis system can be applied not only to traffic demand forecasting and transportation policy analysis but also to analysis of environment (micro dust exposure) impact analysis, location analysis, emergency evacuation plan, and tourist related system planning.

References

1. Hägerstrand, T.: What about people in regional science? Reg. Sci. Assoc. Pap. **24**, 7–21 (1970)
2. Capin, F.S.: Human Activity Patterns in the City. Wiley, New York (1974)
3. Activity-based modeling system for travel demand forecasting, Travel Model Improvement Program, U.S. Department of Transportation (1995)
4. A system of activity-based models for Portland, Oregon, Travel Model Improvement Program, U.S. Department of Transportation (1998)
5. Arentze, T., Timmermans, H.: Albatross: a learning based transportation oriented simulation system. European Institute of Retailing and Service Studies (2000)
6. Bao, Q., Kochan, B., Shen, Y., Creemers, L., Bellemans, T., Janssens, D., Wets, G.: Applying FEATHERS for travel demand analysis: model considerations. Appl. Sci. **211**, 1–12 (2018)

Design and Application of a Visual System for the Supply Chain of Thermal Coal Based on Big Data

Xinyue Zhang[1], Yanmin Han[1(✉)], Wei Ge[2], Daqiang Yan[1],
and Yiming Chen[3]

[1] Big Data Center, State Power Investment Corporation Limited, Beijing, China
hanyanmin@spic.com.cn
[2] Central Research Institute, State Power Investment Corporation Limited,
Beijing, China
[3] State Power Investment Corporation Limited, Beijing, China

Abstract. Big data is now applied to many different fields. The paper will introduce the application of big data in the coal supply chain of the power industry. We designed and implemented a visual system for the Supply Chain of Thermal Coal (SCTC). This system can analyze and predict the coal demand for power generation enterprises. In the system, power companies can easily find suitable coal suppliers by comparing the price of coal, transportation cost, supply cycle, industry status, enterprise credit, etc. So they can reduce power generation cost and storage cost, match power generation plan, and understand regional situation. In addition, the system provides enterprise portrait for each coal company from many aspects, such as credit, risk information, service quality and so on. At the same time, we used actual data to verify the system. It is hoped that the application of this study can provide reference for peers and related industries.

Keywords: Thermal coal supply chain · Enterprise portrait · Big data

1 Introduction

China is a big country of coal production and consumption in the world. Coal is mainly used in the four industries of power, metallurgy, building materials and chemical industry, among which the proportion of the power industry is the largest. The coal and electricity energy supply chain consists of coal production, coal transportation, power generation, power transmission, and power utilization [1].

Since China's electricity is mainly derived from coal power generation, it has important influence on the supply and demand of the whole coal market. At present, there are many researches on the separate part of the thermal coal supply chain at home and abroad, but there is less research on the overall price risk. As a link between electricity and coal, the power coal can truly reflect the relations between the two industries. So it is important to study the thermal coal supply chain and deeply analyze factors that affect the thermal coal demand for power generation companies, including power generation plans, inventory, coal prices, transportation costs, supply cycle, etc. It

© Springer International Publishing AG, part of Springer Nature 2018
F. Y. L. Chin et al. (Eds.): BIGDATA 2018, LNCS 10968, pp. 319–325, 2018.
https://doi.org/10.1007/978-3-319-94301-5_25

can help them predict thermal coal demand, rational arrange fuel purchase plans and optimize business strategy. In addition, the selection of suppliers also plays an important role in supply chain management. It directly affects the competitiveness of power companies in the entire supply chain. Therefore, it is necessary to portray corporate portraits for coal suppliers by big data technology. Through corporate portraits, power companies can find suitable coal suppliers in terms of service, delivery time, coal prices, and transportation costs.

2   System Design

With the reform of organization in electric power industry, the operational mechanism and the management methods of power supply enterprise have great changes. Electric power enterprises are going into market economy. If the power generation enterprises do not obtain and analyze the data rooted in production and management, realize cost reduction and efficiency enhancement, grasp the needs of customers, and understand marketing strategies of competitors, they will lose their initiative and core competitiveness in the reform. Therefore, we designed SCTC based on big data and visualization technology. SCTC can assist power enterprises in predicting their own coal demand within a limited time, and can quickly select suitable partners from neighboring coal companies when purchasing coal. So the system mainly implements the following three functions:

- Coal demand forecast
- Regional situation analysis
- Enterprise portraits of coal suppliers.

3   Technical Route

3.1   System Architecture

In the paper, we designed and implemented a visual system for the supply chain of thermal coal, which is analyzed and displayed from multi-dimensional. It facilitates decision makers to compare and analyze coal suppliers in a short period of time. The system uses Echarts data visualization technology to achieve a flexible, intuitive and interactive chart. At the same time, it also provides functions such as data views, area selection, multi-map linkage, and sub-area maps. The user can mine and integrate data in the system.

The system is based on the user request response framework. It mainly consists of the following three parts: database, server and Web client. The client sends a HTTP request to the server first, and the front controller distributes it to the corresponding page controller according to the request information (such as URL). After receiving the request, the page controller delegates the request object to the business object for processing, interacts with the database through MyBatis, and implements data processing. After the end of the process, the data and view are returned to the front end controller. Then the front controller reclaims control, hands over data and views to Vue.js, and Vue.js binds the data to Echarts. Finally the front controller presents the data and the page to the user [2]. Its framework is shown in Fig. 1.

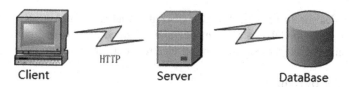

Client   HTTP   Server   DataBase

Fig. 1. The framework of SCTC

3.2   Data Preparation and Processing

The data used in the paper is obtained through network crawlers on many networks. These data are divided into three main categories: coal-related data, coal enterprise related data and power generation enterprise data. Figure 2 shows the classification of a coal network data after processing.

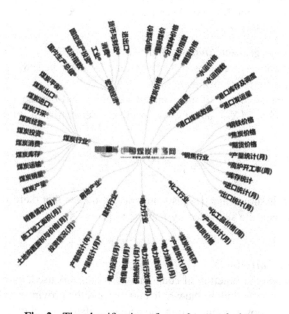

Fig. 2. The classification of a coal network data

We acquired coal-related data such as coal price, coal quality, production volume, inventory, transportation cost, and supply cycle, and related coal enterprise data such as company size, corporate background, cooperation, corporate credit, and relationship network, and the location information of the coal supply enterprise through the web crawler. In addition, we also obtained data related to power companies of the Group's subsidiaries, such as geographical location, power generation plan, inventory, and so on. After preparing the data, the system will store them in the database.

3.3   Coal Demand Forecast

SCTC provides the forecasting function of coal demand, using Auto-Regressive and Moving Average model (ARMA) model and Multivariate Co-integration Analysis model. It can forecast the consumption of electric coal for power companies [3].

3.3.1   Model Introduction

ARMA Model
ARMA model is a combination of Auto-Regressive model (AR) and Moving-Average model (MA). Therefore, it is also called autoregressive moving average model [4]. Its equation is as follows:

$$x_t = c + \emptyset_1 x_{t-1} + \cdots + \emptyset_p x_{t-p} + u_t + \theta_1 u_{t-1} + \cdots + \theta_q u_{t-q}, \ t = 1, 2, \cdots T \qquad (1)$$

In the equation, C is a constant. $\emptyset_i$ and $\emptyset_j$ represent the coefficients of the Auto-Regressive model and the Moving-Average model, respectively. p and q represent the order of the two models.

Multivariate Co-integration Analysis Method
Co-integration theory was proposed by Engle and Granger in 1987. It is the basis for the study of dynamic relationships.

Assume that the sequence of independent variables and the sequence of dependent variables are $\{X_1\}, \cdots \{X_k\}$ and $\{y_t\}$ respectively. Construct a regression model as follows [5].

$$y_t = \beta_0 + \sum_{i=1}^{k} \beta_i x_{it} + \varepsilon_t \qquad (2)$$

If the test result shows that $\{\varepsilon_t\}$ is stationary, it means that there is a cointegration relationship between the sequence $\{y_t\}$ and $\{X_1\}, \cdots \{X_k\}$.

3.3.2   Function Introduction
In the system, when the function of coal demand forecast is used, users can choose the geographical location including the whole country and the provinces and cities. When a user searches for a power plant within the scope of its own authority, the system can display historical data on a monthly basis and predict energy production and consumption of the plant in the coming months.

In SCTC, a two-column display is used for the planned and actual production capacity, and the predicted consumption and actual consumption are displayed with two-color columns of another color. When the mouse is hovered over the two-column column, the proportion prompt will pop up. The function enables business staff to more intuitively understand the plan and the dynamics of actual production, and the ratio of the deviation between the predict consumption and the actual consumption. At the same time, it also helps the system developers to collect data, optimize the algorithm, and improve the accuracy of the system prediction.

3.4 Regional Situation Analysis

The regional situation analysis interface uses the layout of the middle main view and the surrounding auxiliary view. The main view and the auxiliary view can be linked together.

The main view is a GIS map that allows users to select power generation companies within the country or provinces or cities. After selecting a power generation company on the map, the coal suppliers around the power generation company will be immediately displayed on the map, and the distance between the suppliers and the power generation companies and the corresponding transportation costs will be marked.

The auxiliary view includes the overview of the electric power enterprise, the coal price, the supply cycle, the business operation, the coal quality contrast, the enterprise credit and the coal price trend of the top 5 coal enterprises. The overview of the power generation company mainly shows basic information such as the power generation plan, coal inventory, and optimal coal quality of the company. The coal quality section includes information such as calorific value, sulfur content, moisture content, ash content, and comprehensive evaluation of the coal enterprises. In the form of the linkage between the main view and the auxiliary view, the user can select a suitable coal supplier in a short time.

3.5 Enterprise Portraits of Coal Suppliers

In the regional situation analysis interface, the user can grasp the situation of all the coal suppliers in a specific area, and can initially filter out several suitable suppliers in less time. In order to grasp the full range information of coal suppliers, SCTC provides enterprise portraits. When a coal supplier is selected in the regional situation interface, it will automatically jump to the enterprise portraits interface of the coal supplier company.

The portraits of coal companies are mainly described in terms of basic information, evaluation and relations, and business cooperation. The basic information includes the scale of the company, business background, production capacity, coal quality, and coal price. The evaluation and relations include corporate credit evaluation, supplier evaluation, and relationship network. The business cooperation includes contract conditions, cohesiveness, coal purchases, and coal purchases. Three-color labels represent the three states above, equal to, and below the average of all companies for quantifiable indicators. The relationship network is more complex, so it is shown in thumbnails. An individual display window for the relational network pops up when the thumbnail is clicked.

The intelligent analysis part uses big data technology to clean and mine target enterprise data. The comprehensive assessment of coal companies is conducted in terms of transaction volume in the past year, changing trend of historical trading average prices, credit evaluation and so on.

4 The Application

Through the analysis and mining of coal consumption, coal companies and power generation enterprises and other related data in February 2017, the paper forecasts the coal demand of a power generation enterprise, analyzes coal suppliers of the power enterprise, and provides the portrait of a coal supplier company. The results are shown in Figs. 3, 4, and 5.

The top five coal companies that are most suitable for this power generation company to purchase coal in Inner Mongolia, and as shown in Fig. 4.

Fig. 3. A coal demand of a power company

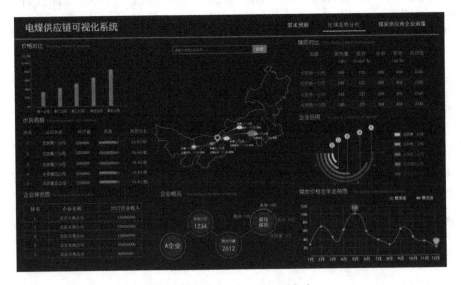

Fig. 4. Regional situation analysis

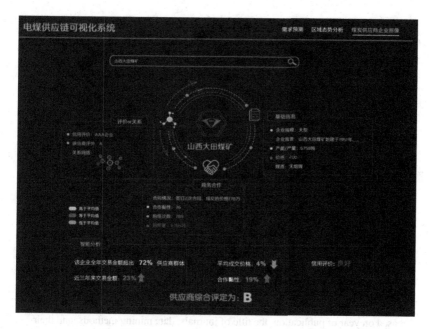

Fig. 5. The enterprise portrait of a coal supplier

5   Conclusion

The paper introduces the design and application of the visual system for the supply chain of thermal coal (SCTC), and implements the functions of coal demand prediction, regional situation analysis and the corporate portrait of the coal supplier. It has been verified by the actual data. The system is simple and friendly. Using the system, users do not have to compare information about coal prices, coal quality, supply time and so on by artificial. SCTC reduces the mistakes that can be made when looking for information subjectively, provides help and support for management and decision-making on purchase of thermal coal.

References

1. Tan, Z., Zhang, H., Liu, W., Wang, S., Zhang, J.: Reviews on risk management of coal and electricity energy supply chain. Modem Electric Power **31**(2), 66–74 (2014)
2. Chen, X., Yang, H.: Design and implementation of medical insurance data visualization system. Softw. Guide **16**(06), 59–62 (2017)
3. Chao, X.: Analysis and prediction of coal demand in electric power industry. North China Electric Power University (2015)
4. Hou, Y., Zhang, W., Zhang, H.: Coal demand combination forecasting. Coal Eng. **11**, 75–78 (2007)
5. Zhu, W.: Research on electricity consumption elasticity in China. Fudan University (2009)

Big Data Analytics on Twitter

A Systematic Review of Applications and Methods

Mudit Pradyumn[1]([✉]), Akshat Kapoor[2], and Nasseh Tabrizi[1]

[1] Department of Computer Science, East Carolina University,
Greenville, NC 27858, USA
pmudit90@gmail.com
[2] Health Services and Information Management, East Carolina University,
Greenville, NC 27858, USA

Abstract. As the amount of digital data is growing at an exponential rate, the emphasis is on forming an insight from the data. Although the new fields of research, including Twitter data analytics, are proven to be fruitful, there is a lack of literature review and classification of the research. Therefore, after segregating 1,025 research papers, we reviewed 29 papers from 20 journals on Twitter data analytics published from 2011 to 2017, and then classified them based on year of publication, the title of journals, data mining methods, and their application. This paper is written with the intent of understanding the trend of research in this field.

Keywords: Big data · Keyword Search · Twitter data · Classification
Analysis

1 Introduction

Presently, a tremendous amount of digital information namely sensor data, social media data, public web, and others are available. To be able to see trends, extract meaningful information, and form insights, from the data accumulating at such rapid rate, it requires specialized methods and techniques. As the technology is evolving, new and effective methods of big data analytics [1] are being developed.

One example of a popular source of big data is Twitter [2] that causes the largest collection of human generated data. Twitter has 330 million active monthly users creating about 500 million tweets, resulting in 200 Billion tweets a year [3]. One advantage of using Twitter data for trend analysis is that, it generates real time data [4], making it possible to gain insights and trending information instantaneously. This paper discusses various methods that are being employed to make use of the data generated on Twitter, as well as the different application areas of that data. These researches are grouped and presented in systematic review on the bases of technique of research such as Sentiment Analysis [5], Linguistic Analysis, Comparison of data, Information System. They can play a game changing role in the areas like Disaster Management, Impact on People, Cybercrime Detection, Public Health Service, Disease Management, Medical Complaints. This paper discusses the results from the analysis of

© Springer International Publishing AG, part of Springer Nature 2018
F. Y. L. Chin et al. (Eds.): BIGDATA 2018, LNCS 10968, pp. 326–333, 2018.
https://doi.org/10.1007/978-3-319-94301-5_26

selected papers and segregation based on the four categories namely, the year of publication, the journal, their field of research, and relevance to the topic.

2   Systematic Review of the Papers

2.1   Sentiment Analysis

The research published in Technological Forecasting and Social Change #iamhappy-because: Gross National Happiness through Twitter analysis and big data [6], was conducted to check the gross national happiness in Turkey. Twenty thousand people tweeted 35 million tweets, where they were analyzed by open source sentiment analysis tool. Data from previous years was compared and based on level of happiness they were categorized as happy, neutral, or negative [6]. Here are some insights.

- There is a relationship between happiness and stock index.
- Study published in detecting suicidality on Twitter [7], showed that people do use Twitter data to express their suicidal feelings among other sentimental expression.
- World Cup 2014 in the Twitter World brought out a variety of public moods [8, 9]. As predicted, emotions of fear and anger peaked after events were not in favor of the U.S. soccer team.
- English composed tweets with geolocation information were collected from March 2014 to December 2014 using Twitter's Streaming API [10]. After cleaning and filtering the tweets, a sample of 146,357 tweets was found using a keyword Search [11] for "cancer". Hedonometric analysis [12] was used to compute the average happiness of each type of cancer on a scale from 1–9.

The tweets with both negative and positive emotions carry many new words common in computer language that enhance the lexicon [13]. The words in the tweets were then tagged using a part-of-speech tagger and the features of the words were calculated. The research presented a new approach for opinion lexicon expansion using data from Twitter. Another study found that the sentiment of a tweet is less useful in terms of prediction than the number of tweets posted by a user [14]. A *klout score* [15], a score that shows the level of influence an individual has, was also calculated for each tweet. An ordinary least square regression and a linear probability model were used to review the relations between the stocks and the sentiments of the tweets. There seems to be a connection between the outcome of games and the sentiment of the fan base on Twitter [16]. The researchers built a Central Sport system, which collects data from Twitter to use in combination with Twitter's streaming API to capture tweets using specific hashtags. It was found that the accuracy of the different models did not prove to be more accurate than a baseline odd only approach. Tweets on company's events can also enhance its market scope and stock value [17]. With the increase in use of Internet and social media, the micro-blog data and blog sentiment provide a useful material which can be used for stock market prediction, its volatility and survey sentiments [18].

2.2 Linguistic Analysis

The study published in Understanding U.S. regional linguistic variation with Twitter data analysis [19], aimed to look at the regional linguistic variation across the continental United States. The study collected geotagged tweets within the continental U.S from October 7, 2013 to October 6, 2014. Lexical alternations were then used to look at the difference in language across the U.S. A variant preference and a mean variant preference were calculated for each county and their alternation. Further research [20] was conducted using linguistic analysis to determine the sarcastic sentences and differentiate them with irony [21]. At times grammatical mistakes are made intentionally to express the views to the end user [22]. The goal of the study published in "Looking for the perfect tweet. The use of data mining techniques to find influencers on twitter" [23] was to investigate and determine the characteristics of influencers on Twitter. IBM SPSS Statistics 23.0 [24] version was used to analyze the different variables of the tweets. Big influencers on Twitter used more hashtags and have more mentions than average users but they use less links. Their tweets were shorter in length and usually express a clear opinion. They also follow many people. They were able to find a clear trend in the ways influencers use Twitter [25]. The shortcoming in the results was that it used Spanish composed tweets and two keywords and data was only collected over a twelve-day span. Further study [26] reveals while gender and income had positive associations with real world opinion leaders, these characteristics had little association with opinion leaders on Twitter.

2.3 Disaster Management

The study titled "Crowd sourcing disaster management: The complex nature of Twitter usage in Padang Indonesia" [27] was designed to analyze whether Twitter could be used in disaster management situations. Five different methods of data collection were examined. The research indicates that situation assessment of densely populated areas can be done using Twitter. Another study titled [28], set out to analyze the way Twitter is used during disaster situations, specifically for Japan's tsunami, and address the current problems and concerns users have.

2.4 Information System

Tweets provide vital information about some important situation or event, etc. The Twitter data in "From Twitter to detector: Real-time traffic incident detection using social media data" was acquired by adaptive data acquisition [29]. From the extracted data, features are extracted based on the keyword. These are then classified and geotagged to gain the information. The tweets provide immediate information, which matches with the information obtained from reliable sources. The data is then compared with HERE [30], which has the time-varying travel time on most of the roads. It was noticed that the data matched closely. But the location of the tweeter is seldom available, which left the information deficient of location. Further, the key words do not pin point the subject of information. In the event of emergency, tweeters provide quick responses from the people [31].

2.5   Cybercrime Detection

Cyberbullying has become a serious problem within the vast amounts of social networks, which was the topic of study in the paper "Computers in Human Behavior Cybercrime detection in online communications: The experimental case of cyberbullying detection in the Twitter network" [32]. Geotagged tweets were collected within the state of California from January - February 2015. Network information, activity information, user information, and tweet content were key features used in the machine learning algorithms to detect cyberbullying. Different combinations of the features were tested with four different classifiers to see which features and classifier would give greater accuracy in detecting cyberbullying. Naive Bayes [33], support vector machine [34], random forest [35], and k-nearest neighbor were the algorithms used in the study. The study's goal was to be able to differentiate cyberbullying tweets from non-cyberbullying tweets using key features and machine learning algorithms. The machine learning algorithms could correctly label a non-cyberbullying tweet 99.4% of the time while it could only correctly label a cyberbullying tweet 71.4% of the time [32].

2.6   Public Health Services

The information on use of marijuana concentrate in different parts of America was gathered in this study "'Time for dabs': Analyzing Twitter data on marijuana concentrates across the U.S." [36]. The tweets were filtered from the drug related tweets using Twitter's API. Keywords used for this research were: tweets: "dabs"; "hash oil"; "butane honey oil"; "smoke/smoking shatter"; "smoke/smoking budder"; "smoke/smoking concentrates." The eDrugTrends system provided a Twitter filtering and aggregation framework [37]. A sample of 125,255 tweets was collected over two-month period, out of which 27,018 of the tweets contained geolocation information. It was found that California, Texas, Florida, and New York had the highest raw number of dabs-related tweets, but after adjusting for different activity levels in each state, Oregon, Colorado, and Washington had the highest proportion of dabs-related tweets. The average adjusted proportion for Status 1, Status 2, and Status 3 states of dabs related tweets was 5.1%, 2.3%, and 1.4% respectively. The study found that dabs-related tweets were more common in states where medical and recreational use of cannabis is legal. Furthermore, the Western region of the United States of America had greater dabs-related tweeting activity [36].

2.7   Disease Management

The study titled "Predicting Flu Trends using Twitter Data" [38], aimed to track flu trends using Twitter and be able to predict influenza like illnesses (ILI). It was shown that there was a Pearson correlation coefficient of 0.9846. A regressive model was built and tested with old CDC data. Using Twitter data improved the model's accuracy in predicting ILI cases and can provide real time analysis of influenza activity. Twitter was shown to be able to effectively track influenza like illnesses and help accurately predict influenza activity.

3 Results

Based on the popularity, importance to topic and citations, we selected following papers (See Table 1) for our review.

Table 1. Selection of papers based on field of research

Field of research	Available papers with keywords	Selection based on relevance
Sentiment analysis	261	10
Linguistic analysis	150	3
Comparison of data	75	2
Influencer identification	54	2
Disaster management	63	2
Information system	34	2
Impact on people	21	1
Detecting relevant topic	4	1
Computer communication	5	1
Cybercrime detection	68	1
Public health service	120	1
Disease management	78	1
Future event prediction	49	1
Medical complaints	43	1
Total	1025	29

Our classification reflects diverse research in the field of Twitter data extraction. The classification is done in four main categories, the year of publication, the journal in which these articles or research papers are published, field of research and relevance to the topic. After completing screening, 29 research papers from 22 journals met the criteria to be included. These classifications are represented below:

3.1 Year of Publication

It is evident from Fig. 1 that the amount of research being done in this field is rapidly increasing over the past few years.

3.2 Journals

The journal named "Computers in Human Behavior" published the maximum number of articles related to our research field. It published 20.69% of all research papers analyzed for classification. See Fig. 2.

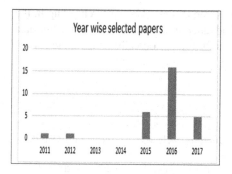

Fig. 1. Year wise selected papers

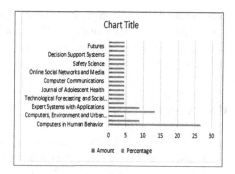

Fig. 2. Selection of papers based on journals

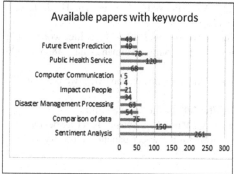

Fig. 3. Available papers with keywords

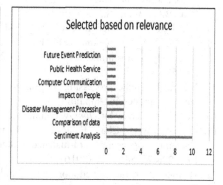

Fig. 4. Selected papers based on relevance

3.3   Research Area

Figures 3 and 4 show available papers with keywords and selected papers based on relevance.

4   Conclusion

The aim of this systematic review was to study the various developments in the field of Twitter data analytics and gain insight to their applications and methods. For better understanding, we describe the idea of research methods and the various steps involved in the data science process. Based on these principles, various research papers were analyzed and classified into four categories based on the selected parameters.

The classification is helpful to understand the basics of big data methods of collecting, processing and finding insights out of the data collected. We also described various ongoing research in the field of Twitter data extraction which utilizes the concept of the data collection by means of big data. This helps in understanding the ongoing research in field of Twitter data analytics. This research paper does not claim to be comprehensive, but we have tried to put some of the major research going in the field of the Twitter data analysis. With rapid advancements in the field of Twitter data analytics, we recommend re-visiting these methods and periodically revising the review to include new developments.

References

1. Data analytics. https://www.techopedia.com/definition/26418/data-a. Accessed 2 Oct 2018
2. Twitter. https://twitter.com/. Accessed 2 Oct 2018
3. Sayce, D.: https://www.dsayce.com/social-media/tweets-day/. Accessed 2 Oct 2018
4. Real time data. https://www.techopedia.com/definition/31256/real-t. Accessed 2 Oct 2018
5. Sentiment analysis. https://www.lexalytics.com/technology/sentiment. Accessed 3 Oct 2018
6. Durahim, A.O., Co, M.: Technological forecasting & social change #iamhappybecause : gross national happiness through twitter analysis and big data, vol. 99, pp. 92–105 (2015)
7. Dea, B.O., Wan, S., Batterham, P.J., Calear, A.L., Paris, C., Christensen, H.: Detecting suicidality on Twitter. Invent 2(2), 183–188 (2015)
8. Yu, Y., Wang, X.: Computers in human behavior world cup 2014 in the twitter world: a big data analysis of sentiments in U.S. sports fans' tweets. Comput. Human Behav. 48, 392–400 (2015)
9. Natural language processing. https://www.sas.com/en_us/insights/analytics/what-is-natural-language-processing-nlp.html
10. Keyword Search. http://www.columbia.edu/cu/lweb/help/clio/keyword.html
11. Crannell, W.C., Clark, E., Jones, C., James, T.A., Moore, J.: Sciencedirect association for academic surgery a pattern-matched twitter analysis of us cancer-patient sentiments. J. Surg. Res. 206(2), 536–542 (2016)
12. Hedonometer. https://hedonometer.org/index.html
13. Bravo-marquez, F., Frank, E., Pfahringer, B.: Knowledge-based systems building a twitter opinion lexicon from automatically-annotated tweets, vol. 108, pp. 65–78 (2016)
14. Corea, F.: Can twitter proxy the investors' sentiment? the case for the technology sector. Big Data Res. 4, 70–74 (2016)
15. Klout. https://klout.com/corp/score
16. Schumaker, R.P., Jarmoszko, A.T., Jr, C.S.L.: Predicting wins and spread in the premier league using a sentiment analysis of twitter. Decis. Support Syst. 88, 76–84 (2016)
17. Daniel, M., Neves, R.F., Horta, N.: Company event popularity for financial markets using twitter and sentiment analysis. Expert Syst. Appl. 71, 111–124 (2016)
18. Pandey, A.C., Rajpoot, D.S., Saraswat, M.: Twitter sentiment analysis using hybrid cuckoo search method. Inf. Process. Manag. 53(4), 764–779 (2017)
19. Oliveira, N., Cortez, P., Areal, N.: The impact of microblogging data for stock market prediction: using Twitter to predict returns, volatility, trading volume and survey sentiment indices. Expert Syst. Appl. 73, 125–144 (2017)
20. Huang, Y., Guo, D., Kasakoff, A., Grieve, J.: Understanding U.S. regional linguistic variation with Twitter data. Comput. Environ. Urban Syst. 59, 244–255 (2016)
21. Sulis, E., Irazú, D., Farías, H., Rosso, P., Patti, V.: Knowledge-Based Systems Figurative messages and affect in Twitter : Differences between #irony, #sarcasm and #not, vol. 108, pp. 132–143 (2016)
22. Oussalah, M., Escallier, B., Daher, D.: An automated system for grammatical analysis of Twitter messages. a learning task application. Knowl. Based Syst. 101, 31–47 (2015)

23. Lahuerta-Otero, E., Cordero-Gutirrez, R.: Looking for the perfect tweet. the use of data mining techniques to find influencers on twitter. Comput. Hum. Behav. **64**, 575–583 (2016)
24. IBM.   https://www.ibm.com/analytics/data-science/predictive-analytics/spss-statistical-software
25. Wilcoxon signed-rank test. https://statistics.laerd.com/spss-tutorials/wilcoxon-signed-rank-test-using-spss-statistics.php
26. Park, C.S., Kaye, B.K.: The tweet goes on: interconnection of twitter opinion leadership, network size, and civic eng. Comput. Hum. Behav. **69**, 174–180 (2017)
27. Carley, K.M., Malik, M., Landwehr, P.M., Pfeffer, J., Kowalchuck, M.: Crowd sourcing disaster management: the complex nature of twitter usage in padang Indonesia. Saf. Sci. **90**, 48–61 (2016)
28. Communities, W.B.: Twitter for crisis communication : lessons learned from Japan' s tsunami disaster Adam Acar * and Yuya Muraki, vol. 7(3), pp. 392–402 (2011)
29. Gu, Y., Sean, Z., Chen, F.: From Twitter to detector: real-time traffic incident detection using social media data. Transp. Res. Part C **67**, 321–342 (2016)
30. HERE. https://www.here.com/en
31. Laylavi, F., Rajabifard, A., Kalantari, M.: Event relatedness assessment of twitter messages for emergency respons. Inf. Process. Manag. **53**(1), 266–280 (2015)
32. Lin, X., Lachlan, K.A., Spence, P.R.: Computers in human behavior exploring extreme events on social media: A comparison of user reposting/retweeting behaviors on Twitter and Weibo. Comput. Human Behav. **65**, 576–581 (2016)
33. Naive Bayesian. http://www.statsoft.com/Textbook/Naive-Bayes-Classifier
34. Support vector machine. http://scikit-learn.org/stable/modules/svm.html
35. Random Forest. http://www.stat.berkeley.edu/~breiman/RandomForest/cc_home.htm
36. Daniulaityte, R., et al.: ' Time for dabs': analyzing twitter data on marijuana concentrates across the U. S. Drug Alcohol Depend. **155**, 307–311 (2015)
37. Kayser, V., Bierwisch, A.: Using twitter for foresight: an opportunity? Futures **84**, 50–63 (2016)
38. Achrekar, H., Lazarus, R., Park, W.C.: Predicting Flu Trends using Twitter Data, pp. 702–707 (2011)

A Survey of Big Data Use in Large and Medium Ecuadorian Companies

Rosa Quelal$^{(\boxtimes)}$ and Monica Villavicencio$^{(\boxtimes)}$

Escuela Superior Politécnica del Litoral (ESPOL),
Facultad de Ingeniería en Electricidad y Computación (FIEC),
Campus Gustavo Galindo, Km 30.5 vía Perimetral, Guayaquil, Ecuador
{rquelal,mvillavi}@espol.edu.ec

Abstract. Big data has become a subject of great interest among a variety of organizations, both from the scientific and business sectors. In this line, it is important to know the focus of attention that companies have on big data. This paper presents a study conducted in Ecuador about big data initiatives among large and medium companies. Results indicate that companies do not have a clear understanding of the implications of big data for their own benefits. Also, higher interest on big data initiatives comes from the private rather than the public sector. And, companies have a preference for contracting big data services from third-parties instead of hiring specialized personnel.

Keywords: Big data · Survey · Large companies · Ecuador

1 Introduction

Big data is a term used to describe very large datasets, with structured or unstructured data, with own and/or external data (i.e. ERP, CRM, legacy systems, web sites, camera's pictures, or devices information); where the data manipulation requires fast processing ways to solve the volume and variety of data problems [1]. Usually, big data is used in Data Science projects, applied in disciplines as biology, medicine, manufacturing, marketing, and others [2]. It is also possible to analyze large volumes of business transactional data [3], especially in large companies, or companies with a big volume of data [4]. The business analytical interest relies on various fields like manufacture of healthcare machines, banking transactions, social media, satellite imaging [5], and analysis for intelligent decision-making [6]. However, credit card, financial, and healthcare providers do not use 80 to 90% of their data [7]. Big data represents an opportunity to companies to increase informed strategic direction, improve customer service, and identify new products, new customers and markets [5]. As T. H. Davenport said, *"Any type of firm in any industry, can participate in the data driven economy. . . can all develop databased offerings for customers, as well as supporting internal decisions with big data"* [3].

Australia, United Kingdom, United States, Japan, Germany, Netherlands, India, Mexico and Brazil are some of the countries which have studied the use of big data in companies [8]. Some Latin American countries like Mexico, Colombia, Chile, Argentina and Brazil have reported an increment of big data initiatives since 2012,

F. Y. L. Chin et al. (Eds.): BIGDATA 2018, LNCS 10968, pp. 334–342, 2018.
https://doi.org/10.1007/978-3-319-94301-5_27

especially in the use of social media and google trends information to analyze different social topics [6, 9, 10].

In this paper, we present the findings of a survey conducted to Ecuadorian companies aimed to get insights about their big data initiatives. The next section presents the research methodology; Sect. 3 shows the obtained results; and Sect. 4 concludes the article.

2 Research Methodology

The survey instrument designed for this study was based on two main sources [11, 12]. The instrument included 49 questions divided into four sections: (1) general information of companies, (2) respondents' information, (3) analysis capabilities and current state of data, and (4) big data initiatives.

The target population were Information Technology (IT) professionals working in medium and large Ecuadorian companies. The survey was administered in person from August to September 2017, collecting a total of 31 responses. To perform the statistical analysis, we used R 3.4.1.

3 Results and Data Analysis

Table 1 summarizes the demographics of respondents. As it can be noticed, respondents mainly occupy a leadership position (52%), and have a master degree. Regarding companies, more than 30% are public (mainly utilities providers), nearly 90% are large, and around 68% have more than 20 years in the market (see Table 2). In this sample, the majority of companies develop Transaction Processing Applications (i.e. inventory, human resources systems) and use Electronic Sheets to analyze data (see Table 3); data aged 10 years old or less (74.19%).

Table 1. Demographics of respondents

Role within the organization	Percentage
CTO	29%
IT project manager	23%
Software architect	13%
IT analyst	29%
Software testing engineer	6%
Age of respondent	**Percentage**
25–34	16,13%
35–44	61,29%
>45	22,58%
Degree of education	**Percentage**
Master	64,51%
Bachelor	29,03%
Other	6,45%

Table 2. Demographics of companies

Economic sectors of companies	Percentage
Public service/Local government/Central government	32,26%
Agro industrial	16,13%
Banking/Finance/Insurance	16,13%
Wholesale/Retail trade	16,13%
Education	6,45%
Manufacturing	3,22%
Military	3,22%
Health	3,22%
Telecommunications	3,22%
Size of companies	**Percentage**
Large (>200 employees)	87,10%
Medium (50–199 employees)	12,90%
Age of companies	**Percentage**
<10 years	19,35%
10–20 years	12,90%
>20 years	67,75%
Type of companies	**Percentage**
Private	54,84%
Public	38,71%
Public & Private	6,45%
Type of customers	**Percentage**
Retail customers	61,29%
Wholesale customers	35,48%
Wholesale & Retail customers	3,23%

Table 3. IT General information

What type of applications have been developed in your organization?	Percentage
Transaction processing applications	80,65%
Real-time applications	67,74%
Management information systems	64,52%
Web applications or shopping carts	41,94%
Applications on mobile devices	35,48%
What type of applications have been developed in your organization?	**Percentage**
Multimedia Systems	32,26%
Tools used to analyze data	**Percentage**
Electronic sheets	74,19%
Business intelligence	58,06%
Statistical software	51,61%
Dashboard	41,94%

Regarding big data understanding, 29% of respondents self-rated them as con-noisseurs (i.e. understand what big data means in detail), 55% as novices (i.e. under-stand but not in detail), and the rest with no understanding at all. Table 4 shows the actual level of understanding of big data concepts. With respect to big data initiatives, only 16% of organizations are already working in a project, 23% are in the planning stage, and the rest has not started yet. When respondents were asked about how the big data projects are managed, 56% mentioned that big data initiatives are led in collab-oration between business and IT areas. Fifty percent of the big data projects are performed by third parties under direction and supervision of the company, using the methodologies commonly applied for developing software, like Scrum, Kanban, and Extreme programming. Also, 71% of companies decided to use tools developed by third parties. In addition, we found that 83% do not plan to hire data scientists, instead to work with product sellers, consultants, and PhD interns.

Table 4. Actual level of understanding of big data concepts

Big Data is a term used to describe very large datasets of internal and external origin of organizations	64.52%
Big Data is a term used to describe complex data sets such as those from social media, video, photos, unstructured text, or those collected by devices	61,29%
Manipulation of very large datasets often requires applications running in parallel on tens, hundreds, or even thousands of servers	45,16%
Big Data is used when fast processing is required to solve volume and variety of data problems	41,94%
Big Data is used when you need to improve processing time	35,48%

The business functions currently analyzed are similar to the ones that are planned with big data projects, except from digital marketing, which notably increases with big data initiatives (See Table 5). The main reasons for adopting these initiatives are: to get new insights about the business (82%), to have information and ideas about the market (65%), and to analyze data in less time (53%). The initiatives are mostly oriented to analyze data from internal and/or external sources (71%), data of unknown size (63%), and new data types (41%). The expectation is to analyze real time data and data integrated with social networks to provide real-time alerts and perform relationship analysis. The desired analytical functions are presented in Table 6, and the data types and structures in Table 7. Transactional and geospatial data are the two most currently used in big data projects; nevertheless, voice, audio and social network data are expected to increase and use in the next 3 years. Relational databases and data ware-houses are still predominant in ongoing and planned big data projects (Table 7).

Table 5. Analyzed business functions

Business function	Current data analysis	Focus of big data initiatives
Operation	67,74%	62,50%
Market analysis	61,29%	56,25%
Customer service	51,62%	43,75%
Information technology	51,61%	37,50%
Product management	41,94%	37,50%
Risk management	25,81%	18,75%
Fraud management	25,81%	12,50%
E-Commerce	16,13%	12,50%
Digital marketing	3,23%	43,75%

Table 6. Desired analytical functions

Data visualization	82,35%
Social networking analysis	41,18%
Advanced analysis algorithms	41,18%
Run algorithms faster	41,18%
Analysis of feelings	29,41%
Automatic learning	23,53%
Text analysis	23,53%
Run algorithms with more data	17,65%
Modulation of topics	11,76%

Table 7. Data types and structures

What kind of data are you considering analyzing with Big Data technologies?	Currently	In 3 years
Transactional data	76,47%	0,00%
Geospatial data	52,94%	23,53%
Image data	41,18%	17,65%
Data of sensors machinery and devices	35,29%	35,29%
Social networking data	23,53%	47,06%
Mail/documents (Unstructured)	23,53%	41,18%
What kind of data are you considering analyzing with Big Data technologies?	**Currently**	**In 3 years**
Clickstream	17,65%	35,29%
Scientific data	12,50%	29,41%
Voice/Audio	0,00%	41,18%

(Continued)

Table 7. (*Continued*)

Which data structures are of particular interest in your Big Data initiatives?	Percentage
Relational database/Data warehouse	94,12%
Images	47,06%
XML	41,18%
Flat file	35,29%
Non-structured texts	17,65%
Json	17,65%
Semantic web	11,76%
Audio video	11,76%
Series-Logs	6,25%

Next, we illustrate the characteristics of the companies participating in this study by using spider plots. We observe from Fig. 1 that companies from the public, agro industrial and wholesale/retail trade sectors have mainly up to 5 people analyzing data. Also, we see in Fig. 2 that the public and wholesale/retail trade sectors have 5 or less years of information to analyze, while banking/finance/Insurance and agro industrial sectors 11 to 15 years of information. Additionally, private companies like agro industrial, banking/finance/insurance and wholesale/retail trade sectors reported a greater number of ongoing or planned big data initiatives compared to public (see Fig. 3). Finally, we observed that companies with ongoing big data initiatives, also reported a better understanding of big data concepts (see Fig. 4).

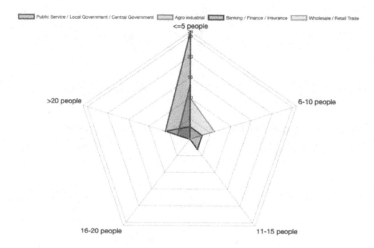

Fig. 1. Spider plot of the number of people required to analyze data by economic sectors

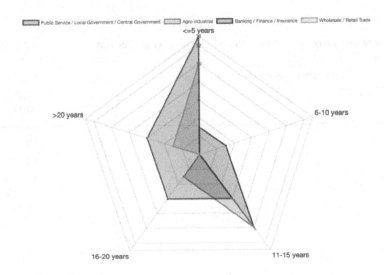

Fig. 2. Spider plot showing the age of analyzed data by economic sectors

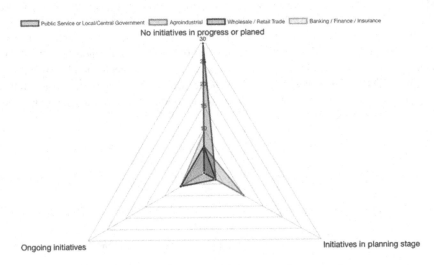

Fig. 3. Spider plot of Big Data initiatives stage by economic sectors

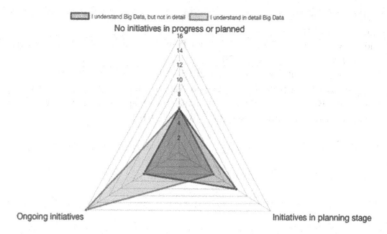

Fig. 4. Spider plot of initiatives of big data based on the understanding of concepts

4   Conclusions

This article summarizes the findings of the state of adoption of big data initiatives in Ecuador. We found that although most of the surveyed professionals have middle and high management positions and a master degree, their understanding of big data concepts is rather low; most of them mostly know that big data is applicable to large and complex datasets, but few believe that it can solve problems of processing time.

Almost 40% of respondents has big data initiatives in place, 71% of them plan to use tools developed by third parties and 83% are not planning to hire data scientists. The main reason of having big data initiatives is to get new insights of the business to innovate. People expect to analyze data from different sources in less time, and process data in real Time.

References

1. Isitor, E., Stanier, C.: Defining Big Data (2016)
2. Grill, E., Müller, M., Mansmann, U.: Health—exploring complexity: an interdisciplinary systems approach HEC2016. Eur. J. Epidemiol. **31**(1), 1–239 (2016)
3. Larsona, D., Chang, V.: A review and future direction of agile, business intelligence, analytics and data science. Int. J. Inf. Manag. **36**, 700–710 (2016)
4. Davenport, T.H., Dyché, J.: Big Data in Big Companies. Thomas H. Davenport and SAS Institute (2013)
5. Khan, N., Yaqoob, I.: Big data: survey, technologies, opportunities, and challenges. Sci. World J. **2014**, 18 (2014). ID 712826
6. Hilbert, M.: Big Data for Development: A Review of Promises and Challenges. Overseas Development Institute (2016)
7. Zikopoulos, P., Eaton, C., deRoos, D., Deutsch, T.: Understanding Big Data: Analytics for Enterprise Class Hadoop and Streaming Data. McGraw-Hill, New York (2012)

8. Siddiqui, S., Gupta, D.: Big data process analytics: a survey. Int. J. Emerg. Res. Manage. Technol. **3**(7) (2014)
9. Fernández, A., Gómez, A., Lecumberry, F.: Pattern Recognition in Latin America in the "Big Data" Era. Pattern Recognit. (2014)
10. Rigby, D.: Management Tools & Trends 2013. Bain & Company (2013)
11. Report of the 2015 Big Data Survey. United Nations Statistics Division (2016)
12. Frankováa, P., Balco, P.: Agile project management approach and its use in big data management. Procedia Comput. Sci. **83**, 576–583 (2016)

Short Paper Track: BigData Modeling

K-mer Counting for Genomic Big Data

Jianqiu Ge[1,2], Ning Guo[1], Jintao Meng[1], Bingqiang Wang[1],
Pavan Balaji[3], Shengzhong Feng[1], Jiaxiu Zhou[4], and Yanjie Wei[1(✉)]

[1] Shenzhen Institutes of Advanced Technology, Chinese Academy of Sciences,
Shenzhen 518055, China
yj.wei@siat.ac.cn
[2] University of Science and Technology of China, Hefei 230041, China
[3] Argonne National Laboratory, Lemont, IL 60439, USA
[4] Shenzhen Children's Hospital, Shenzhen 518038, China
shirleyzjx@163.com

Abstract. Counting the abundance of all the k-mers (substrings of length k) in sequencing reads is an important step of many bioinformatics applications, including de novo assembly, error correction and multiple sequence alignment. However, processing large amount of genomic dataset (TB range) has become a bottle neck in these bioinformatics pipelines. At present, most of the k-mer counting tools are based on single node, and cannot handle the data at TB level efficiently. In this paper, we propose a new distributed method for k-mer counting with high scalability. We test our k-mer counting tool on Mira supercomputer at Argonne National Lab, the experimental results show that it can scale to 8192 cores with an efficiency of 43% when processing 2 TB simulated genome dataset with 200 billion distinct k-mers (graph size), and only 578 s is used for the whole genome statistical analysis.

Keywords: K-mer counting · Genome sequence analysis
Performance and scalability

1 Introduction

With the rapid development of Next-Generation Sequencing (NGS) technology, the growth rate of genomic data is even faster than Moore's law. Pre-analyzing the genomic data has become an essential component in many bioinformatics applications and it poses a great challenge, especially in genome assembly [1, 2]. Most of the popular parallel genome assemblers are based on de Bruijn Graph strategy, a throughout overview of the number and the distribution of all distinct k-mers provides detailed information about the size of de Bruijn graph. For this reason, it becomes increasingly important to build the histogram of frequency of each distinct k-mer so as to meet the demand of these critical bioinformatics pipelines, including genome assembly, error correction [3], multiple sequence alignment [4], metagenomic data classification and clustering.

State-of-art k-mer counting tools can be classified into two categories: one is based on hard disk and shared memory environment, the other is based on distributed memory environment.

© Springer International Publishing AG, part of Springer Nature 2018
F. Y. L. Chin et al. (Eds.): BIGDATA 2018, LNCS 10968, pp. 345–351, 2018.
https://doi.org/10.1007/978-3-319-94301-5_28

Shared Memory Tools. Tools developed in early days are mainly classified into this category. These k-mer counting tools rely on different techniques. Firstly, the most well-known k-mer counting tool is Jellyfish [5], which uses a lock-free hash table and lock-free queues for communication so that several threads can update the hash table at the same time. KMC2 [6], MSPKmerCounter [7] and DSK [8] use disk-based partitioning technique to enable a low memory footprint with huge data processing. Methods such as BFCounter [9], Turtle [10] and KHMer [11] filter low coverage k-mers with Bloom filter to save memory consumption.

Distributed Memory Tools. As the size of genomic data increases dramatically in recent years, shared memory tools have failed to handle these data. Using high performance cluster to accelerate this procedure and breaking the limit of memory usage is a tendency. Representative tools include Kmerind [12] and Bloomfish [13]. Kmerind is the first k-mer counting and indexing library for distributed memory environment, it is implemented over MPI and contains many optimizations on efficient SIMD implementation and data structures. Bloomfish integrated Jellyfish into a MapReduce framework called Mimir [14], and 24 TB data were processed in 1 h.

In this paper, we propose a distributed k-mer counter with a higher scalability than distributed counting tools such as Bloomfish and Kmerind. The experimental results on Mira supercomputer show that, with 8192 cores, it can process 2 TB simulated genomic data with 200 billion distinct k-mers (graph size) in approximately 578 s.

2   System Design and Implementation

Our proposed k-mer counter contains three components, parallel I/O, k-mer extraction and distribution, counting. These components are illustrated in Fig. 1 and will be introduced in the following. For speeding up the pipeline during implementation, these three phases are further overlapped at a high degree with data streaming technology.

Parallel I/O. Improving the loading speed for TB or even PB level of sequencing data into memory from hard drive with multiple processes faces great challenges. Many traditional genome assembler and k-mer counting tools use one thread to load data, which usually takes several hours as the data size scales to TB level.

In our work, we adopt a similar approach as SWAP2 [15] and HipMer [16], parallel I/O module is used to speed up the loading process. This module supports both FASTQ and FASTA format sequence data. Firstly, the data are divided into n virtual fragments, where n is the number of processes. Since splitting data may break the DNA reads in the middle, a location function is used to check the beginning and ending point, so that each process can quickly locate the position information and send it to neighboring process.

K-mer Extraction and Distribution. After reading the sequences from the data files, each process will extract k-mers from sequences and then distribute k-mers to their corresponding processes according to a given hash function. In the extraction phase, a sliding window of length k will be applied to break short reads to k-mers. In the distribution phase, the generated k-mers are collected and grouped with a given hash

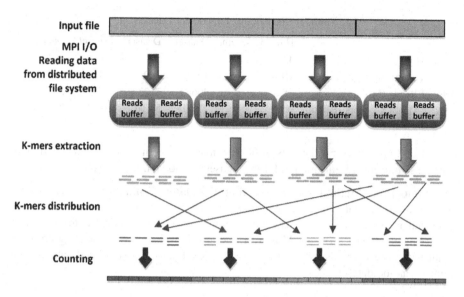

Fig. 1. Pipeline of distributed k-mer counting

function, and then each group of k-mers will be delivered to its corresponding processes with all_to_all collective communication protocols.

In this step, two other techniques are applied to further improve the computation and communication efficiency. The first one is data pre-compression and instruction level optimization on k-mer extractions. In this method, the bit operation and SSE instruction is used to further compress the computing time used in this phase. The second is to overlap the computing time in k-mer extraction phase and the communication time in k-mer distribution phases. In this method, the non-blocking all_to_all communication optimization helps to improve the efficiency.

Counting. Two types of results are counted and analyzed. One is the number of distinct k-mers, the other is frequency of each distinct k-mer. By using these basic results, almost all the other counting results can be further computed. Since the direction of DNA sequences is rarely known, genome assemblers, k-mer counting tools and many other bioinformatics applications usually treat the k-mers and their reverse complements as equivalent, we also use the same technique for our k-mer counter.

3 Experiment Setup

3.1 Dataset

To analyze the performance of k-mer counting tools, we selected two types of the datasets, real sequencing data from 1000 Genomes, and four other simulated datasets (ranging from 250G to 2 TB) generated from human reference (Downloaded from NCBI: GRCh38,

Table 1. Statistics of Experiment Datasets

	Human Genome	Dataset1	Dataset2	Dataset3	Dataset4
Size (GB)	4.5	250	500	1000	2000
Read Length (bp)	100	100	100	100	100
No. of Reads (Million)	28.8	3066.8	7133.6	14267.2	28534.4
No. of Distinct k-mers (Billion)	0.3	27.69	54.16	105.36	203.49

Data size: 3 GB). The test datasets in fasta format have both good coverage and large genome size. Detailed information for the datasets is shown in Table 1.

The Simulation Method for Generating Simulated Data. We selected human genome as the initial reference sequences in the experiment. Mutation is randomly generated in human reference sequences, simulation coverage is 50X, the length of each sequence read is 100 bp. Moreover, we also introduce 1% sequencing errors in simulated sequence reads. See Table 1 for the details of the 4 simulated datasets and human genome dataset.

3.2 Platform

In this section, we examine the performance of k-mer counting tools both on single node and high performance computing clusters. Single-node tests were conducted on a HPC server, which has Ivy Bridge-E Intel Xeon E5-2692 processor with 20 M L3 cache, 64 GB of DDR3 RAM. Distributed memory experiments were performed on IBM Blue Gene Q-Mira supercomputer at Argonne National Laboratory. Mira contains 48 cabinets, 49,152 compute nodes. Each compute node is equipped with 16 IBM PowerPC A2 processors and 16 GB RAM. All the compute nodes are connected by Infiniband, and the network of Mira follows 5D-torus architecture, the communication bandwidth is 0.9 GB/s per node. In addition, the I/O of Mira uses IBM GPFS system; it supports parallel I/O and the theoretical peak performance of its bandwidth is 32 GB per node.

4   Performance Evaluation

4.1   Single Node Test

In this section, we conducted a performance evaluation for several tools. We selected three tools, KMC2, Jellyfish and MSPKmerCounter. Both the real human sequencing data and simulated datasets are used in the experiments. During the test, we assigned one thread per core when running Jellyfish, KMC2 and MSPKmerCounter. Experimental results are shown in Tables 2 and 3.

Table 2. The time usage for human dataset using 24 cpu cores

	KMC2	Jellyfish	MSKPKmerCounter
Execution time (seconds)	27	37	62

Table 3. The time usage for simulated datasets using 24 cpu cores

	Dataset1	Dataset2	Dataset3	Dataset4
MSPKmerCounter	130 m	>6 h	out of memory	out of memory
KMC2	23 m	93 m	260 m	out of memory
Jellyfish	150 m	354 m	771 m	1495 m

As can be seen, KMC2 is the fastest k-mer counter. For GB range genomic data, MSPKmerCounter is the slowest. Even if the running time of these tools shows slight difference, all the tools can finish the run in relatively less time (minutes range).

We also tested the tools on four bigger simulated datasets. From Table 3, we can see that KMC2 is the fastest, but for the 2 TB dataset, it runs out of memory. As the dataset increased from 250 GB to 1 TB, the time consumption increased by 11.3 times, and the efficiency is only about 35%. For MSPKmerCounter, it cannot run the simulated dataset larger than 500 GB. Only Jellyfish can handle 2 TB simulated dataset, but it takes more than one day to count all the k-mers. Our proposed k-mer counter is designed for distributed environment with more memories, thus it is tested in the following step with more cores.

4.2 Scaling Test

To evaluate the scalability of the proposed tool, we conducted both the weak scaling and strong scaling tests. The size of simulated datasets ranges from 250 GB to 2 TB, and the number of CPU cores increases correspondingly from 1024 to 8192. The k value is set to 31. Results are shown in Figs. 2 and 3.

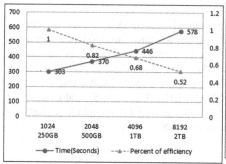

Fig. 2. Execution time and efficiency for weak scaling

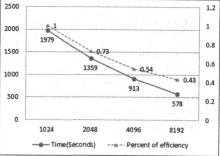

Fig. 3. Execution time and efficiency for strong scaling

Weak Scaling Test: For weak scaling test, we performed an experiment with 4 simulated datasets generated in the previous section. The data size increases from 250G to 2 TB as the number of cores increases from 1024 to 8192. From Fig. 2, we observe that when the cores and the data size doubled, the execution time shows a slightly increase. The ideal situation is a straight line. While the real human sequencing dataset has a fixed distinct k-mer number, and the rising number of distinct k-mers in simulated datasets accounts for this increase. When the number of cores doubles, the rising communication complexity will also affect the performance.

Strong Scaling Test: To analyze the strong scaling performance, we conduct a test on 2 TB simulated dataset with a fix problem size and double the number of cores in each round. Figure 3 shows that the proposed k-mer counter can scale to 8192 cores on 2 TB simulated dataset with more than 200 billion distinct k-mers, the execution time is 578 s. We note that both Bloomfish and kmerind can scale to 1536 cores on 384 GB sequence data from 1000 Genome Data on Comet at San Diego Supercomputer Center. Additionally, Bloomfish can scale to 3072 cores on 24 TB dataset in Tianhe-2A Supercomputer, while the scalability is still lower than our proposed tool. Besides, the strong scaling test results of our proposed tool show a linear decrease in execution time. Similar to the weak scaling test, the communication becomes more complex when the number of cores doubled, and finally a parallel efficiency of 43% for 8192 cores is obtained (1024 cores as a baseline).

5   Conclusions

K-mer counting has become an essential component in bioinformatics, which provides much important information for other bioinformatics pipelines. In this paper, we present a new distributed k-mer counter, which can take advantage of high performance clusters to count k-mers on thousands of cores with limited time. Moreover, it also has a higher scalability than Bloomfish and Kmerind.

Acknowledgements. This work is supported by Guangdong Provincial Department of Science and Technology under grant No. 2016B090918122, National Key Research and Development Program of China under grant No. 2016YFB0201305, the Science Technology and Innovation Committee of Shenzhen under grant No. JCYJ20160331190123578 and No. GJHZ20170314154722613, National Science Foundation of China under grant no. U1435215, 61702494 and 61433012, Shenzhen Basic Research Fund under grant no. JCYJ20170413093358429, Special Program for Applied Research on Super Computation of the NSFC-Guangdong Joint Fund under grant No. U1501501, and Youth Innovation Promotion Association, CAS to Yanjie Wei.

References

1. Meng, J., Wang, B., Wei, Y., Feng, S., Balaji, P.: SWAP-assembler: scalable and efficient genome assembly towards thousands of cores. BMC Bioinform. **15**, S2 (2014)
2. Simpson, J.T., Wong, K., Jackman, S.D., Schein, J.E., Jones, S.J., Birol, I.: Abyss:a parallel assembler for short read sequence data. Genome Res. **19**(6), 1117–1123 (2009)

3. Kelley, D.R., Schatz, M.C., Salzberg, S.L.: Quake: quality-aware detection and correction of sequencing errors. Genome Biol. **11**(11), R116 (2010)

4. Kent, W.J.: Blatthe blast-like alignment tool. Genome Res. **12**(4), 656–664 (2002)

5. Marcais, G., Kingsford, C.: A fast, lock-free approach for efficient parallel counting of occurrences of k-mers. Bioinformatics **27**(6), 764–770 (2011)

6. Deorowicz, S., Kokot, M., Grabowski, S., Debudaj-Grabysz, A.: Kmc 2: fast and resource-frugal k-mer counting. Bioinformatics **31**(10), 1569–1576 (2015)

7. Li, Y., et al.: Mspkmercounter: a fast and memory efficient approach for k-mer counting. arXiv preprint arXiv:1505.06550 (2015)

8. Rizk, G., Lavenier, D., Chikhi, R.: Dsk: k-mer counting with very low memory usage. Bioinformatics **29**(5), 652–653 (2013)

9. Melsted, P., Pritchard, J.K.: Efficient counting of k-mers in dna sequences using a bloom filter. BMC Bioinform. **12**(1), 333 (2011)

10. Roy, R.S., Bhattacharya, D., Schliep, A.: Turtle: identifying frequent k-mers with cache-efficient algorithms. Bioinformatics **30**(14), 1950–1957 (2014)

11. Zhang, Q., Pell, J., Caninokoning, R., Howe, A., Brown, C.T.: These are not the k-mers you are looking for: efficient online k-mer counting using a probabilistic data structure. PLOS ONE **9**(7), e101271 (2014)

12. Pan, T., Flick, P., Jain, C., Liu, Y., Aluru, S.: Kmerind: a flexible parallel library for k-mer indexing of biological sequences on distributed memory systems. IEEE/ACM Trans. Comput. Biol. Bioinform. (2017)

13. Gao, T., Guo, Y., Wei, Y., Wang, B., Lu, Y., Cicotti, P., Balaji, P., Taufer, M.: Bloomfish: a highly scalable distributed k-mer counting framework. In: ICPADS IEEE International Conference on Parallel and Distributed Systems. IEEE (2017). http://www.futurenet.ac.cn/icpads2017/?program-Gid_33.html

14. Gao, T., Guo, Y., Zhang, B., Cicotti, P., Lu, Y., Balaji, P., Taufer, M.: Mimir: Memory-efficient and scalable mapreduce for large supercomputing systems. In: 2017 IEEE International Parallel and Distributed Processing Symposium (IPDPS), pp. 1098–1108. IEEE (2017)

15. Meng, J., Seo, S., Balaji, P., Wei, Y., Wang, B., Feng, S.: SWAP-assembler 2: optimization of de novo genome assembler at extreme scale. In: 2016 45th International Conference on Parallel Processing (ICPP), pp. 195–204. IEEE (2016)

16. Georganas, E., Buluc, A., Chapman, J., Hofmeyr, S., Aluru, C., Egan, R., Oliker, L., Rokhsar, D., Yelick, K.: Hipmer: an extreme-scale de novo genome assembler. In: Proceedings of the International Conference for High Performance Computing, Networking, Storage and Analysis, p. 14. ACM (2015)

Ensemble Learning Based Gender Recognition from Physiological Signals

Huiling Zhang[1], Ning Guo[1], Guangyuan Liu[2], Junhao Hu[3],
Jiaxiu Zhou[4], Shengzhong Feng[1], and Yanjie Wei[1(✉)]

[1] Shenzhen Institutes of Advanced Technology, Chinese Academy of Sciences,
Shenzhen 518055, China
yj.wei@siat.ac.cn
[2] College of Electronic and Information Engineering, Southwest University,
Chongqing 400715, China
liugy@swu.edu.cn
[3] Chongqing Optoelectronics Research Institute, Chongqing 400060, China
[4] Shenzhen Children's Hospital, Shenzhen 518000, China

Abstract. Gender recognition based on facial image, body gesture and speech has been widely studied. In this paper, we propose a gender recognition approach based on four different types of physiological signals, namely, electrocardiogram (ECG), electromyogram (EMG), respiratory (RSP) and galvanic skin response (GSR). The core steps of the experiment consist of data collection, feature extraction and feature selection & classification. We developed a wrapper method based on Adaboost and sequential backward selection for feature selection and classification. Through the data analysis of 234 participants, we obtained a recognition accuracy of 91.1% with a subset of 12 features from ECG/EMG/RSP/GSR, 82.3% with 11 features from ECG only, 80.8% with 5 features from RSP only, indicating the effectiveness of the proposed method. The ECG, EMG, RSP, GSR signals are collected from human wrist, face, chest and fingers respectively, hence the method proposed in this paper can be easily applied to wearable devices.

Keywords: Gender recognition · Physiological signal · Feature selection
Ensemble learning · Wearable devices

1 Introduction

Gender contains a wide range of information regarding the characteristics difference between male and female. Automated gender recognition has numerous applications, including gender medicine [1, 2], video surveillance [3, 4], human machine interaction [5, 6]. Recently, with the development of social networks and mobile devices such as smartphones, gender recognition applications become more and more important. The research contents include facial image [7–9], speech [10, 11], body gesture [12, 13], and physiological signal [14] based gender recognition, among which gender recognition using physiological signals is more reliable but more difficult for data acquisition and analysis.

F. Y. L. Chin et al. (Eds.): BIGDATA 2018, LNCS 10968, pp. 352–359, 2018.
https://doi.org/10.1007/978-3-319-94301-5_29

In practice, automatic gender recognition is a two-class classification problem. With little prior knowledge, massive number of features will be extracted from the raw data. Searching for an optimal feature subset from a high dimensional feature space is known to be an NP-complete problem. As a key issue in machine learning and related fields, feature selection (FS) is used to select a better feature combination from many solutions, the essence of which is combinatorial optimization. Wrapper feature selection method, which utilizes the learning machine of interest as a black box to score subsets of feature according to their predictive power [15], has shown its superior performance in various machine learning applications.

In this paper, we propose a gender recognition method from multiple physiological signals, in particular, we developed a wrapper algorithm based on Adaboost.M1 [16] and sequential backward selection (SBS) for physiological feature selection. Through the data acquisition, feature extraction and feature selection & gender recognition procedure, we obtained a prediction accuracy of 91.1% on a dataset of 234 participants, and we also find a subset of 12 features which can best represent our gender recognition model.

2   Materials and Methods

The proposed physiological-signal based gender recognition system is composed of three core components: Data Collection module, Feature Extraction module and Feature Selection & Classification module.

2.1   Data Acquisition

234 students from Southwest University with no history of cardiac disease and mental disease voluntarily participated in the test. The electrocardiogram (ECG), electromyogram (EMG), respiratory (RSP) and galvanic skin response (GSR) signals are collected with BIOPAC System MP150 from the subject's wrist, facial muscle, chest and fingers, respectively. The sampling rates are 200 Hz for ECG, 1000 Hz for EMG, 100 Hz for RSP, and 20 Hz for GSR.

234 groups (154 female vs. 80 male samples) of valid data were obtained, and each signal record is an 80-s fragment. Figure 1 illustrates the raw signals of ECG, EMG, RSP and GSR from one participant.

2.2   Feature Extraction

The raw physiological signals are firstly preprocessed using wavelet transform. A bunch of statistical features such as maximum, minimum, mean and standard deviation are then extracted from the preprocessed signals as well as different transformations of the signals. The raw features are extracted mainly by the AuBT Biosignal Toolbox [17]. The details of the features can be found on our website http://hpcc.siat. ac.cn/~hlzhang/GR/193_features.html.

We have 234 samples with 84 ECG features, 21 EMG features, 67 RSP features and 21 GSR features, resulting in a raw data matrix of size 234 * 193. The value for

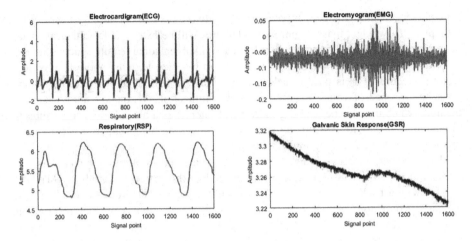

Fig. 1. Examples of raw ECG, EMG, RSP and GSR signals

each feature position is then normalized by Z-score to enable faster convergence of our feature selection algorithm.

2.3   Feature Selection and Gender Classification

The feature selection and classification algorithm, a wrapper method combining adaboost and SBS, is outlined in Fig. 2.

In this paper, classification and regression trees (CART) [18] is used as weak classifier of Adaboost. Assuming that the data record number is m and the feature dimension is n, the time complexity of CART is $O(nm\log m)$ ($\log m$ is the depth of tree and $O(nm)$ is the computational complexity of each layer), the iteration number of the SBS procedure (*while* loop in Algorithm Boost_FS) is n-c (c is a constant determined by line 17 in the algorithm), and the time complexity of quick-sort (feature importance as indicated by line 12 in the algorithm) is $O(n\log n)$. For a v-fold (v is a constant, which is 20 in this paper) cross validation, the computational complexity of the proposed algorithm is:

$$O((n - c) * (v * O(knm\log m) + O(n\log n) + n - c)) \approx O(kn^2 m\log m) \qquad (1)$$

where k is the number of trees used.

As seen from Eq. (1), the computational complexity of Algorithm Boost_FS grows as a quadratic function of feature dimension and as a $m\log m$ function of data record numbers, which demonstrates the scalability of the algorithm for big data applications.

In this paper, we call the feature subset found by Boost_FS algorithm with highest prediction accuracy as the best feature subset.

Algorithm Boost_FS

Input: Data Set S with N features

Output: Optimal feature subset

Steps:

01    $ACC_{Glob\_Max} = 0$

02    $ft\_num = N$

03   **while** $(ft\_num > 0)$

04        $S[1,..,v] \leftarrow$ Divide(S, v) /*Randomly divide S into v equal parts*/

05        $ACC_{Loc\_Max} = 0$

06        $ACC_{Loc}[1 : v] \leftarrow 0$

07   **for** $(i$ in $1 : v)$

08   Model$\leftarrow$Adaboost.M1.train$(S[, -i])$

09            $Acc\_Loc[i] \leftarrow$ Adaboost.M1.test$(S[i])$

10   **if** $(ACC_{Loc\_Max} <= ACC_{Loc}[i])$

11                    $ACC_{Loc\_Max} = ACC_{Loc}[i]$

12                    $Feat\_Order \leftarrow$ Feature_Importance$(Model)$

13        $ACC_{Loc\_Mean} \leftarrow$ Mean(ACC_{Loc})

14   **if** $(ACC_{Glob\_Max} <= ACC_{Loc\_Mean})$

15            $ACC_{Glob\_Max} = ACC_{Loc\_Mean}$

16            $Feat\_Order\_Glob = Feat\_Order$

17        $(S, n) \leftarrow$ Remove_feature$(S, Feat\_Order)$ /*Remove n least important feature(s) with the same score */

18        $ft\_num = ft\_num - n$

19   **return** $Feat\_Order\_Glob$

Fig. 2. The wrapper algorithm Boost_FS

2.4 Evaluation Metrics

We use various metrics, such as accuracy, precision, recall (also known as sensitivity), specificity, F1 score and ROC curve [19], to measure the quality of the prediction results. Shown as below are the definitions of accuracy, precision, recall and specificity values:

$$Accuracy = (TP + TN)/(TP + TN + FP + FN) \tag{2}$$

$$Precision = TP/(TP + FP) \tag{3}$$

$$Recall = Sensitivity = TP/(TP + FN) \tag{4}$$

$$Specificity = TN/(TN + FP) \tag{5}$$

where TP, TN, FP and FN represent the number of true positives, true negatives, false positives and false negatives.

F1 is the harmonic average of the precision and recall calculated as follows:

$$F1 = 2 * Precision/(Precision + Recall) \tag{6}$$

We also draw the Receiver Operating Characteristic (ROC) curve and calculate the area under this curve (AUC) for performance evaluation of our gender recognition models. The ROC curve, which is defined as a plot of test Sensitivity as the y coordinate versus its 1 - Specificity as the x coordinate, is an effective method of evaluating the performance of classification models. The AUC value, ranging from 0 to 1, shows the stability and performance of a model. An AUC value of 0 indicates a perfectly inaccurate test and a value of 1 reflects a perfectly accurate test.

3 Results and Analysis

In order to evaluate the performance of the proposed method, we employ a 20-fold cross-validation scheme. The dataset is divided into 20 folds with approximately 11/12 samples in each fold. 19 folds are used for training the gender recognition model, and the remaining fold is used for testing. We have 5 individual runs of BOOST_FS, with different input feature matrix extracted from different physiological signals. Each iteration (*while* loop in Algorithm BOOST_FS in Fig. 2) in the run generates a feature subset and the corresponding evaluation metrics. The best feature subset in each run is the subset with the highest prediction accuracy.

Figures 3, 4 and Table 1 show the overall performance of different models with or without the feature selection. From left to right, Fig. 3 illustrates the prediction accuracies, precisions, recalls, specificities and F1_scores using different gender recognition models. The model from 4 signals (ECG/EMG/RSP/GSR) with FS achieves the highest performance for all metrics: 91.1% accuracy, 92.4% precision, 94.2% recall, 85% precision and 99.0% F1_socre. The GSR based model without FS shows the worst performance for all metrics except for the F1_socre. Figure 4 shows the ROC curves of tests with and without using BOOST_FS. 4_signals_with_FS shows the highest AUC value of 0.951, followed by 4_signals_without_FS of 0.921, ECG_with_FS of 0.880 and RSP_with_FS of 0.833, which are consistent with the results and analysis in Fig. 3.

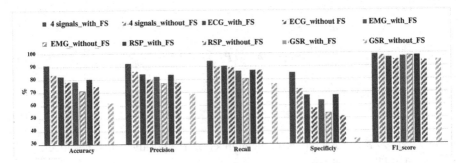

Fig. 3. Performance of accuracy, precision, recall, specificity, F1_socre for 10 different recognition models

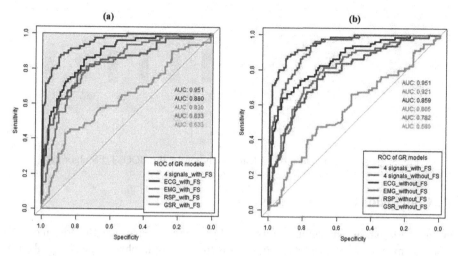

Fig. 4. ROC curve for (a) recognition models with FS; (b) recognition models without FS

Table 1. Prediction accuracies and feature numbers for different gender recognition models

Physiological signal	Accuracy without FS	Acc with FS	Total feature number	Best feature number
ECG/EMG/RSP/GSR	83.7%	91.1%	193	12
ECG	78.2%	82.3%	84	11
EMG	71.4%	78.8%	21	12
RSP	74.8%	80.8%	67	5
GSR	61.8%	69.1%	21	2

The feature number of the best subset for each type of signal and the corresponding prediction accuracy are tabulated in Table 1. For comparison, we also list the prediction accuracies and original feature numbers without FS. For 5 group of physiological signals shown in Table 1, the prediction accuracies using FS are increased by 7.4%/4.1%/7.4%/6%/7.3% compared with those without FS, correspondingly, feature numbers using FS are reduced by 181/73/9/62/19. The highest prediction accuracy from BOOST_FS is 91.1% with 12 features from the combined ECG/EMG/RSP/GSR data.

Figure 5 shows the feature names and feature importances in the best subset determined by the BOOST_FS algorithm. Detail information of the features can be found in Sect. 2.2 and on our website. 7 out of the 12 features selected from massive computational efforts are ECG features, and 3 are RSP features, which are reasonable since previous studies have reported the physiological difference between men and women in cardiac [20] and thoraco-abdominal [21] functions.

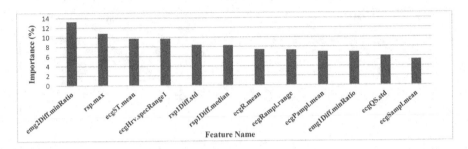

Fig. 5. Feature Importance of 12 features chosen by the BOOST_FS algorithm

4 Conclusions

In this work, we introduced an automated physiological signal based gender recognition system. We observed that a model built from multiple physiological signals can outperform model based on a single physiological signal. We further showed that recognition performance can be improved obviously through using a wrapper feature selection procedure. Finally, we analyzed the best feature subset which can best represent gender differences. The future work would concentrate on developing more effective feature selection algorithm, taking the effect of human age into account and applying our gender recognition system to human machine interface.

Acknowledgement. This work is supported by National Science Foundation of China under grant no. U1435215 and 61433012, Guangdong Provincial Department of Science and Technology under grant No. 2016B090918122, the Science Technology and Innovation Committee of Shenzhen Municipality under grant No. JCYJ20160331190123578, and No. GJHZ201703141 54722613, Special Program for Applied Research on Super Computation of the NSFC-Guangdong Joint Fund under Grant No. U1501501, and Youth Innovation Promotion Association, CAS to Yanjie Wei.

References

1. Ludwig, S., Oertelt-Prigione, S., Kurmeyer, C., Gross, M., Grüters-Kieslich, A., Regitz-Zagrosek, V., Peters, H.: A successful strategy to integrate sex and gender medicine into a newly developed medical curriculum. J. Women's Health **24**, 996–1005 (2015)
2. Canevelli, M., Quarata, F., Remiddi, F., Lucchini, F., Lacorte, E., Vanacore, N., Bruno, G., Cesari, M.: Sex and gender differences in the treatment of Alzheimer's disease: a systematic review of randomized controlled trials. Pharmacol. Res. **115**, 218–223 (2017)
3. Bonetto, M., Korshunov, P., Ramponi, G., Ebrahimi, T.: Privacy in mini-drone based video surveillance. In: 11th IEEE International Conference and Workshops on Automatic Face and Gesture Recognition (FG), Slovenia, pp. 1–6. IEEE (2015)
4. Venetianer, P.L., Lipton, A.J., Chosak, A.J., Frazier, M.F., Haering, N., Myers, G.W., Yin, W., Zhang, Z., Cutting, R.: Video surveillance system employing video primitives. Google Patents (2018)
5. Rukavina, S., Gruss, S., Hoffmann, H., Tan, J.-W., Walter, S., Traue, H.C.: Affective computing and the impact of gender and age. PLoS ONE **11**, e0150584 (2016)

6. Han, H., Otto, C., Liu, X., Jain, A.K.: Demographic estimation from face images: Human vs. machine performance. IEEE Trans. Pattern Anal. Mach. Intell. **37**, 1148–1161 (2015)
7. Bekios-Calfa, J., Buenaposada, J.M., Baumela, L.: Robust gender recognition by exploiting facial attributes dependencies. Pattern Recogn. Lett. **36**, 228–234 (2014)
8. Levi, G., Hassner, T.: Age and gender classification using convolutional neural networks. In: Proceedings of the IEEE Conference on Computer Vision and Pattern Recognition Workshops, Boston, pp. 34–42. IEEE (2015)
9. Dantcheva, A., Brémond, F.: Gender estimation based on smile-dynamics. IEEE Trans. Inf. Forensics Secur. **12**, 719–729 (2017)
10. Pahwa, A., Aggarwal, G.: Speech feature extraction for gender recognition. Int. J. Image Graph. Sig. Process. **8**, 17 (2016)
11. Li, M., Han, K.J., Narayanan, S.: Automatic speaker age and gender recognition using acoustic and prosodic level information fusion. Comput. Speech Lang. **27**, 151–167 (2013)
12. Lu, J., Wang, G., Moulin, P.: Human identity and gender recognition from gait sequences with arbitrary walking directions. IEEE Trans. Inf. Forensics Secur. **9**, 51–61 (2014)
13. Cao, L., Dikmen, M., Fu, Y., Huang, T.S.: Gender recognition from body. In: Proceedings of the 16th ACM International Conference on Multimedia, Vancouver, Canada, pp. 725–728. ACM (2008)
14. Hu, J.: An approach to EEG-based gender recognition using entropy measurement methods. Knowl.-Based Syst. **140**, 134–141 (2018)
15. Guyon, I., Elisseeff, A.: An introduction to variable and feature selection. J. Mach. Learn. Res. **3**, 1157–1182 (2003)
16. Schapire, R.E., Freund, Y.: Boosting: Foundations and Algorithms. Adaptive Computation and Machine Learning Series. The MIT Press, Cambridge (2012)
17. AuBT. https://www.informatik.uni-augsburg.de/lehrstuehle/hcm/projects/tools/aubt/. Accessed 25 Feb 2013
18. Steinberg, D., Colla, P.: CART: classification and regression trees. In: The Top Ten Algorithms in Data Mining, vol. 9, p. 179 (2009)
19. Powers, D.M.: Evaluation: from precision, recall and F-measure to ROC, informedness, markedness and correlation. Int. J. Mach. Learn. Technol. **2**, 37–63 (2011)
20. Alfakih, K., Walters, K., Jones, T., Ridgway, J., Hall, A.S., Sivananthan, M.: New gender-specific partition values for ECG criteria of left ventricular hypertrophy: recalibration against cardiac MRI. Hypertension **44**, 175–179 (2004)
21. Romei, M., Mauro, A.L., D'angelo, M., Turconi, A., Bresolin, N., Pedotti, A., Aliverti, A.: Effects of gender and posture on thoraco-abdominal kinematics during quiet breathing in healthy adults. Respir. Physiol. Neurobiol. **172**, 184–191 (2010)

Developing a Chinese Food Nutrient Data Analysis System for Precise Dietary Intake Management

Xiaowei Xu[1], Li Hou[1], Zhen Guo[1], Ju Wang[2], and Jiao Li[1(✉)]

[1] Institute of Medical Information, Chinese Academy of Medical Sciences/Peking Union Medical College, Beijing 100020, China
{xu.xiaowei,hou.li,guo.zhen,li.jiao}@imicams.ac.cn
[2] The University of Texas Health Science Center at Houston, 77030 Houston, USA
373353139@qq.com

Abstract. A big mount of dietary data can be recorded in the daily life with the development of Internet of Things (e.g., RFID-equipped food carriers and food vending machines). Via monitoring and analyzing of personal dietary, it can provide valuable information for disease diagnosis, body weight control, and dietary habit management. The big data analysis benefits for patients, dieters, nutritionists and individuals who concern their health. While various techniques have been used for dietary monitoring in clinical trials and user studies, they are not ready for daily use. Existing solutions either require tedious manual recording or may impede normal daily activities. In this paper, we designed a smart big data framework using RFID technology to analyze the nutrition intake from dietary every day. The framework is capable to record Chinese food dietary information efficiently and effectively. It is promising for individuals and dietarians to set up personalized nutrient plan in the future.

Keywords: Big data · Dietary record · Nutrition analysis · Chinese food

1 Introduction

Dietary intake is important for health management and disease treatment. It is an important factor when evaluating group or an individual's health condition. Unbalance diet may cause a variety of health problems including the overweight, cardiovascular disease, liver disease, and type 2 diabetes [1–4].

However, accurate collection and assessment of food consumption is challenging. In previous work, some traditional self-report methods of dietary assessment, including weighted diet records, 24-hour dietary recall, food frequency questionnaire relying on the respondent's memory and ability to estimate portion sizes [5–7]. The populations using these methods tend to underreport the total energy intake by as much as 30% [8–12]. With the rapid development of information technology, many emerging innovations involving camera and mobile telephone technology to capture food and meal images for dietary recording before and after meal are available for use [13–16]. However, it is time-consuming in that dedicated time is required to enter dietary information [17].

© Springer International Publishing AG, part of Springer Nature 2018
F. Y. L. Chin et al. (Eds.): BIGDATA 2018, LNCS 10968, pp. 360–366, 2018.
https://doi.org/10.1007/978-3-319-94301-5_30

Radio Frequency Identification (RFID) is a technology mainly implemented in the form of tags that contain information that can be read by specific RFID tag readers [18–20]. With RFID, information interaction can be achieved without human intervention and awareness. The RFID technology has been applied in dietary recording [20, 21]. The wireless interaction between the tags and the readers facilitates automatic assessment of product with minimal human interference. In this study, we used the iPlate system equipped with RIFD, recording the dietary intake of individuals automatically.

Despite several relevant studies, such as dietary management using RFID technology [20, 21] and validation of the existing electronic diet recording methods [17], dietary intake management for Chinese food covering both food consumption and nutrient analysis is still an unsolved problem.

In this study, we designed a framework to collect and analyze the dishes consumed and estimate the nutrients (carbohydrate, fat, fiber and etc.) of individuals automatically.

2 Design of Analytics Framework

In this study, we designed a dietary analytic framework for Chinese food (see Fig. 1). It includes two key components: the RFID-based dietary data collection module, the Chinese food data representation module.

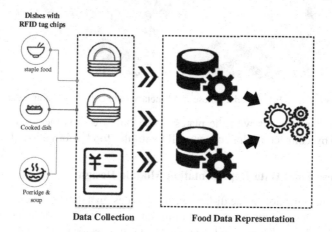

Fig. 1. Nutrition analytics framework of dietary intake

2.1 RFID-Based Dietary Data Collection Module Design

The iPlate is developed by the Sovell Science and Technology Limited company [22]. A standard set of iPlate includes three components: three types of plates embedded with RFID chips which could be written food names in the plates and corresponding weight, a production bench which can erase and rewrite the RFID chips embedded in the plates, and a settlement bench with RFID reading function.

Foods are classified into three categories and are put in three types of plates, staple food (e.g., rice) is in Plate I, cooked dish (e.g., cooked tomato with eggs) in Plate II, and soup (e.g., egg drop soup) in Plate III. A diner collects all the plates with food he/she wants to eat on one tray, and puts the tray on the settlement bench, then the RFID reader reads the information written into RFID chips of each plate including the recipe name and its weight. The price and the diet information will be obtained and stored in the central repository. When a diner checks out after choosing plates of foods,

Fig. 2. RFID-based Chinese food dietary data collection equipment

the RFID reader interacts with the plates, meanwhile, records the food consumption information by connecting the consumer ID with the food information (Fig. 2).

2.2 Chinese Food Data Representation Module Design

To calculate the nutrition of the dishes, we have to design a data representation for food nutrition first. The data representation in this study is capable to formulize five objects including completed dishes, food ingredients, food classification, cooking steps and nutrition composition. And also we build relationship between the above five data objects, which could be used in the further calculation. The relationship of the above objects is shown in Fig. 3.

2.3 Basic Object Design

(1) COOK_FOOD: This table is used to store the Chinese dishes data, which defines the name of the dish (cookfood_name), the cooking method description (cookfood_desc).

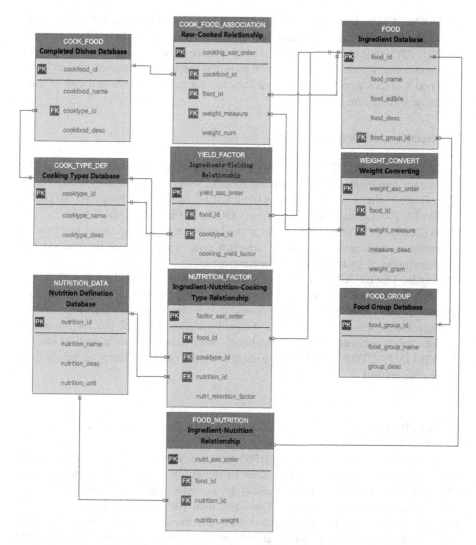

Fig. 3. The relationship of the databases

(2) FOOD: This table is used to store the common ingredients of Chinese foods, whose data source is "China Food Composition" [23, 24]. It defines the name of the ingredients (food_name), description of the ingredients (food_desc), the edible part of the ingredients (food_edible).

(3) FOOD_GROUP: This table is used to store the data of food group, whose data source is also the "China Food Composition" [23, 24]. And it defines the unique id of the food group (food_group_id), name of each group (food_group_name), and description of each group(group_desc).

(4) COOK_TYPE_DEF: This table is used to store the cooking method of Chinese food, which defines the name and description of each cooking method (cooktype_name).

(5) NUTRITION_DATA: This table is used to store the nutrition information of each ingredients, whose data source is the common nutrition information of the "China Food Composition" [23, 24]. It defines the name, description and unit of each nutrient (nutrition_name).

2.4   Relationship Design

(1) NUTRITION_FACTOR: This table is used to store the nutrient retention factor [25] of each ingredient in different cooking method, which is influenced by the data fields including: the food_id from FOOD TABLE, nutrition_id from NUTRITION_DATA TABLE and nutrition_id from COOK_TYPE_DEF TABLE,field.

(2) WEIGHT_CONVERT: This table is used to store converting factor between the common measurements and the standard unit, which defines the method and description of each ingredient' measurement.

(3) COOK_FOOD_ASSOCIATION: This table is used to store the ingredients consisted in cooked food and the corresponding weight information (weight_num).

(4) FOOD_NUTRITION: This table is used to store the nutrients information of each ingredients, whose data source is "China Food Composition" [23, 24]. It defines how much nutrients(nutrition_weight) of 100 g of edible part ingredient has.

(5) YIELD_FACTOR: This table is used to store the changing factor of weight from raw to cooked

3   Conclusions

In this study, the iPlate system using RFID technology is used to track and record the dishes and the corresponding weight eaten by the group. Meanwhile, we designed the data representation from completed dishes to nutrients, which could be used for the future calculation of the nutrients. The equipment for collecting data could obtain consumption data without human intervention, and the data representation is expendable for other model applications. Based on the above design, we could provide decision support for the restaurant in the future.

In future study, we will realize the data representation in a real-world environment and accomplish the whole calculation of dietary nutrition. Based on the experiments, we will take more factors into nutrition calculation to minify the limitations above and make the results more accurate.

Acknowledgements. This study was supported by the Key Laboratory of Medical Information Intelligent Technology Chinese Academy of Medical Sciences, The National Key Research and Development Program of China (Grant No. 2016YFC0901901), the National Population and Health Scientific Data Sharing Program of China, and the Knowledge Centre for Engineering Sciences and Technology (Medical Centre).

References

1. Saslow, L.R., Mason, A.E., Kim, S., Goldman, V., Ploutz-Snyder, R., Bayandorian, H., Daubenmier, J., Hecht, F.M., Moskowitz, J.T.: An online intervention comparing a very low-carbohydrate ketogenic diet and lifestyle recommendations versus a plate method diet in overweight individuals with type 2 diabetes: a randomized controlled trial. J. Med. Internet Res. 19(2), e36 (2017). https://doi.org/10.2196/jmir.5806. PMID: 28193599
2. Jospe, M.R., Fairbairn, K.A., Green, P., Perry, T.L.: Diet app use by sports dietitians: a survey in five countries. JMIR mHealth uHealth 3(1), e7 (2015)
3. Desroches, S., Lapointe, A., Ratté, S., Gravel, K., Légaré, F., Turcotte, S.: Interventions to enhance adherence to dietary advice for preventing and managing chronic diseases in adults. Cochrane Database Syst. Rev. 2, CD008722 (2013). https://doi.org/10.1002/14651858. cd008722.pub2. Medline: 23450587
4. Dhurandhar, N.V., Thomas, D.: The link between dietary sugar intake and cardiovascular disease mortality: an unresolved question. JAMA 313(9), 959–960 (2015). https://doi.org/10.1001/jama.2014.18267. Medline: 25734737
5. Mahabir, S., Baer, D.J., Giffen, C., Subar, A., Campbell, W., Hartman, T.J., et al.: Calorie intake misreporting by diet record and food frequency questionnaire compared to doubly labeled water among postmenopausal women. Eur. J. Clin. Nutr. 60(4), 561–565 (2006). https://doi.org/10.1038/sj.ejcn.1602359. Medline: 16391574
6. Probst, Y.C., Tapsell, L.C.: Overview of computerized dietary assessment programs for research and practice in nutrition education. J. Nutr. Educ. Behav. 37(1), 20–26 (2005). Medline: 15745652
7. Forster, H., Walsh, M.C., Gibney, M.J., Brennan, L., Gibney, E.R.: Personalised nutrition: the role of new dietary assessment methods. Proc. Nutr. Soc. 75(1), 96–105 (2016). https://doi.org/10.1017/s0029665115002086. Medline: 26032731
8. Subar, A.F., Kipnis, V., Troiano, R.P., Midthune, D., Schoeller, D.A., Bingham, S., et al.: Using intake biomarkers to evaluate the extent of dietary misreporting in a large sample of adults: the OPEN study. Am. J. Epidemiol. 158(1), 1–13 (2003). Medline: 12835280
9. Champagne, C.M., Baker, N.B., DeLany, J.P., Harsha, D.W., Bray, G.A.: Assessment of energy intake underreporting by doubly labeled water and observations on reported nutrient intakes in children. J. Am. Diet. Assoc. 98(4), 426–433 (1998). https://doi.org/10.1016/s0002-8223(98)00097-2. Medline: 9550166
10. Gersovitz, M., Madden, J.P., Smiciklas-Wright, H.: Validity of the 24-hr. dietary recall and seven-day record for group comparisons. J. Am. Diet. Assoc. 73(1), 48–55 (1978). Medline: 659761
11. Australian Bureau of Statistics. Australian Health Survey: Users' Guide, 2011–2013: Under-Reporting in Nutrition Surveys (2014). http://www.abs.gov.au/ausstats/abs@.nsf/Lookup/4363.0.55.001Chapter651512011-13. Accessed 09 Aug 2016. WebCite Cache ID 6jcWW3HQR
12. Moshfegh, A.J., Rhodes, D.G., Baer, D.J., Murayi, T., Clemens, J.C., Rumpler, W.V., et al.: The US department of agriculture automated multiple-pass method reduces bias in the collection of energy intakes. Am. J. Clin. Nutr. 88(2), 324–332 (2008). FREE Full text. Medline: 18689367
13. Anton, S.D., LeBlanc, E., Allen, H.R., Karabetian, C., Sacks, F., Bray, G., et al.: Use of a computerized tracking system to monitor and provide feedback on dietary goals for calorie-restricted diets: the POUNDS LOST study. J. Diabetes Sci. Technol. 5, 1216–1225 (2012). FREE Full text. Medline: 23063049

14. Springvloet, L., Lechner, L., Oenema, A.: Planned development and evaluation protocol of two versions of a web-based computer-tailored nutrition education intervention aimed at adults, including cognitive and environmental feedback. BMC Public Health **14**, 47 (2014). https://doi.org/10.1186/1471-2458-14-47. FREE Full text. Medline: 24438381

15. Charney, P., Peterson, S.J.: Practice paper of the academy of nutrition and dietetics abstract: critical thinking skills in nutrition assessment and diagnosis. J. Acad. Nutr. Diet. **113**(11), 1545 (2013). https://doi.org/10.1016/j.jand.2013.09.006

16. Daugherty, B.L., Schap, T.E., Ettienne-Gittens, R., Zhu, F.M., Bosch, M., Delp, E.J., et al.: Novel technologies for assessing dietary intake: evaluating the usability of a mobile telephone food record among adults and adolescents. J. Med. Internet Res. **14**(2), e58 (2012). https://doi.org/10.2196/jmir.1967. Medline: 22504018

17. Raatz, S.K., Scheett, A.J., Johnson, L.K., Jahns, L.: Validity of electronic diet recording nutrient estimates compared to dietitian analysis of diet records: randomized controlled trial. J. Med. Internet Res. **17**(1), e21 (2015). https://doi.org/10.2196/jmir.3744. http://www.jmir.org/2015/1/e21/. PMID: 25604640

18. Roy, W.: An introduction to RFID technology. IEEE Pervasive Comput. **5**(1), 25–33 (2010)

19. Shepard, S.: RFID: radio frequency identification. McGraw-Hill, New York (2005). ISBN 9780071442992

20. Chen, P.H., Liang, Y.H., Lin, T.C.: Using E-Plate to implement a custom dietary management system. In: International Symposium on Computer, Consumer and Control, pp. 978–981 (2014)

21. Chen, P.H., Liang, Y.H., Chou, W.C.: E-tag plate application for dietary management. In: International Symposium on Computer, Consumer and Control. pp. 223–226 (2014)

22. The Sovell Science and Technology Limited Company (2017). http://www.sovell.com.cn/. Accessed 7 Jul 2017. WebCite Cache ID 6tIHI2Guf (in Chinese)

23. Yang, Y.X., Wang, G.Y., Pan, X.C.: China food composition. Peking University Medical Press, Beijing (2009). ISBN 9787811167276

24. Yang, Y.X.: China food composition. Peking University Medical Press, Beijing (2004). ISBN 9787810716789

25. Matthews, R., Garrison, Y.: Agriculture handbook No. 102: Food yields summarized by different stages of preparation. USDA Agricultural Research Service, Washington, DC (1975)

A Real-Time Professional Content Recommendation System for Healthcare Providers' Knowledge Acquisition

Lu Qin, Xiaowei Xu, and Jiao Li[(✉)]

Institute of Medical Information, Chinese Academy of Medical Sciences/Peking Union Medical College, Beijing 100020, China
{qin.lu, xu.xiaowei, li.jiao}@imicams.ac.cn

Abstract. Lifelong learning has become an essential component in a professional's career path. With the wide use of mobile devices in working environments, professionals can acquire knowledge anytime and anywhere. For healthcare providers, their clinical knowledge acquisition in a timely manner can significantly improve patient treatment. In this paper, we present a real-time professional content recommendation system for healthcare providers' knowledge acquisition. The system includes five layers: Data Layer (healthcare provider profile, learning behavior, social network and etc.), Algorithm Layer (clinical/medical knowledge classification, medical expertise similarity measurement, professional-knowledge matching algorithms and etc.), Service Layer (click/browsing behavior monitor, healthcare provider feedback collection and etc.), Application Layer (medical content recommendation, retrieval optimization and etc.) and Management Layer (clinical scenario configuration, medical terminology management and etc.). The system has been applied in a knowledge service applications targeting clinicians in mobile Health (mHealth) scenario.

Keywords: Personalization · Recommendation system · Healthcare providers

1 Introduction

The rapid development of Internet results in the big problem of information overload. Recommendation technology has been widely used in Internet applications to help people find their favorite content from the huge amount of information. For example, the famous e-commerce platforms such as amazon.com and taobao.com recommend items to users based on their browsing and purchase history [1]; Netflix.com uses recommendation to help people find their interested films and videos using collaborative filtering [2]; Youtube.com builds personalized homepage to show recommended videos for every user according to the playing history [3]. Recommendation system already becomes one of the most important and effective tools to help reduce the cost of information navigation.

In the medical and healthcare field, Internet is becoming a primary source of medical knowledge acquisition and dissemination for healthcare providers. Medical professionals are open-minded to the medical applications [4]. According to the investigation [5], the top Chinese mobile health applications consists largely of medical education applications

© Springer International Publishing AG, part of Springer Nature 2018
F. Y. L. Chin et al. (Eds.): BIGDATA 2018, LNCS 10968, pp. 367–371, 2018.
https://doi.org/10.1007/978-3-319-94301-5_31

which focus on knowledge acquisition. XingShuLin [6], YiXueJie [7], DingXiangYuan [8] are all applications focus on Continuing Medical Education (CME). How to provide information effectively from the huge amount of information to satisfy the healthcare providers' taste has become the urgent problem to solve.

This paper presents a recommendation system with multi-dimensional recommendation algorithms which takes the healthcare providers' specific information into account, and comprehensive modeling method of user in the medical and healthcare field, which aims to extract user-specific information from their historical behaviors to represent their interest profile. Besides, demographic and background information including gender, age, title, department and other specific information of healthcare providers are also used to better reveal theirs potential choice. With this system, healthcare providers will get a better user experience and get a comprehensive view of the information with cheaper time cost.

2 Related Work

2.1 Recommendation Algorithm

Recommendation systems have gained a lot of attention since the advent of the internet. Various algorithms including neighborhood based [9] such as user based collaborative filtering and item based collaborative filtering, latent interest models such as matrix factorization [10] have been used in web applications. As personality is becoming more effective in the field of personalized recommendation, recent research on recommender services has been interested in personality-based approaches. For example, Rana and Jain emphasized the potential use of personality information in recommendation systems in their current overview [11]; Hu and Pu proved that personality-based recommendation system was not only more accurate than the rating-based one, but also it was much easier to use [12].

2.2 Personality and User Behavior

When it comes to the personalization, how to define the characteristics of the user becomes important. As the basis of recommendation, behavior modeling is also becoming the focus of current research. Ngai et al. emphasized that "personality traits are often taken to be one of the fundamental theories explaining the characteristics affecting users' subsequent behavior" [13]; Rezarta et al. worked to understand the PubMed user behaviors through the searching log analysis, and according the analysis to improve the information retrieval of the website [14]; as the mobile phones occupies the majority of the market and it could know a lot of user behaviors through the sensors embedded, some researchers put emphasis on the mobile user behaviors, Bent et al. proposed a method of user behavior modeling which taking the personalized education and some other different activities into consideration to better analyze the uses of the mobile applications [15].

Compared with previous studies, the main contribution of our work is that we design a personalized recommendation framework which exploits algorithms to cover different scenarios. We also proposed a common user modeling method using information from various sources including demographics, specific information about healthcare providers and social relationship among them, which aims to provide more effective information to the medical staff.

3 Framework Design

In this study, the framework we proposed consists of five layers (see Fig. 1). We will introduce each layer in the following sections.

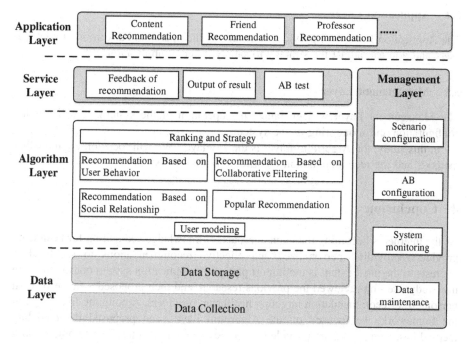

Fig. 1. Framework of the personalized recommendation system.

3.1 Data Layer

As the basis of the framework, Data Layer is responsible for the data collection and the data storage. The data source could be the log of medicine website or the mobile applications, which will be collected and stored for further analysis.

3.2   Algorithm Layer

In the Algorithm Layer, we focus on the user modeling. With the user model above, we proposed four kinds of algorithms to cover the scenarios in the recommendation. At the same time, we proposed ranking and strategy to present the result of the algorithms.

3.3   Service Layer

The Service Layer acts as a connecting link between the Algorithm Layer and the Application Layer. It provides recommendation results to the Application Layer and recycles the feedback of the users to improve the performance of the framework. In order to guarantee the stability of the online service, sometimes AB Test is needed to verify the accuracy and applicability of the algorithms.

3.4   Application Layer

The Application Layer will provide the results of the operation background to the real users and you can adjust the strategy to satisfy different applications.

3.5   Management Layer

The Management Layer will guarantee the operation of the system. It will provide configuration and monitoring of the system.

In this study, the Algorithm Layer is the key layer. We will focus on the modeling methods and the recommendation algorithms in the next section.

4   Conclusion

As the internet is becoming the main approach of knowledge acquisition for healthcare providers, providing personalized and effective recommendations to satisfy their interests to the most extent is the import part of the application system construction. In this study, we first reviewed the previous research, and then proposed a framework on personalized recommendation targeting healthcare providers. Among the five layers of the framework, we focused on the Algorithm Layer, and proposed the modeling method based on healthcare providers' specific behaviors and the four complementary recommendation algorithms.

In the future, we will build an online system to run our model and to improve its performance according to the online tests. Meanwhile, other user behaviors e.g. video behaviors will also be taken into consideration.

Acknowledgement. This study was supported by the Key Laboratory of Medical Information Intelligent Technology Chinese Academy of Medical Sciences, The National Key Research and Development Program of China (Grant No. 2016YFC0901901), the National Population and Health Scientific Data Sharing Program of China, and the Knowledge Centre for Engineering Sciences and Technology (Medical Centre).

References

1. Linden, G., Smith, B., York, J.: Amazon.com recommendations: item-to-item collaborative filtering. IEEE Internet Comput. **7**(1), 76–80 (2003)
2. Bennett, J., Lanning, S.: The netflix prize. In: Proceedings of KDD Cup and Workshop, pp. 3–6 (2007)
3. Davidson, J., Liebald, B., Liu, J., Nandy, P., Vleet, T.V., Gargi, U., et al.: The YouTube video recommendation system. In: ACM Conference on Recommender Systems, pp. 293–296. ACM (2010)
4. Ponce, L. B.: Analysis of certificated mobile application for medical education purposes. In: International Conference on Technological Ecosystems for Enhancing Multiculturality, pp. 13–17. ACM (2014)
5. Jeffrey, H., Liu, D., Yu, Y.M., Zhao, H.T., Chen, Z.R., Li, J., et al.: The top chinese mobile health apps: a systematic investigation. J. Med. Internet Res. **18**(8), e222 (2016)
6. XingShuLin Homepage. https://www.xingshulin.com/. Accessed 11 Mar 2018
7. YiXueJie Homepage. http://test1.yxj.org.cn/. Accessed 11 Mar 2018
8. DingXiangYuan Homepage. http://www.dxy.cn/. Accessed 11 Mar 2018
9. Resnick, P., Iacovou, N., Suchak, M., Bergstrom, P., Riedl, J.: GroupLens: an open architecture for collaborative filtering of netnews. In: ACM Conference on Computer Supported Cooperative Work, pp. 175–186. ACM (1994)
10. Koren, Y., Bell, R., Volinsky, C.: Matrix factorization techniques for recommender systems. Computer **42**(8), 30–37 (2009)
11. Rana, C., Jain, S.K.: A study of the dynamic features of recommender systems. Artif. Intell. Rev. **43**(1), 141–153 (2015)
12. Hu, R., Pu, P.: Acceptance issues of personality-based recommender systems. In: ACM Conference on Recommender Systems, pp. 221–224. ACM (2009)
13. Ngai, E.W.T., Tao, S.S.C., Moon, K.K.L.: Social media research: theories, constructs, and conceptual frameworks. Int. J. Inf. Manage. **35**(1), 33–44 (2015)
14. Dogan, R.I., Murray, G.C., Névéol, A., Lu, Z.: Understanding pubmed® user search behavior through log analysis. Database J. Biol. Databases Curation **2009**, bap018 (2009)
15. Bent, O., Dey, P., Weldemariam, K., Mohania, M.K.: Modeling user behavior data in systems of engagement. Future Gener. Comput. Syst. **68**, 456–464 (2017)

Study on Big Data Visualization of Joint Operation Command and Control System

Gang Liu[✉] and Yi Su

Academy of Art and Design, Tsinghua University,
Haidian District, Beijing 100084, China
309799496@qq.com

Abstract. As more and more weapons and equipment are connected to the joint operational command and control system, the amount of information generated by the weapons and equipment is also increasing, which has brought tremendous challenges to the commanders and soldiers' ability to make decisions. Based on the concept of "user-centered design", this paper studies the information visualization method of the operational command and control system. The implementation of visualization is divided into three levels: user behavior, interaction architecture, and visual performance. Through the user's task investigation, the logic of interactive architecture, and visual element coding, a joint air defense command and control system interface was designed. After a user test of the interface, the results show that the new design scheme is better than the original program. In this study, task research, design development, and user testing are used to propose a method of information visualization for command and control systems.

Keywords: Joint operation · Command and control
Human-machine interaction · Big data · Visualization · C4ISR

1 Research Background

1.1 The Summary of the Research Background

The number of weapons and equipment connected to the operational command and control system is increasing, and the amount of information generated by the weaponry and equipment is increasing (Lan et al. 2015). This has brought great challenges to the commanders and soldiers' ability to make decisions. The visual design research of command and control system (It is also called C4ISR System) aims to improve the ease of use and aesthetics of the operation interface and reduce the visual burden and cognitive load of the commanders (Lei 2016). By designing and innovating the system's interactive interface, commanders can be better able to command different military types, complete operational tasks, and adapt to multiple usage environments.

The human-machine interaction of traditional weapon systems is a unilateral interaction between man and machine. The new operational command and control system pays more attention to joint operations and needs to combine all types of military, various types of weapon systems, and different battlefield areas. With the

© Springer International Publishing AG, part of Springer Nature 2018
F. Y. L. Chin et al. (Eds.): BIGDATA 2018, LNCS 10968, pp. 372–380, 2018.
https://doi.org/10.1007/978-3-319-94301-5_32

development of new sensing and communication equipment, joint operations have become synonymous with modern warfare. Joint combat systems combine widely distributed combat units such as sensors, accusation nodes, and weapon platforms. In the process of command and control systems linking combat units, the acquisition, integration, transmission and distribution of various types of information will generate a large amount of data. The results of these data analyses will be presented on the computer screen of the command center. The commander will make judgments based on the data, so the quality of the data visual design will directly affect the efficiency of the commander in making decisions (Liao et al. 2017) (Fig. 1).

Fig. 1. Structure diagram of air defense system (Picture Resource: http://blog.sina.com.cn/s/blog_856c53b20101fa1f.html)

The most important task of data visualization design of command and control system is to combine information presentation technology with human behavior requirements. The visual design of the command and control system is divided into two main parts: the first part is the design of the digital interface interaction architecture, and the other part is the visual design of the text diagram and the battlefield situation.

1.2 Domestic and Foreign Research Overview

The application of data visualization methods in the field of military operations has a long history. Prior to the invention of the computer, the maps, situation charts, and sand tables used by the commanders were the manifestations of information visualization. In the digital era where computers are widely used, both combat command and control and combat training simulation require the support of various types of data on the battlefield. At the same time, the visualization of large amounts of data has become an important part of combat informationization.

The United States is the proponent of the theory of joint operations in the world. It first built a joint combat command and control system. The National Military Command Center (NMCC) is a strategic command and control system built and delivered by the United States in 1959. The United States has established the main theoretical basis for joint operations and has continuously improved relevant construction goals through practice. At the beginning of the 21st century, the United States proposed the concept of Global Information Grid and C2 Constellation. One important subsystem related to the visualization of big data information is the defense communication system. The system is a defense communication network composed of three subsystems, the Defense Switched Network (DSN), the Defense Data Network (DDN), and the Defense Satellite Communications System (DSCS). It has a variety of information transmission capabilities such as images, data, video, and text. The information communication on the display terminal is mainly through visual methods (Fig. 2).

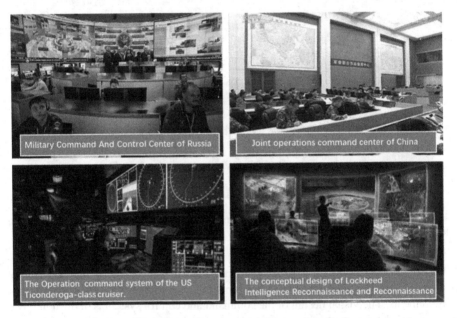

Fig. 2. Command and control system usage scenarios

Russia officially launched the National Defense Command Center at the end of 2014. In the period of peace, it is responsible for monitoring the national security dynamics. Once a war breaks out, the National Defense Command Center will take over the armed forces of the entire country. On April 20, 2016, China exposed the newly established Joint Operations Command Center of the Central Military Commission on CCTV and proposed the requirements for establishing a joint operational command system. Lockheed Martin of the United States demonstrated advanced interactive and display technologies—"21st century intelligence, surveillance and reconnaissance systems".

The human-computer interaction and interface design research of the command and control system not only needs to solve the system's functional problems, but also needs to improve the experience of the commander and the soldier more, reduce the operational pressure and cognitive burden, and increase the "know-decision-operation" of the commander-in-command. The accuracy and comfort of the process.

2   Interaction Architecture and Information Hierarchy

2.1   Cognition and Design

The design goal of the system interaction architecture is to improve system availability, improve user experience satisfaction, and focus on the fluency and comfort of system commanders during use. The design of the system needs to adhere to the design concept of "centered on the commander and the combatant", and the system design is based on human cognition and human behavior.

The presentation of information visualization content in the interface is constrained by human cognitive ability. Broadbent proposed treating people as a source of a series of processing information, using communication system metrics to consider information flow (Jeff 2014). The information is contained in a series of (n) alternative stimuli. Each stimulus (i) has a probability of occurrence (Pi). By simulating these stimuli, information can be quantified.

$$\text{Information content} = \sum_{i=0}^{n} -P_i \log_2\left(\frac{1}{P_i}\right) \text{ bits}$$

The amount of information the person's brain is concerned about at the same time is about 125 bits per second. People can deliver about 40 bits of information per second during the conversation. Human short-term memory can only remember 7 ($\pm$) 2 chunks at a time (Gu 2016). Human cognitive ability determines the boundaries of information presented by the user interface. Information that exceeds the cognitive range in a period will result in a decrease in cognitive efficiency. If people are in an emergency, their cognitive ability is still lower than normal. Therefore, data visualization design should give priority to the display of important information. One solution is to use artificial intelligence inference engines to reduce useless information in advance before the results of data visualization are presented. The inference engine can sort the information urgently through algorithm simulation and machine learning, and recommend the most important information to the area that is most likely to be noticed by the commander (Fig. 3).

The system research tasks with "commander as the center" can be divided into three levels: user behavior, interaction architecture, and visual presentation. They correspond to the user's "know-decision-operation" interaction model. This research proposes a research framework for the man-machine interface design of the command and control system, which includes three aspects: user behavior investigation, interaction logic framework design, and graphical user interface design (Shneiderman 2017).

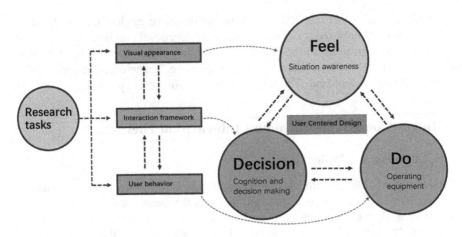

Fig. 3. Command and control system usage scenarios

2.2 Information Layered

Joint operations are the main features of modern warfare. Joint operations are the unified combat forces in a command and control system. Therefore, the interactive architecture of this system is very complicated. Joint operations require the support of information networks, which requires the organization and visualization of large amounts of data, and the conversion of complex big data into information that can be quickly interpreted. Data connection between people and systems, as well as people and equipment, data presentation requires a lot of human-computer interaction design and interface visualization design. Through the classification of information, the commander can accurately find the visual battlefield situation information (Fig. 4).

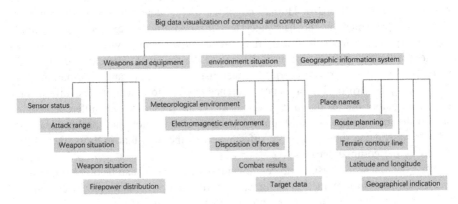

Fig. 4. Data visualization of command and control system

The human-computer interaction design goal of the joint operations command and control system is to integrate image processing, streaming media, data mining, information processing, and visualization technologies, and to graphically design and express elements such as content, relationship, and time in command information. Visualization of the battlefield includes visualizing the battlefield environment, visualizing operational operations, and visualizing intelligence data, which together constitute a command decision support system.

The information visualization method of the joint operations command and control system can be applied to the three stages of the information analysis process, the information collation stage, the information analysis stage and the information presentation stage. The purpose of the collation and analysis of the early stage is to stratify the information and arrange the analysis results according to the importance and urgency. For example, alarm information and security situation information should be placed in an important position, and map situation information should not be blocked by a popup box (Xue 2015).

3 Design Case of Air Defense System

3.1 Modular Design of Interactive Interfaces

The user interface is the space where visual information is carried and is the key content of the human-computer interaction of informationized combat systems. The joint operation command and control system includes various types of software and hardware. The human-computer interaction interface needs to have enough flexibility to customize the functional areas in a modular manner. The human-computer interaction interface should include the system default form and custom layout form. The system allows the user to personalize and divide the role of function settings, the interface is divided into special seat function area, common function area, custom function area and other areas.

The command and control system needs to adapt to different display contents, and it needs to have normative standards and compatibility. The overall system consists of modular, standardized and generalized components. This study used interviews to survey 12 experienced combat system design engineers (Han 2016). According to the conclusion of the investigation and the suggestion of the engineer, the system interface layout is divided into several large modules, and the position between the module and the module can be replaced. Users access the system according to their job rights, and the user's seat will present different modules. At the same time, the system can display a personalized and customized user interface, and the personalized and customized interface can be adjusted according to the task and improve the information utilization efficiency (Fig. 5).

The command and control system should adapt to different display contents, need to have normative standards, and have sufficient compatibility, with modularization, standardization and generalization. In this study, we divided the interface layout into several large modules by investigating the function of the system. A feature plane replaced each module, and the position between the module and the module could be

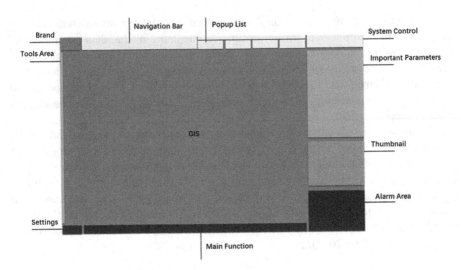

Fig. 5. User interface design framework analysis

replaced. Users of different functions access the system according to the permission, and the user seats will also display different modules. At the same time, the system can display the customized user interface, and the personalized customization interface can use the operational command personnel to layout functional modules according to specific tasks and improve the efficiency of information utilization.

With the increasing importance of air supremacy in the war, air defense command and control systems (Air Defense Operations Center) are becoming more and more important in the national defense system. Many countries place air defense systems on the priority of military development. In fact, the operational command and control automation system is originated from the air defense field. The SAGE Semi-Automatic Ground Environment is recognized as the world's first semi-automatic air defense command and control network (Fig. 6).

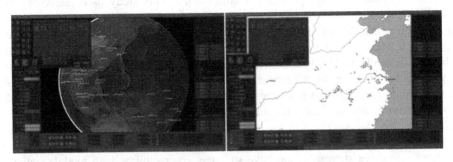

Fig. 6. The concept user interface design of command and control system

In this study, we designed a new interactive interface and situational display for a concept air defense command and control system. We have visualized the battlefield situation based on geographic information systems. The new system uses image processing, streaming media and other technologies to build a new system interface with "functional modularity and data visualization". The new command and control system enables efficient display of data content such as battlefield environments, decision actions, and intelligence information.

This air defense system uses a two-dimensional and three-dimensional map design method. In this system, three-dimensional models have been established for heavy priority areas, weapons and equipment, target buildings, and radar envelope maps. The simulation and rendering of the three-dimensional model enhances the display effect and forms a dynamic, highly simulated scenario that visually makes it easier for users to identify various types of targets. After testing 10 ordinary users, the new system improved the interface quality, reduced the user's cognitive load, and improved the operational command efficiency.

3.2 Battlefield Data Visualization

Situation data mainly includes integrated management of multiple sources of data, real-time modeling of battlefield environmental factors, organization and presentation of various military icons, and display of various weapons and equipment parameters and models. The trend display has multi-dimensional, multi-variable, and multi-type features. The visual expression should use visual coding methods to express the information with the minimum visual element content. The human-computer interaction of the command and control system is also developing toward intelligence. The current system has integrated content screening algorithms that can help commanders optimize the battlefield situation to assist in decision-making.

Plot icons are the main information visualization indicators in the situation. The battlefield visualization system displays battlefield situation information by plotting a variety of military graphic symbols or labels with specific meaning on the map. Plotting icons are widely used in battle deduction and auxiliary command decisions. Situation plotting is an important part of information visualization. The use of military plots needs to be accurate and standardized. Plotting icons requires innovation based on the diversity of tasks. On the two-dimensional map, there are standard military icons that express most of the information. On three-dimensional maps, military plots are rendered using three-dimensional modeling and graphic rendering.

4 Conclusion

The command and control system is the core of multi-arms joint operations. The most important human-machine interaction between the system and the commanders and officers is information communication. The presentation of massive information generated by the operation of various types of equipment requires efficient and accurate visual design. This study applies the "user-centered" design concept. The goal is to improve the user experience of the command and control system. This study divides the visualization into three levels: user behavior, interaction architecture, and visual performance. Based on the user's behavioral characteristics, operating habits, and psychological characteristics, we designed an innovative air defense command and control system design, and proposes a design method of user interface of command and control system. This study tested the method of designing a modular, combinable framework-aided command and control system. Through the three parts of task definition, design development and user test, this study has determined an effective command and control system design process.

References

Shneiderman, B.: Designing the User Interface: Strategies for Effective Human-Computer Interaction. Publishing House of Electronics industry, Beijing (2017)

Lei, C.: C4ISR System, 2nd edn. National Defense Industrial Press, Beijing (2016)

Gu, Z.: Principles & Processes of Interaction Design. Tsinghua University Press, Beijing (2016)

Han, T.: User Research and Experience Design. Shanghai Jiao Tong University Press, Shanghai (2016)

Jeff, J.: Designing with the Mind in Mind Simple Guide to Understanding User Interface Design Guidelines. Elsevier, Singapore (2014)

Lan, Y., Zhao, K., Guo, C.: The system framework of future command and control an information processing system. J. Command Control **1**, 1 (2015)

Liao, Z., et al.: Application of Human Factors Engineering in Command and Control Information System. Publishing House of Electronics Industry (2017)

Xue, C.: The Design Method and Application of Digitalized Human-Computer Interface on Complex Information System. Southeast University Press, Nanjing (2015)

Author Index

Printed in the United States
By Bookmasters